STUDENT SOLUTIONS MANUAL

Richard N. Aufmann
Palomar College

Vernon C. Barker
Palomar College

Joanne S. Lockwood
Plymouth State College

Christine S. Verity

BASIC COLLEGE MATHEMATICS: AN APPLIED APPROACH

SEVENTH EDITION

Aufmann/Barker/Lockwood

HOUGHTON MIFFLIN COMPANY   BOSTON NEW YORK

Senior Sponsoring Editor: Lynn Cox
Senior Development Editor: Dawn Nuttall
Editorial Assistant: Melissa Parkin
Senior Manufacturing Coordinator: Florence Cadran
Marketing Manager: Ben Rivera

Printed in the U.S.A.

ISBN: 0-618-20287-0

3456789 – PAT – 06 05 04 03

Contents

Chapter 1: Whole Numbers — 1

Chapter 2: Fractions — 19

Chapter 3: Decimals — 44

Chapter 4: Ratio and Proportion — 61

Chapter 5: Percents — 72

Chapter 6: Applications for Business and Consumers — 85

Chapter 7: Statistics and Probability — 103

Chapter 8: U.S. Customary Units of Measurement — 124

Chapter 9: The Metric System of Measurement — 135

Chapter 10: Rational Numbers — 146

Chapter 11: Introduction to Algebra — 159

Chapter 12: Geometry — 183

STUDENT SOLUTIONS
MANUAL

Chapter 1: Whole Numbers

PREP TEST and GO FIGURE

1. 8

2. 1 2 3 4 5 6 7 8 9 10

3. a and D; b and E; c and A; d and B; e and F; f and C

Go Figure

On the first trip, the two children row over. The second trip, one child returns with the boat. The third trip, one adult rows to the other side. The fourth trip, one child returns with the boat. At this point, one adult has crossed the river. Repeat the first four trips an additional four times, one time for each adult. After completing the fifth time, there have been twenty trips taken, and the only people waiting to cross the river are the two children. On the twenty-first trip, the two children cross the river. So the minimum number of trips is twenty-one.

SECTION 1.1

Objective A Exercises

1.
3.

5. $37 < 49$

7. $101 > 87$

9. $245 > 158$

11. $0 < 45$

13. $815 < 928$

Objective B Exercises

15. millions

17. hundred-thousands

19. Three thousand seven hundred ninety

21. Fifty-eight thousand four hundred seventy-three

23. Four hundred ninety-eight thousand five hundred twelve

25. Six million eight hundred forty-two thousand seven hundred fifteen

27. 357

29. 63,780

31. 7,024,709

Objective C Exercises

33. $6000 + 200 + 90 + 5$

35. $400,000 + 50,000 + 3000 + 900 + 20 + 1$

37. $300,000 + 1000 + 800 + 9$

39. $3,000,000 + 600 + 40 + 2$

Objective D Exercises

41. 850

43. 4000

45. 53,000

47. 250,000

Applying the Concepts

49. 999; 10,000

SECTION 1.2

Objective A Exercises

1. 28

3. 125

5. 102

7. 154

9. 1489

11. 828

13.
$$\begin{array}{r} \overset{1}{859} \\ +725 \\ \hline 1584 \end{array}$$

15.
$$\begin{array}{r} \overset{1}{470} \\ +749 \\ \hline 1219 \end{array}$$

17.
$$\begin{array}{r} \overset{1\ 1\ \ 1}{36,925} \\ +65,392 \\ \hline 102,317 \end{array}$$

19.
$$\begin{array}{r} \overset{1\ \ \ 1}{50,873} \\ +28,453 \\ \hline 79,326 \end{array}$$

21.
$$\begin{array}{r} \overset{2\,2}{878} \\ 737 \\ +189 \\ \hline 1804 \end{array}$$

23.
$$\begin{array}{r} \overset{1}{319} \\ 348 \\ +\ 912 \\ \hline 1579 \end{array}$$

25.
$$\begin{array}{r} \overset{1\,2}{9409} \\ 3253 \\ +\ 7078 \\ \hline 19,740 \end{array}$$

27.
$$\begin{array}{r} \overset{12}{2038} \\ 2243 \\ + \ 3139 \\ \hline 7420 \end{array}$$

29.
$$\begin{array}{r} \overset{11 \ 11}{67,428} \\ 32,171 \\ + 20,971 \\ \hline 120,570 \end{array}$$

31.
$$\begin{array}{r} \overset{11 \ \ 1}{76,290} \\ 43,761 \\ + 87,402 \\ \hline 207,453 \end{array}$$

33.
$$\begin{array}{r} \overset{1 \ 11}{20,958} \\ 3,218 \\ + \ \ \ 42 \\ \hline 24,218 \end{array}$$

35.
$$\begin{array}{r} \overset{1 \ 11}{392} \\ 37 \\ 10,924 \\ + \ \ 621 \\ \hline 11,974 \end{array}$$

37.
$$\begin{array}{r} \overset{122}{294} \\ 1029 \\ 7935 \\ + \ \ 65 \\ \hline 9323 \end{array}$$

39.
$$\begin{array}{r} \overset{11 \ 21}{97} \\ 7,234 \\ 69,532 \\ + \ \ 276 \\ \hline 77,139 \end{array}$$

41. 14,383

43. 9473

45. 33,247

47. 5058

49. 1992

51. 68,263

53.
$$\begin{array}{rclr} 1234 & \approx & & 1200 \\ 9780 & \approx & & 9800 \\ + \ 6740 & \approx & + & 6700 \\ \hline \end{array}$$
Cal.: 17,754   Est.: 17,700

55.
$$\begin{array}{rclr} 241 & \approx & & 200 \\ 569 & \approx & & 600 \\ 390 & \approx & & 400 \\ + \ 1672 & \approx & + & 1700 \\ \hline \end{array}$$
Cal.: 2872   Est.: 2900

57.
$$\begin{array}{rclr} 32,461 & \approx & & 32,000 \\ 9,844 & \approx & & 10,000 \\ + \ 59,407 & \approx & + & 59,000 \\ \hline \end{array}$$
Cal.: 101,712   Est.: 101,000

59.
$$\begin{array}{rclr} 25,432 & \approx & & 25,000 \\ 62,941 & \approx & & 63,000 \\ + \ 70,390 & \approx & + & 70,000 \\ \hline \end{array}$$
Cal.: 158,763   Est.: 158,000

61.
$$\begin{array}{rclr} 67,421 & \approx & & 70,000 \\ 82,984 & \approx & & 80,000 \\ 66,361 & \approx & & 70,000 \\ 10,792 & \approx & & 10,000 \\ + \ 34,037 & \approx & + & 30,000 \\ \hline \end{array}$$
Cal.: 261,595   Est.: 260,000

63.
$$\begin{array}{rclr} 281,421 & \approx & & 280,000 \\ 9,874 & \approx & & 10,000 \\ 34,394 & \approx & & 30,000 \\ 526,398 & \approx & & 530,000 \\ + \ 94,631 & \approx & + & 90,000 \\ \hline \end{array}$$
Cal.: 946,718   Est.: 940,000

65.
$$\begin{array}{rclr} 28,627,052 & \approx & & 29,000,000 \\ 983,073 & \approx & & 1,000,000 \\ + \ 3,081,496 & \approx & + & 3,000,000 \\ \hline \end{array}$$
Cal.: 32,691,621   Est.: 33,000,000

67.
$$\begin{array}{rclr} 12,377,491 & \approx & & 12,000,000 \\ 3,409,723 & \approx & & 3,000,000 \\ 7,928,026 & \approx & & 8,000,000 \\ + \ 10,705,682 & \approx & + & 11,000,000 \\ \hline \end{array}$$
Cal.: 34,420,922   Est.: 34,000,000

Objective B Application Problems

69. Strategy   To find the total number of multiple births, add the four amounts (110,670, 6919, 627, and 79).

Solution
$$\begin{array}{r} 110,670 \\ 6919 \\ 627 \\ + \ \ \ 79 \\ \hline 118,295 \end{array}$$

The total amount of multiple births was 118,295.

71. Strategy   To estimate the total income from the four Star Wars movies, estimate each movie to the nearest hundred million and then add the estimates.

Solution
$$\begin{array}{rclr} 461,000,000 & \approx & & 500,000,000 \\ 290,200,000 & \approx & & 300,000,000 \\ 309,100,000 & \approx & & 300,000,000 \\ 431,100,000 & \approx & + & 400,000,000 \\ \hline & & & \$1,500,000,000 \end{array}$$

The estimated income from the four Star Wars movies was $1,500,000,000.

73a. Strategy   To find the total income from the two with the lowest box-office incomes, add the incomes from Episode V ($290,200,000) and Episode VI ($309,100,000).

Solution

$$\begin{array}{r} 290,200,000 \\ + 309,100,000 \\ \hline \$599,300,000 \end{array}$$

The income from the two with the lowest box-office returns is $599,300,000.

73b. The income from the 1977 Star Wars production is $461,000,000. The income from the two with the lowest box-office returns is $599,300,000 (from exercise 73a). Yes this income exceeds the income from 1977 Star Wars production.

75a. Strategy   To find the total number of miles driven during the three days, add the three amounts (515, 492, and 278 miles).

Solution

$$\begin{array}{r} 515 \\ 492 \\ + 278 \\ \hline 1285 \end{array}$$

The number of miles that will be driven during the three days is 1285.

75b. Strategy   To find what the odometer reading will be by the end of the trip, add the total number of miles driven during the three days (1285) to the original odometer reading (68,692).

Solution

$$\begin{array}{r} 1,285 \\ + 68,692 \\ \hline 69,977 \end{array}$$

At the end of the trip, the odometer will read 69,977 miles.

77. Strategy   To find the total average amount invested for Americans ages 16 to 34, add the amounts in checking accounts ($375), savings accounts ($1155), and U.S. Savings Bonds ($266).

Solution

$$\begin{array}{r} \$375 \\ 1155 \\ + 266 \\ \hline \$1796 \end{array}$$

Americans ages 16 to 34 have invested $1796 in these three investments.

79. Strategy   To find the greater amount invested in home equity and retirement:
- Add $43,070 and $9016 to find the amount invested by all Americans
- Add $17,184 and $4298 to find the amount invested by Americans 16 to 34.

Solution

$$\begin{array}{r} \$43,070 \\ + 9016 \\ \hline \$52,086 \end{array} \qquad \begin{array}{r} \$17,184 \\ + 4298 \\ \hline \$21,482 \end{array}$$

$52,086 > $21,482

Since 52,086 is greater than 21,482, All Americans invest more than Americans 16 to 34.

Applying the Concepts

81. There are 6 possible outcomes for each die (1, 2, 3, 4, 5, and 6).
The smallest sum on two dice is $1 + 1 = 2$.
The largest sum on two dice is $6 + 6 = 12$.
There are 11 different sums from 2 to 12 (2, 3, 4, 5, 6, 7, 8, 9, 10, 11, and 12).

83. No; $0 + 0 = 0$

85. Ten numbers that are less than 100 end in a 7. They are 7, 17, 27, 37, 47, 57, 67, 77, 87, and 97.

SECTION 1.3

Objective A Exercises

1. 4

3. 4

5. 10

7. 4

9. 9

11. 22

13. 60

15. 66

17. 31

19. 901

21. 791

23. 1125

25. 3131

27. 47

29. 925

31. 4561

33. 3205

35. 1222

37. 3021

39. 3022

41. 3040

43. 212

45. 60,245

Objective B Exercises

47.
$$\begin{array}{r} {}^{8\ 13}\!\!\not{9}\,\not{3} \\ -\ 2\,8 \\ \hline 6\,5 \end{array}$$

49.
$$\begin{array}{r} {}^{3\ 14}\!\!\not{4}\,\not{4} \\ -\ 2\,7 \\ \hline 1\,7 \end{array}$$

51.
$$\begin{array}{r} {}^{4\ 10}\!\!\not{5}\,\not{0} \\ -\ 2\,7 \\ \hline 2\,3 \end{array}$$

53.
$$\begin{array}{r} {}^{8\ 13}9\,\not{9}\,\not{3} \\ -\ 5\,3\,7 \\ \hline 4\,5\,6 \end{array}$$

55.
$$\begin{array}{r} {}^{13}\\{}^{7\ \not{8}\ 10}\!\!\not{8}\,\not{4}\,\not{0} \\ -\ 7\,8\,3 \\ \hline 5\,7 \end{array}$$

57.
$$\begin{array}{r} {}^{6\ 16\ 10}\!\!\not{7}\,\not{7}\,\not{0} \\ -\ 3\,9\,5 \\ \hline 3\,7\,5 \end{array}$$

59.
$$\begin{array}{r} {}^{11}\\{}^{4\ \not{1}\ 16}3\,\not{5}\,\not{2}\,6 \\ -\ \ 3\,8\,7 \\ \hline 3\,1\,3\,9 \end{array}$$

61.
$$\begin{array}{r} {}^{3\ 13\ 4\ 10}\!\!\not{4}\,\not{3}\,\not{5}\,\not{0} \\ -\ \ 7\,2\,9 \\ \hline 3\,6\,2\,1 \end{array}$$

63.
$$\begin{array}{r} {}^{15\ 9}\\{}^{0\ \not{5}\ 10\ 17}\!\!\not{1}\,\not{6}\,\not{0}\,\not{7} \\ -\ \ \ 8\,6\,9 \\ \hline 7\,3\,8 \end{array}$$

65.
$$\begin{array}{r} {}^{6\ 12\ 8\ 13}\!\!\not{7}\,\not{2}\,\not{9}\,\not{3} \\ -\ 3\,7\,4\,8 \\ \hline 3\,5\,4\,5 \end{array}$$

67.
$$\begin{array}{r} {}^{16\ 9}\\{}^{2\ 6\ 10\ 16}\!\!\not{3}\,\not{7}\,\not{0}\,\not{6} \\ -\ 2\,9\,5\,7 \\ \hline 7\,4\,9 \end{array}$$

69.
$$\begin{array}{r} {}^{7\ 10\ 4\ 12}\!\!\not{8}\,\not{0}\,\not{5}\,\not{2} \\ -\ 2\,7\,0\,9 \\ \hline 5\,3\,4\,3 \end{array}$$

71.
$$\begin{array}{r} {}^{9}\\{}^{6\ 10\ 6\ 10\ 12}\!\!\not{7}\,\not{0}\,.\,\not{7}\,\not{0}\,\not{2} \\ -\ \ \ 4,2\,3\,9 \\ \hline 6\,6,4\,6\,3 \end{array}$$

73.
$$\begin{array}{r} {}^{7\ 10\ 10\ 10}\!\!\not{8}\,\not{0},\not{0}\,\not{0}\,9 \\ -\ 6\,3,4\,1\,9 \\ \hline 1\,6,5\,9\,0 \end{array}$$

75.
$$\begin{array}{r} {}^{9}\\{}^{7\ 10\ 10\ 4\ 13}\!\!\not{8}\,\not{0},\not{0}\,\not{5}\,\not{3} \\ -\ 2\,7,6\,4\,9 \\ \hline 5\,2,4\,0\,4 \end{array}$$

77.
$$\begin{array}{r} {}^{9}\\{}^{7\ 10\ 7\ 10\ 10}\!\!\not{8}\,\not{0},\not{8}\,\not{0}\,\not{0} \\ -\ 4\,2,0\,2\,3 \\ \hline 3\,8,7\,7\,7 \end{array}$$

79.
$$\begin{array}{r} {}^{13\ 9}\\{}^{7\ \not{3}\ 10\ 10}\!\!\not{8}\,\not{4}\,\not{0}\,\not{0} \\ -\ 3\,7\,6\,2 \\ \hline 4\,6\,3\,8 \end{array}$$

81.
$$\begin{array}{r} {}^{9}\\{}^{5\ 10\ 10}\!\!\not{6}\,\not{0}\,\not{0}\,4 \\ -\ 2\,3\,9\,2 \\ \hline 3\,6\,1\,2 \end{array}$$

83.
$$\begin{array}{r} {}^{6\ 10\ 4\ 10}\!\!\not{7}\,\not{0}\,\not{5}\,\not{0} \\ -\ 4\,1\,3\,7 \\ \hline 2\,9\,1\,3 \end{array}$$

85.
$$\begin{array}{r} {}^{11}\\{}^{3\ \not{1}\ 10}4\,\not{2}\,\not{0}\,7 \\ -\ 1\,6\,2\,4 \\ \hline 2\,5\,8\,3 \end{array}$$

87.
$$\begin{array}{r} {}^{9\ 9}\\{}^{7\ 10\ 10\ 13}\!\!\not{8}\,\not{0}\,\not{0}\,\not{3} \\ -\ 2\,7\,3\,5 \\ \hline 5\,2\,6\,8 \end{array}$$

89.
$$\begin{array}{r} {}^{9\ 9\ 9}\\{}^{7\ 10\ 10\ 10\ 14}\!\!\not{8}\,\not{0},\not{0}\,\not{0}\,\not{4} \\ -\ \ 8,2\,3\,7 \\ \hline 7\,1,7\,6\,7 \end{array}$$

91.
$$\begin{array}{r} {}^{6\ \ 9\ 12\ 11}1\,\not{7},\not{0}\,\not{3}\,\not{1} \\ -\ \ 5,7\,9\,2 \\ \hline 1\,1,2\,3\,9 \end{array}$$

93.
$$\begin{array}{r} {}^{7\ 17}2\,9,\not{8}\,\not{7}\,4 \\ -\ 2\,1,3\,9\,2 \\ \hline 8,4\,8\,2 \end{array}$$

95.
$$\begin{array}{r} {}^{6\ 9\ 9\ 9\ 14}\!\!\not{7}\,\not{0},\not{0}\,\not{0}\,\not{4} \\ 6\,9,3\,7\,9 \\ \hline 6\,2\,5 \end{array}$$

97.
$$\begin{array}{r} {}^{7\ 15\ \ 16\ 9\ 11} \\ 8\,6\,,\,7\,0\,1 \\ -\ 9\,,\,9\,7\,6 \\ \hline 7\,6\,,\,7\,2\,5 \end{array}$$

99. **Strategy**   To find the amount that completes the statement, subtract the addend (67) from sum (90).

     Solution
$$\begin{array}{r} 90 \\ -67 \\ \hline 23 \end{array}$$
Therefore 23 completes the statement, $67 + 23 = 90$.

101. **Strategy**   To find the amount that completes the statement, subtract the addend (253) from sum (4901).

     Solution
$$\begin{array}{r} 4901 \\ -253 \\ \hline 4648 \end{array}$$
Therefore 4648 completes the statement, $253 + 4648 = 4901$.

103.
$$\begin{array}{rcr} 90,765 & \approx & 90,000 \\ -\ 60,928 & \approx & -\ 60,000 \\ \hline \text{Cal.}: 29,837 & & \text{Est.}: 30,000 \end{array}$$

105.
$$\begin{array}{rcr} 96,430 & \approx & 100,000 \\ -\ 59,762 & \approx & -\ 60,000 \\ \hline \text{Cal.}: 36,668 & & \text{Est.}: 40,000 \end{array}$$

107.
$$\begin{array}{rcr} 300,712 & \approx & 300,000 \\ -\ 198,714 & \approx & -\ 200,000 \\ \hline \text{Cal.}: 101,998 & & \text{Est.}: 100,000 \end{array}$$

Objective C Application Problems

109. **Strategy**   To find the difference in the number of patents, subtract the number that Canon received (1934) from the number that Canon received (2682).

     Solution
$$\begin{array}{r} 2682 \\ -1934 \\ \hline 748 \end{array}$$
IBM was granted 748 more patents than Canon.

111. **Strategy**   To find the amount that remains to be paid, subtract the down payment ($950) from the cost ($11,225).

     Solution
$$\begin{array}{r} \$11,225 \\ -\ \ \ 950 \\ \hline \$10,275 \end{array}$$
The amount that remains to be paid is $4775.

113. **Strategy**   To find the difference in maximum heights between the two geysers, subtract the height of the Valentine (75 feet) from the height of the Great Fountain (90 feet).

     Solution
$$\begin{array}{r} 90 \\ -75 \\ \hline 15 \end{array}$$
The Great Fountain geyser erupts 15 feet more than the Valentine geyser.

115a. **Strategy**   To find how much more the Rolling Stones grossed, subtract the gross from the Grateful Dead ($285,000,000) from the gross of the Rolling Stones ($751,000,000).

     Solution
$$\begin{array}{r} \$751,000,000 \\ -\ 285,000,000 \\ \hline \$466,000,000 \end{array}$$
The Rolling Stones grossed $466,000,000 more than the Grateful Dead.

115b. **Strategy**   To find how much more the Neil Diamond grossed, subtract the gross from the Jimmy Buffet ($167,000,000) from the gross of the Neil Diamond ($183,000,000).

     Solution
$$\begin{array}{r} \$183,000,000 \\ -167,000,000 \\ \hline \$16,000,000 \end{array}$$
The Neil Diamond grossed $16,000,000 more than the Jimmy Buffet.

117a. Strategy To find which 2 year period had the smallest increase, find the difference in each of the 2 year periods and determine which is the smallest difference.

Solution For 2010 − 2012
$$\begin{array}{r} 146,000 \\ -129,000 \\ \hline 17,000 \end{array}$$

For 2012 − 2014
$$\begin{array}{r} 166,000 \\ -146,000 \\ \hline 20,000 \end{array}$$

For 2014 − 2016
$$\begin{array}{r} 187,000 \\ -166,000 \\ \hline 21,000 \end{array}$$

For 2016 − 2018
$$\begin{array}{r} 208,000 \\ -187,000 \\ \hline 21,000 \end{array}$$

For 2018 − 2020
$$\begin{array}{r} 235,000 \\ -208,000 \\ \hline 27,000 \end{array}$$

The smallest 2 year increase was 17,000 between 2010 − 2012.

117b. Strategy To find which 2 year period had the greatest increase, find the difference in each of the 2 year periods and determine which is the greatest difference.

Solution Using the calculations from 117a, the greatest 2 year increase was 27,000 between 2018 − 2020.

Applying the Concepts

119. Answers will vary. For example:
Pat has earned 15 college credits, and Leslie has earned 8 college credits. How many more college credits has Pat earned? 7 college credits

SECTION 1.4

Objective A Exercises

1. 6×2 or $6 \cdot 2$

3. 4×7 or $4 \cdot 7$

5. 12

7. 35

9. 25

11. 0

13. 72

15.
$$\begin{array}{r} \overset{1}{66} \\ \times\ 3 \\ \hline 198 \end{array}$$

17.
$$\begin{array}{r} \overset{3}{67} \\ \times\ 5 \\ \hline 335 \end{array}$$

19.
$$\begin{array}{r} \overset{1}{623} \\ \times\ 4 \\ \hline 2492 \end{array}$$

21.
$$\begin{array}{r} \overset{6}{607} \\ \times\ 9 \\ \hline 5463 \end{array}$$

23.
$$\begin{array}{r} 600 \\ \times\ 7 \\ \hline 4200 \end{array}$$

25.
$$\begin{array}{r} \overset{2}{703} \\ \times\ 9 \\ \hline 6327 \end{array}$$

27.
$$\begin{array}{r} 632 \\ \times\ 3 \\ \hline 1896 \end{array}$$

29.
$$\begin{array}{r} \overset{2\,1}{632} \\ \times\ 8 \\ \hline 5056 \end{array}$$

31.
$$\begin{array}{r} \overset{1\,3}{337} \\ \times\ 5 \\ \hline 1685 \end{array}$$

33.
$$\begin{array}{r} \overset{4\ \ 6}{6709} \\ \times\ 7 \\ \hline 46,963 \end{array}$$

35.
$$\begin{array}{r} \overset{3\,4\,5}{8568} \\ \times\ 7 \\ \hline 59,976 \end{array}$$

37.
$$\begin{array}{r} \overset{3\,3}{4780} \\ \times\ 4 \\ \hline 19,120 \end{array}$$

39.
$$\begin{array}{r} \overset{1\,1\,1}{9895} \\ \times\ 2 \\ \hline 19,790 \end{array}$$

41. $6 \times 2 \times 9\ =\ 108$

43.
$$\begin{array}{r} 458 \\ \times\ 8 \\ \hline 3664 \end{array}$$

45.
$$\begin{array}{r} 5009 \\ \times\ 4 \\ \hline 20,036 \end{array}$$

47.
$$\begin{array}{r} 8957 \\ \times\ 8 \\ \hline 71,656 \end{array}$$

49.
$$\begin{array}{r} 18 \\ \times\ 24 \\ \hline 72 \\ 36\ \ \\ \hline 432 \end{array}$$

51.
$$\begin{array}{r} 27 \\ \times\ 72 \\ \hline 54 \\ 189\ \ \\ \hline 1944 \end{array}$$

53.
$$\begin{array}{r} 581 \\ \times\ 72 \\ \hline 1162 \\ 4067\ \ \\ \hline 41,832 \end{array}$$

55.
$$\begin{array}{r} 727 \\ \times\ 60 \\ \hline 43,620 \end{array}$$

57.
$$\begin{array}{r} 9577 \\ \times\ 35 \\ \hline 47885 \\ 28731\ \ \\ \hline 335,195 \end{array}$$

59.
$$\begin{array}{r} 8875 \\ \times\ 67 \\ \hline 62125 \\ 53250\ \ \\ \hline 594,625 \end{array}$$

61.   6702
     × 48
    53616
   26808
  321,696

63.   6003
     × 57
   42021
  30015
  342,171

65.   607
    × 460
   36420
    2428
  279,220

67.   700
    × 274
    2800
    4900
    1400
  191,800

69.   688
    × 674
    2752
    4816
    4128
  463,712

71.   423
    × 427
    2961
    846
    1692
  180,621

73.   684
    × 700
  478,800

75.   758
    × 209
    6822
   15160
  158,422

77.   5207
    × 902
   10414
  468630
  4,696,714

79.   6327
    × 876
   37962
   44289
  50616
  5,542,452

81.   7349
    × 27
   51443
   14698
  198,423

83.   6 × 73 = 438

       438
      × 43
    1314
   1752
  18,834

85.   842
    × 309
    7578
   2526
  260,178

87.   34,985
   × 9007
  244895
  314865
  315,109,895

89.   4732 ≈   5000
     × 93 ≈   × 90
Cal. : 440,076   Est. : 450,000

91.   8941 ≈   9000
     × 726 ≈   × 700
Cal. : 6,491,166   Est. : 6,300,000

93.   6379 ≈   6000
     × 2936 ≈   × 3000
Cal. : 18,728,744   Est. : 18,000,000

95.   62,504 ≈   60,000
     × 923 ≈   × 900
Cal. : 57,691,192   Est. : 54,000,000

Objective C Application Problems

97.   Strategy   To find the number of gallons of fuel used on a 6-hour flight, multiply the number of gallons used in 1 hour (865) by 6.

     Solution   865
             × 6
          5190

          The plane used 5190 gallons of fuel in a 6-hour flight.

99.   Strategy   To find the area, multiply the length of one side (16 miles) by itself (16 miles).

     Solution   16
            × 16
          96
          16
         256

          The area is 256 square miles.

101.   Company A   Company B
     43 × 2 = 86   43 × 3 = 129
     15 × 6 = 90   15 × 4 = 60
     20 × 12 = 240   20 × 11 = 220
     1 × 998 = 998   1 × 1089 = 1089
     Total = $1414   Total = $1498

     Company A offers the lower total price.

103. Total cost = no. of   ×   no. of hours   ×   wages
              electricians   each works   per hour
              $= 3 \times 50 \times 34$
              $= \$5100$

105.
     Electrician= $1 \times 30 \times \$34 =$ $\$1020$
      Plumber= $1 \times 33 \times \$30 =$ $\$990$
         Clerk= $1 \times 3 \times \$16 =$ $\$48$
  Bookkeeper= $1 \times 4 \times \$20 =$ $\$80$
                          Total = $\$2138$

Applying the Concepts

107.   There is one accidental death every 5 minutes.
       There are 60 minutes in an hour.
       $60 \div 5 = 12$
       There are 12 accidental deaths in an hour.

       There are 12 deaths in an hour.
       There are 24 hours per day.
       $12 \times 24 = 288$
       There are 288 accidental deaths in a day.

       There are 288 deaths in a day.
       There are 365 days in a year.
       $288 \times 365 = 105,120$
       There are 105,120 accidental deaths in a year.

109. $3 \times 37,037 = 111,111$
       By the Multiplication Property of One, a number
       times 1 equals the number. Because the product of
       3 and 37,037 is 111,111, the product of 111,111
       and a number between 1 and 9 will be a six-digit
       number in which all digits equal the number
       between 1 and 9.

SECTION 1.5

Objective A Exercises

1. 2

3. 6

5. 7

7.
```
    16
 6)96
   -6
    36
   -36
     0
```

9.
```
   210
 4)840
   -8
    04
    -4
    00
    -0
     0
```

11.
```
    44
 7)308
   -28
    28
   -28
     0
```

13.
```
    703
 9)6327
   -63
     02
     -0
     27
    -27
      0
```

15.
```
    910
 8)7280
   -72
     08
     -8
     00
     -0
      0
```

17.
```
    21,560
 3)64,680
   -6
    04
    -3
     16
    -15
     18
    -18
     00
     -0
      0
```

19.
```
    3,580
 6)21,480
   -18
     34
    -30
     48
    -48
     00
     -0
      0
```

21.
```
    482
 3)1446
   -12
     24
    -24
     06
     -6
      0
```

23.
```
  1075
7)7525
 -7
  05
  -0
  52
 -49
  35
 -35
   0
```

25.
```
   52
7)364
 -35
  14
 -14
   0
```

27.
```
     5
34)170
  -170
     0
```

29.
```
  2 r1
4)9
 -8
  1
```

31.
```
  5 r2
5)27
-25
  2
```

33.
```
  13 r1
3)40
 -3
 10
 -9
  1
```

35.
```
  10 r3
8)83
 -8
 03
 -0
  3
```

37.
```
  90 r2
7)632
-63
 02
 -0
  2
```

39.
```
  230 r1
4)921
 -8
 12
-12
 01
 -0
  1
```

41.
```
  204 r3
8)1635
 -16
  03
  -0
  35
 -32
   3
```

43.
```
  1347 r3
7)9432
 -7
 24
-21
 33
-28
 52
-49
  3
```

45.
```
  1720 r2
3)5162
 -3
 21
-21
 06
 -6
 02
 -0
  2
```

47.
```
   409 r2
8)3274
 -32
  07
  -0
  74
 -72
   2
```

49.
```
   6,214 r2
7)43,500
 -42
  1 5
 -14
  10
  -7
  30
 -28
   2
```

51.
```
   8,708 r2
5)43,542
 -40
  3 5
 -3 5
  0 4
  -0
  42
 -40
   2
```

53.
```
  1080 r2
8)8642
 -8
 06
 -0
 64
-64
 02
-0
 2
```

55.
```
   4,210 r6
9)37,896
 -36
  1 8
 -18
  09
  -9
  06
  -0
   6
```
Round to 4200.

57.
```
   19,586 r1
4)78,345
 -4
 38
-36
 23
-20
 34
-32
 25
-24
  1
```
Round to 19,600.

Objective C Exercises

59.
```
   1 r38
44)82
 -44
  38
```

61.
```
   1 r26
67)93
 -67
  26
```

63.
```
   21 r21
32)693
 -64
  53
 -32
  21
```

65.
```
   30 r22
25)772
 -75
  22
  -0
  22
```

67.
```
    5 r40
92)500
 -460
   40
```

69.
```
      9  r17
50)467
  -450
    17
```

71.
```
    200 r21
44)8821
  -88
   02
   -0
   21
   - 0
   21
```

73.
```
    303 r1
32)9697
  -96
   09
   -0
   97
  -96
    1
```

75.
```
     67 r13
92)6177
  -552
   657
  -644
    13
```

77.
```
    176 r13
27)4765
  -27
   206
  -189
   175
  -162
    13
```

79.
```
   1,086 r7
77)83,629
  -77
   66
   -0
   662
  -616
   469
  -462
     7
```

81.
```
    403
78)31,434
  -312
    23
    -0
   234
  -234
     0
```

83.
```
     12 r456
504)6504
   -504
    1464
   -1008
     456
```

85.
```
      4 r160
546)2344
  -2184
    160
```

87.
```
    160 r27
53)8507
  -53
   320
  -318
    27
```

89.
```
   1,669 r14
46)76,788
  -46
   307
  -276
   318
  -276
   428
  -414
    14
```

91.
```
   7,948  r17
43)341,781
  -301
   407
  -387
   208
  -172
   361
  -344
    17
```

Round to 7950.

93. Cal.: $53\overline{)117{,}925}$ → $2{,}225$ Est.: $50\overline{)100{,}000}$ → $2{,}000$

95. Cal.: $67\overline{)738{,}072}$ → $11{,}016$ Est.: $70\overline{)700{,}000}$ → $10{,}000$

97. Cal.: $34\overline{)906{,}304}$ → $26{,}656$ Est.: $30\overline{)900{,}000}$ → $30{,}000$

99. Cal.: $642\overline{)323{,}568}$ → 504 Est.: $600\overline{)300{,}000}$ → 500

101. Cal.: $614\overline{)332{,}174}$ → 541 Est.: $600\overline{)300{,}000}$ → 500

103. Cal.: $374\overline{)7{,}712{,}254}$ → $20{,}621$ Est.: $400\overline{)8{,}000{,}000}$ → $20{,}000$

Objective D Application Problems

105. Strategy   To find the monthly expense for housing, divide annual housing expense ($11,713) by the number of months (12).

Solution

$$\begin{array}{r} 976 \\ 12\overline{)11,713} \\ -108 \\ \hline 91 \\ -84 \\ \hline 73 \\ -72 \\ \hline 1 \end{array}$$

The average monthly expense for housing is $976.

107. Strategy   To find the average monthly claim for theft, divide the annual claim for theft ($300,000) by the number of months (12).

Solution

$$\begin{array}{r} 25,000 \\ 12\overline{)300,000} \\ -24 \\ \hline 60 \\ -60 \\ \hline 00 \\ -0 \\ \hline 00 \\ -0 \\ \hline 00 \\ -0 \\ \hline 0 \end{array}$$

The average monthly claim for theft is $25,000.

109. Strategy   To find the average hours worked by employees in Britain, divide annual hours worked (1731) by the number of weeks (50).

Solution

$$\begin{array}{r} 35 \\ 50\overline{)1713} \\ -150 \\ \hline 213 \\ -200 \\ \hline 13 \end{array}$$

Since 13 is less than half of 50, the average number of hours worked by employees in Britain is 35 hours.

111. Strategy   To find the approximate number of pennies per person divide the number of pennies in circulation (114,000,000,000) by the number of people (280,000,000).

Solution

$$\begin{array}{r} 407 \\ 280,000,000\overline{)114\,000\,000\,000} \\ -112\,000\,000\,0 \\ \hline 2\,000\,000\,00 \\ -\qquad 0 \\ \hline 2\,000\,000\,000 \\ -1\,960\,000\,000 \\ \hline 40\,000\,000 \end{array}$$

Because 40,000,000 is less than half of 280,000,000, round to 407. Approximately 407 pennies are in circulation for each person.

Applying the Concepts

113. The smallest four-digit palindromic number that is divisible by 8 is 2112.
Consider the list of palindromic numbers:
1001, 1111, 1221, 1331, ... 1991.
2002, 2112, 2222, 2332, ... 2992.
The first ten numbers have an odd ending digit, which is not divisible by 4.

SECTION 1.6

Objective A Exercises

1. 2^3

3. $6^3 \cdot 7^4$

5. $2^3 \cdot 3^3$

7. $5 \cdot 7^5$

9. $3^3 \cdot 6^4$

11. $3^3 \cdot 5 \cdot 9^3$

13. $2 \cdot 2 \cdot 2 = 8$

15. $2 \cdot 2 \cdot 2 \cdot 2 \cdot 5 \cdot 5 = 16 \cdot 25 = 400$

17. $3 \cdot 3 \cdot 10 \cdot 10 = 9 \cdot 100 = 900$

19. $6 \cdot 6 \cdot 3 \cdot 3 \cdot 3 = 36 \cdot 27 = 972$

21. $5 \cdot 2 \cdot 2 \cdot 2 \cdot 3 = 5 \cdot 8 \cdot 3 = 120$

23. $2 \cdot 2 \cdot 3 \cdot 3 \cdot 10 = 4 \cdot 9 \cdot 10 = 360$

25. $0 \cdot 0 \cdot 4 \cdot 4 \cdot 4 = 0 \cdot 64 = 0$

27. $3 \cdot 3 \cdot 10 \cdot 10 \cdot 10 \cdot 10 = 9 \cdot 10,000 = 90,000$

29. $2 \cdot 2 \cdot 3 \cdot 3 \cdot 3 \cdot 5 = 4 \cdot 27 \cdot 5 = 540$

31. $2 \cdot 3 \cdot 3 \cdot 3 \cdot 3 \cdot 5 \cdot 5 = 2 \cdot 81 \cdot 25 = 4050$

33. $5 \cdot 5 \cdot 3 \cdot 3 \cdot 7 \cdot 7 = 25 \cdot 9 \cdot 49 = 11,025$

35. $3 \cdot 3 \cdot 3 \cdot 3 \cdot 2 \cdot 2 \cdot 2 \cdot 2 \cdot 2 \cdot 2 \cdot 5 = 81 \cdot 64 \cdot 5$
$= 25,920$

37. $4 \cdot 4 \cdot 3 \cdot 3 \cdot 3 \cdot 10 \cdot 10 \cdot 10 \cdot 10 = 16 \cdot 27 \cdot 10,000$
$= 4,320,000$

Objective B Exercises

39. $6 - 3 + 3 = 3 + 2 = 5$

41. $8 \div 4 + 8 = 2 + 8 = 10$

43. $5 \cdot 9 + 2 = 45 + 2 = 47$

45. $5^2 - 17 = 25 - 17 = 8$

47. $3 + (4 + 2) \div 3 = 3 + 6 \div 3$
$= 3 + 2 = 5$

49. $8 - 2^2 + 4 = 8 - 4 + 4$
$= 4 + 4 = 8$

51. $12 \cdot (1 + 5) \div 12 = 12 \cdot 6 \div 12$
$= 72 \div 12 = 6$

53. $5 \cdot 3^2 + 8 = 5 \cdot 9 + 8$
$= 45 + 8 = 53$

55. $12 + 4 \cdot 2^3 = 12 + 4 \cdot 8$
$= 12 + 32 = 44$

57. $7 + (9 - 5) \cdot 3 = 7 + 4 \cdot 3$
$= 7 + 12 = 19$

59. $3^3 + 5 \cdot (8 - 6)^3 = 3^3 + 5 \cdot 2^3$
$= 27 + 5 \cdot 8$
$= 27 + 40 = 67$

61. $4 \cdot 6 + 3^2 \cdot 4^2 = 4 \cdot 6 + 9 \cdot 16$
$= 24 + 9 \cdot 16$
$= 24 + 144 = 168$

63. $12 + 3 \cdot 5 = 12 + 15 = 27$

65. $5 \cdot (8 - 4) - 6 = 5 \cdot 4 - 6 = 20 - 6 = 14$

67. $12 - (12 - 4) \div 4 = 12 - 8 \div 4 = 12 - 2 = 10$

69. $10 + 1 - 5 \cdot 2 \div 5 = 10 + 1 - 10 \div 5$
$= 10 + 1 - 2$
$= 11 - 2 = 9$

71. $(7 - 3)^2 \div 2 - 4 + 8 = 4^2 \div 2 - 4 + 8$
$= 16 \div 2 - 4 + 8$
$= 8 - 4 + 8$
$= 4 + 8 = 12$

73. $12 \div 3 \cdot 2^2 + (7 - 3)^2 = 12 \div 3 \cdot 2^2 + 4^2$
$= 12 \div 3 \cdot 4 + 16$
$= 4 \cdot 4 + 16$
$= 16 + 16 = 32$

75. $18 - 2 \cdot 3 + (4 - 1)^3 = 18 - 2 \cdot 3 + 3^3$
$= 18 - 2 \cdot 3 + 27$
$= 18 - 6 + 27$
$= 12 + 27 = 39$

Applying the Concepts

77. $2^{10} = 1024$

79. Yes; 30 is a product of factors of 6 and 10 $(2 \cdot 3 \cdot 5)$.

SECTION 1.7

Objective A Exercises

1. $4 \div 1 = 4$
$4 \div 2 = 2$
Factors are 1, 2, and 4.

3. $10 \div 1 = 10$
$10 \div 2 = 5$
$10 \div 5 = 2$
Factors are 1, 2, 5, and 10.

5. $7 \div 1 = 7$
$7 \div 7 = 1$
Factors are 1 and 7.

7. $9 \div 1 = 9$
$9 \div 3 = 3$
Factors are 1, 3, and 9.

9. $13 \div 1 = 13$
$13 \div 13 = 1$
Factors are 1 and 13.

11. $18 \div 1 = 18$
$18 \div 2 = 9$
$18 \div 3 = 6$
$18 \div 6 = 3$
Factors are 1, 2, 3, 6, 9, and 18.

13. $56 \div 1 = 56$
$56 \div 2 = 28$
$56 \div 4 = 14$
$56 \div 7 = 8$
$56 \div 8 = 7$
Factors are 1, 2, 4, 7, 8, 14, 28, and 56.

15. $45 \div 1 = 45$
$45 \div 3 = 15$
$45 \div 5 = 9$
Factors are 1, 3, 5, 9, 15, and 45.

17. $29 \div 1 = 29$
$29 \div 29 = 1$
Factors are 1 and 29.

19. $22 \div 1 = 22$
$22 \div 2 = 11$
$22 \div 11 = 2$
Factors are 1, 2, 11, and 22.

21. $52 \div 1 = 52$
$52 \div 2 = 26$
$52 \div 4 = 13$
$52 \div 13 = 4$
Factors are 1, 2, 4, 13, 26, and 52.

23. $82 \div 1 = 82$
$82 \div 2 = 41$
$82 \div 41 = 2$
Factors are 1, 2, 41, and 82.

25. $57 \div 1 = 57$
$57 \div 3 = 19$
$57 \div 19 = 3$
Factors are 1, 3, 19, and 57.

27. $48 \div 1 = 48$
$48 \div 2 = 24$
$48 \div 3 = 16$
$48 \div 4 = 12$
$48 \div 6 = 8$
$48 \div 8 = 6$
Factors are 1, 2, 3, 4, 6, 8, 12, 16, 24, and 48.

29. $95 \div 1 = 95$
$95 \div 5 = 19$
$95 \div 19 = 5$
Factors are 1, 5, 19, and 95.

31. $54 \div 1 = 54$
$54 \div 2 = 27$
$54 \div 3 = 18$
$54 \div 6 = 9$
$54 \div 9 = 6$
Factors are 1, 2, 3, 6, 9, 18, 27, and 54.

33. $66 \div 1 = 66$
$66 \div 2 = 33$
$66 \div 3 = 22$
$66 \div 6 = 11$
$66 \div 11 = 6$
Factors are 1, 2, 3, 6, 11, 22, 33, and 66.

35. $80 \div 1 = 80$
$80 \div 2 = 40$
$80 \div 4 = 20$
$80 \div 5 = 16$
$80 \div 8 = 10$
$80 \div 10 = 8$
Factors are 1, 2, 4, 5, 8, 10, 16, 20, 40, and 80.

37. $96 \div 1 = 96$
$96 \div 2 = 48$
$96 \div 3 = 32$
$96 \div 4 = 24$
$96 \div 6 = 16$
$96 \div 8 = 12$
$96 \div 12 = 8$
Factors are 1, 2, 3, 4, 6, 8, 12, 16, 24, 32, 48, and 96.

39. $90 \div 1 = 90$
$90 \div 2 = 45$
$90 \div 3 = 30$
$90 \div 5 = 18$
$90 \div 6 = 15$
$90 \div 9 = 10$
$90 \div 10 = 9$
Factors are 1, 2, 3, 5, 6, 9, 10, 15, 18, 30, 45, and 90.

Objective B Exercises

41. $2\overline{)6}$ with quotient 3
$6 = 2 \cdot 3$

43. 17 is prime.

45. $2\overline{)6}$ with quotient 3
$2\overline{)12}$
$2\overline{)24}$
$24 = 2 \cdot 2 \cdot 2 \cdot 3$

47. $3\overline{)9}$ with quotient 3
$3\overline{)27}$
$27 = 3 \cdot 3 \cdot 3$

49. $3\overline{)9}$ with quotient 3
$2\overline{)18}$
$2\overline{)36}$
$36 = 2 \cdot 2 \cdot 3 \cdot 3$

51. 19 is prime.

53. $3\overline{)15}$ with quotient 5
$3\overline{)45}$
$2\overline{)90}$
$90 = 2 \cdot 3 \cdot 3 \cdot 5$

55. $5\overline{)115}$ with quotient 23
$115 = 5 \cdot 23$

57. $3\overline{)9}$ with quotient 3
$2\overline{)18}$
$18 = 2 \cdot 3 \cdot 3$

59. $2\overline{)14}$ with quotient 7
$2\overline{)28}$
$28 = 2 \cdot 2 \cdot 7$

61. 31 is prime.

63. $2\overline{)62}$ with quotient 31
$62 = 2 \cdot 31$

65. $2\overline{)22}$ with quotient 11
$22 = 2 \cdot 11$

67. 101 is prime.

69. $3\overline{)33}$ with quotient 11
$2\overline{)66}$
$66 = 2 \cdot 3 \cdot 11$

71.
$$2)\overline{74} \quad \overset{37}{}$$
$$74 = 2 \cdot 37$$

73. 67 is prime.

75.
$$5)\overline{55} \quad \overset{11}{}$$
$$55 = 5 \cdot 11$$

77.
$$3)\overline{15} \quad \overset{5}{}$$
$$2)\overline{30}$$
$$2)\overline{60}$$
$$2)\overline{120}$$
$$120 = 2 \cdot 2 \cdot 2 \cdot 3 \cdot 5$$

79.
$$2)\overline{10} \quad \overset{5}{}$$
$$2)\overline{20}$$
$$2)\overline{40}$$
$$2)\overline{80}$$
$$2)\overline{160}$$
$$160 = 2 \cdot 2 \cdot 2 \cdot 2 \cdot 2 \cdot 5$$

81.
$$3)\overline{9} \quad \overset{3}{}$$
$$3)\overline{27}$$
$$2)\overline{54}$$
$$2)\overline{108}$$
$$2)\overline{216}$$
$$216 = 2 \cdot 2 \cdot 2 \cdot 3 \cdot 3 \cdot 3$$

83.
$$5)\overline{25} \quad \overset{5}{}$$
$$5)\overline{125}$$
$$5)\overline{625}$$
$$625 = 5 \cdot 5 \cdot 5 \cdot 5$$

Applying the Concepts

85. 3, 5; 5, 7; 11, 13; 41, 43; 71, 73 and (17, 19) are all the twin primes less than 100.

87. 2 is the *only* even prime number because all other even numbers have 2 as a factor and thus cannot be prime.

CHAPTER REVIEW

1. $3 \cdot 2^3 \cdot 5^2 = 3 \cdot 8 \cdot 25$
$$= 24 \cdot 25 = 600$$

2. $10,000 + 300 + 20 + 7$

3. $18 \div 1 = 18$
$18 \div 2 = 9$
$18 \div 3 = 6$
$18 \div 6 = 3$
Factors are 1, 2, 3, 6, 9, and 18.

4.
$$\overset{1\;1\;1}{5894}$$
$$6301$$
$$+\;298$$
$$\overline{12,493}$$

5.
$$\overset{8\;\overset{11}{1}\;16}{49\,2\,6}$$
$$-\;3\,1\,7\,7$$
$$\overline{1\,7\,4\,9}$$

6.
$$7)\overline{14,945} \quad 2,135$$
$$\underline{-14}$$
$$09$$
$$\underline{-7}$$
$$24$$
$$\underline{-21}$$
$$35$$
$$\underline{-35}$$
$$0$$

7. $>$

8. $5 \cdot 5 \cdot 7 \cdot 7 \cdot 7 \cdot 7 \cdot 7 = 5^2 \cdot 7^5$

9.
$$\overset{6}{2019}$$
$$\times\;\;\;307$$
$$\overline{14133}$$
$$60570$$
$$\overline{619,833}$$

10.
$$\overset{0\;\overset{10}{10}\;11\;2\;14}{10,1\,3\,4}$$
$$-4,7\,2\,5$$
$$\overline{5,4\,0\,9}$$

11.
$$\overset{1\;1}{298}$$
$$461$$
$$+\;322$$
$$\overline{1081}$$

12. $2^3 - 3 \cdot 2 = 8 - 6 = 2$

13. 45,700

14. Two hundred seventy-six thousand fifty-seven

15.
$$84)\overline{109,763} \quad 1306\;r59$$
$$\underline{-84}$$
$$257$$
$$\underline{-252}$$
$$56$$
$$\underline{-0}$$
$$563$$
$$\underline{-504}$$
$$59$$

16. 2,011,044

17.
$$\begin{array}{r} 488\,r2 \\ 8\overline{)3906} \\ \underline{-32} \\ 70 \\ \underline{-64} \\ 66 \\ \underline{-64} \\ 2 \end{array}$$

18. $3^2 + 2^2 \cdot (5 - 3) = 3^2 + 2^2 \cdot (2)$
$ = 9 + 4 \cdot 2$
$ = 9 + 8 = 17$

19. $8 \cdot (6 - 2)^2 \div 4 = 8 \cdot 4^2 \div 4$
$ = 8 \cdot 16 \div 4$
$ = 128 \div 4 = 32$

20. $72 = 2 \cdot 2 \cdot 2 \cdot 3 \cdot 3$
$$\begin{array}{r} 3 \\ 3\overline{)9} \\ 2\overline{)18} \\ 2\overline{)36} \\ 2\overline{)72} \end{array}$$

21. 2133

22.
$$\begin{array}{r} {}^{3\,2} \\ 843 \\ \times\ 27 \\ \hline 5901 \\ \underline{1686} \\ 22{,}761 \end{array}$$

23. Strategy To find the total pay for last week's work:
• Multiply the overtime rate ($24) by the number of hours worked (12).
• Add the total earned as overtime to the assistant's salary ($480).

Solution
$$\begin{array}{r} \$24 \\ \times\ 12 \\ \hline 48 \\ \underline{24} \\ \$288 \end{array} \qquad \begin{array}{r} \$480 \\ +\ 288 \\ \hline \$768 \end{array}$$

The total pay for last week's work is $768.

24. Strategy To find the number of miles driven per gallon of gasoline, divide the total number of miles driven (351) by the number of gallons used (13).

Solution
$$\begin{array}{r} 27 \\ 13\overline{)351} \\ \underline{-26} \\ 91 \\ \underline{-91} \\ 0 \end{array}$$

The number of miles driven per gallon of gasoline is 27.

25. Strategy To find the monthly car payment:
• Subtract the down payment ($3000) from the cost of the car ($17,880) to find the balance.
• Divide the balance by the number of equal payments (48).

Solution
$$\begin{array}{r} \$17{,}880 \\ -\ 3{,}000 \\ \hline \$14{,}880 \end{array} \qquad \begin{array}{r} 310 \\ 48\overline{)14{,}880} \\ \underline{-144} \\ 48 \\ \underline{-48} \\ 00 \\ \underline{-0} \\ 0 \end{array}$$

Each monthly car payment is $310.

26. Strategy To find the total income from commissions, add the amounts received for each of the 4 weeks ($723, $544, $812, and $488).

Solution
$$\begin{array}{r} \$723 \\ 544 \\ 812 \\ +\ 488 \\ \hline \$2567 \end{array}$$

The total income from commissions is $2567.

27a. Strategy To find the total amount deposited, add the two deposits ($88 and $213).

Solution
$$\begin{array}{r} \$88 \\ +\ 213 \\ \hline \$301 \end{array}$$

The total amount deposited is $301.

27b. Strategy To find the new checking account balance, add the total amount deposited ($301) to the original balance ($516).

Solution
$$\begin{array}{r} \$301 \\ +\ 516 \\ \hline \$817 \end{array}$$

The new checking balance is $817.

28. Strategy To find the total of the car payments over a 12-month period, multiply the amount of each payment ($246) by the number of payments (12).

Solution
$$\begin{array}{r} \$246 \\ \times\ 12 \\ \hline 492 \\ \underline{246} \\ \$2952 \end{array}$$

The total of the car payments is $2952.

29. Strategy To find the salary difference, subtract the salary of a woman with an associate's degree ($21,057) from the salary of a woman with an advanced degree ($50,546).

Solution $50,546 (advanced)
−$21,057 (associate's)
$29,489

A woman with an advanced degree earns $29,489 more than a woman with an associate's degree.

30. Strategy To find the salary difference, subtract the salary of a woman with a bachelor's degree ($30,119) from the salary of a man with a bachelor's degree ($50,056).

Solution $50,056 (man)
−$30,119 (woman)
$19,937

A man with a bachelor's degree earns $19,937 more than a woman with a bachelor's degree.

31. Strategy To find how much more the man earns in five years:
• Subtract the salary of a woman with an associate's degree ($21,057) from the salary of a man with an associate's degree ($33,830).
• Multiply the difference by the number of years (5).

Solution $33,830 $12,773
−21,057   ×   5
$12,773 $63,865

The man earns $63,865 more.

CHAPTER TEST

1. $3^3 \cdot 4^2 = 27 \cdot 16 = 432$

2. Two hundred seven thousand sixty-eight

3.
$$\overset{0\ 17}{17,495}$$
$$-8,162$$
$$9,333$$

4. $20 \div 1 = 20$
$20 \div 2 = 10$
$20 \div 4 = 5$
$20 \div 5 = 4$
Factors are 1, 2, 4, 5, 10, and 20.

5.
$$9736$$
$$\times\ \ \ 704$$
$$38,944$$
$$681,520$$
$$6,854,144$$

6. $4^2 \cdot (4-2) \div 8 + 5 = 4^2 \cdot (2) \div 8 + 5$
$= 16 \cdot (2) \div 8 + 5$
$= 32 \div 8 + 5$
$= 4 + 5 = 9$

7. $900,000 + 6000 + 300 + 70 + 8$

8. $75,000$

9.
$$\begin{array}{r} 1121\,\text{r}27 \\ 97)\overline{108,764} \\ \underline{-97} \\ 117 \\ \underline{-97} \\ 206 \\ \underline{-194} \\ 124 \\ \underline{-97} \\ 27 \end{array}$$

10. $3 \cdot 3 \cdot 3 \cdot 7 \cdot 7 = 3^3 \cdot 7^2$

11.
$$\overset{2\ \ \ 21}{8,756}$$
$$9,094$$
$$+37,065$$
$$54,915$$

12. $2 \cdot 2 \cdot 3 \cdot 7 = 84$
$$\begin{array}{r} 7 \\ 3)\overline{21} \\ 2)\overline{42} \\ 2)\overline{84} \end{array}$$

13. $16 \div 4 \cdot 2 - (7-5)^2 = 16 \div 4 \cdot 2 - 2^2$
$= 16 \div 4 \cdot 2 - 4$
$= 4 \cdot 2 - 4$
$= 8 - 4 = 4$

14.
$$\overset{6\ 52}{90,763}$$
$$\times\ \ \ \ \ 8$$
$$726,104$$

15. $1,204,006$

16.
$$\begin{array}{r} 8710\,\text{r}2 \\ 7)\overline{60972} \\ \underline{-56} \\ 49 \\ \underline{-49} \\ 07 \\ \underline{-7} \\ 02 \\ \underline{-0} \\ 2 \end{array}$$

17. $>$

18.
$$
\begin{array}{r}
703 \\
8\overline{)5624} \\
-56 \\
\hline
02 \\
-0 \\
\hline
24 \\
-24 \\
\hline
0
\end{array}
$$

19.
$$
\begin{array}{r}
25{,}492 \\
+\ 71{,}306 \\
\hline
96{,}798
\end{array}
$$

20.
$$
\begin{array}{r}
{}^{1\ \ 18\ 17} \\
2\,9{,}7\,3\,6 \\
-\ 9{,}8\,1\,4 \\
\hline
19{,}9\,2\,2
\end{array}
$$

21. **Strategy**   To find the difference in total enrollment between 2000 and projected for 2005:
• Find the total enrollment for 2000 by adding the enrollment for K through grade 8 (39,1529,000) to the enrollment for grades 9 through 12 (15,250,000).
• Find the total projected enrollment for 2005 by adding the projected enrollment for K through grade 8 (39,437,000) to the enrollment for grades 9 through 12 (16,434,000).
• Subtract the total enrollment for 2000 from the total projected enrollment for 2005.

Solution
$$
\begin{array}{r}
39{,}152{,}000 \\
+\ 15{,}250{,}000 \\
\hline
54{,}402{,}000
\end{array}
\qquad
\begin{array}{r}
39{,}437{,}000 \\
+\ 16{,}434{,}000 \\
\hline
55{,}871{,}000
\end{array}
$$

$$
\begin{array}{r}
55{,}871{,}000 \\
-\ 54{,}402{,}000 \\
\hline
1{,}469{,}000
\end{array}
$$

The difference between the total enrollment in 2000 and the total projected enrollment in 2005 is 1,469,000 students.

22. **Strategy**   To find the average projected enrollment in each grade of 9 through 12, divide the projected enrollment in 2005 (16,434,000) by the number of grades (4).

Solution
$$
\begin{array}{r}
4{,}108{,}500 \\
4\overline{)16{,}434{,}000} \\
-16 \\
\hline
04 \\
-4 \\
\hline
03 \\
-0 \\
\hline
34 \\
-32 \\
\hline
20 \\
-20 \\
\hline
00 \\
-0 \\
\hline
00 \\
-0 \\
\hline
0
\end{array}
$$

The average enrollment for each of grades 9 through 12 in 2005 is 4,108,500 students.

23. **Strategy**   To find how many boxes were needed to pack the lemons:
• Find the total number of lemons harvested by adding the amounts harvested from the two groves (48,290 and 23,710 pounds).
• Divide the total number of pounds harvested by the number of pounds of lemons that can be packed in each box (24).

Solution
$$
\begin{array}{r}
48{,}290 \\
+\ 23{,}710 \\
\hline
72{,}000
\end{array}
$$

$$
\begin{array}{r}
3000 \\
24\overline{)72{,}000} \\
-72 \\
\hline
00 \\
-0 \\
\hline
00 \\
-0 \\
\hline
00 \\
-0 \\
\hline
0
\end{array}
$$

The number of boxes needed to pack the lemons is 3000.

24. Strategy To find the amount that the investor receives over the period, multiply the amount she receives each month ($237) by the number of months (12).

Solution

$$\begin{array}{r} \$237 \\ \times\ 12 \\ \hline 474 \\ 237 \\ \hline \$2844 \end{array}$$

The investor receives $2844 over the 12-month period.

25a. Strategy To find the total number of miles driven during the 3 days, add the amounts driven each day (425, 187, and 243 miles).

Solution

$$\begin{array}{r} 425 \\ 187 \\ +\ 243 \\ \hline 855 \end{array}$$

The total number of miles driven during the 3 days was 855.

25b. Strategy To find the odometer reading at the end of the 3 days, add the number of miles driven during the 3 days (855) to the odometer reading at the start of the vacation (47,626).

Solution

$$\begin{array}{r} 47{,}626 \\ +\ \ \ \ 855 \\ \hline 48{,}481 \end{array}$$

The odometer reading at the end of the 3 days is 48,481 miles.

Chapter 2: Fractions

PREP TEST and GO FIGURE

1. 20

2. 120

3. 9

4. 10

5. 7

6.
$$\begin{array}{r} 2\ \text{r3} \\ 30\overline{)63} \\ -60 \\ \hline 3 \end{array}$$

7. 1, 2, 3, 4, 6, 12

8. $8 \cdot 7 + 3 = 56 + 3 = 59$

9. 7

10. <

Go Figure

One lap is down to the end of the pool and back. If you swim one lap every four minutes and your friend swims one lap every five minutes, then a visual time reference might look like this

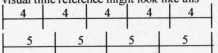

where each section represents one lap. As you can see, you meet up with your friend after you swim 5 laps and your friend swims 4 laps. Five four-minute laps are 20 minutes ($5 \times 4 = 20$). From the visual reference, you can see that you pass each other 4 times during the 20 minutes for each lap. Since one lap is down and back, you pass each other twice as often. So you pass each other 8 times ($4 \times 2 = 8$).

SECTION 2.1

Objective A Exercises

1.
$5 =$ [| 5]
$8 =$ [2·2·2 |]
LCM $= 2 \cdot 2 \cdot 2 \cdot 5 = 40$

3.
$3 =$ [| 3]
$8 =$ [2·2·2 |]
LCM $= 2 \cdot 2 \cdot 2 \cdot 3 = 24$

5.
$5 =$ [| | 5]
$6 =$ [2 | 3 |]
LCM $= 2 \cdot 3 \cdot 5 = 30$

7.
$4 =$ [2·2 |]
$6 =$ [2 | 3]
LCM $= 2 \cdot 2 \cdot 3 = 12$

9.
$8 =$ [2·2·2 |]
$12 =$ [2·2 | 3]
LCM $= 2 \cdot 2 \cdot 2 \cdot 3 = 24$

11.
$5 =$ [| | 5]
$12 =$ [2·2 | 3 |]
LCM $= 2 \cdot 2 \cdot 3 \cdot 5 = 60$

13.
$8 =$ [2·2·2 |]
$14 =$ [2 | 7]
LCM $= 2 \cdot 2 \cdot 2 \cdot 7 = 56$

15.
$3 =$ [3]
$9 =$ [3·3]
LCM $= 3 \cdot 3 = 9$

17.
$8 =$ [2·2·2]
$32 =$ [2·2·2·2·2]
LCM $= 2 \cdot 2 \cdot 2 \cdot 2 \cdot 2 = 32$

19.
$9 =$ [| 3·3]
$36 =$ [2·2 | 3·3]
LCM $= 2 \cdot 2 \cdot 3 \cdot 3 = 36$

21.
$44 =$ [2·2 | | | 11]
$60 =$ [2·2 | 3 | 5 |]
LCM $= 2 \cdot 3 \cdot 5 \cdot 11 = 660$

23.
$102 =$ [2 | 3 | 17 |]
$184 =$ [2·2·2 | | | 23]
LCM $= 2 \cdot 2 \cdot 2 \cdot 3 \cdot 17 \cdot 23 = 9384$

25.
$4 =$ [2·2 |]
$8 =$ [2·2·2 |]
$12 =$ [2·2 | 3]
LCM $= 2 \cdot 2 \cdot 2 \cdot 3 = 24$

27.
$3 =$ [| 3 |]
$5 =$ [| | 5]
$10 =$ [2 | | 5]
LCM $= 2 \cdot 3 \cdot 5 = 30$

29.
$3 =$ [| 3]
$8 =$ [2·2·2 |]
$12 =$ [2·2 | 3]
LCM $= 2 \cdot 2 \cdot 2 \cdot 3 = 24$

31.

	2	3
9 =		3·3
36 =	2·2	3·3
64 =	2·2·2·2·2·2	

LCM = 2·2·2·2·2·2·3·3 = 576

33.

	2	3	5	7
16 =	2·2·2·2			
30 =	2	3	5	
84 =	2·2	3		7

LCM = 2·2·2·2·3·5·7 = 1680

Objective B Exercises

35.

	3	5
3 =	3	
5 =		5

GCF = 1

37.

	2	3
6 =	2	3
9 =		3·3

GCF = 3

39.

	3	5
15 =	3	5
25 =		5·5

GCF = 5

41.

	2	5
25 =		5·5
100 =	2·2	5·5

GCF = 5·5 = 25

43.

	2	3	17
32 =	2·2·2·2·2		
51 =		3	17

GCF = 1

45.

	2	3	5
12 =	2·2	3	
80 =	2·2·2·2		5

GCF = 2·2 = 4

47.

	2	5	7
16 =	2·2·2·2		
140 =	2·2	5	7

GCF = 2·2 = 4

49.

	2	3	5
24 =	2·2·2	3	
30 =	2	3	5

GCF = 2·3 = 6

51.

	2	3	11
44 =	2·2		11
96 =	2·2·2·2·2	3	

GCF = 2·2 = 4

53.

	3	5	11
3 =	3		
5 =		5	
11 =			11

GCF = 1

55.

	2	7
7 =		7
14 =	2	7
49 =		7·7

GCF = 7

57.

	2	3	5
10 =	2		5
15 =		3	5
20 =	2·2		5

GCF = 5

59.

	2	3	5
24 =	2·2·2	3	
40 =	2·2·2		5
72 =	2·2·2	3·3	

GCF = 2·2·2 = 8

61.

	3	17	31
17 =		17	
31 =			31
81 =	3·3·3·3		

GCF = 1

63.

	5
25 =	5·5
125 =	5·5·5
625 =	5·5·5·5

GCF = 25

65.

	2	5	7
28 =	2·2		7
35 =	2	5	7
70 =	2	5	7

GCF = 7

67.

	2	3	7
32 =	2·2·2·2·2		
56 =	2·2·2		7
72 =	2·2·2	3·3	

GCF = 2·2·2 = 8

Applying the Concepts

69. Two composite numbers are relatively prime if they do not have any common factor. Examples: 4 and 5, 8 and 9, 16 and 21.

71. The LCM of 2 and 3 is 6. The LCM of 5 and 7 is 35. The LCM of 11 and 19 is 209. The LCM of two prime numbers is the product of the two numbers. The LCM of three prime numbers is the product of the three numbers.

73. Yes, the LCM of the two numbers is always divisible by the GCF of the two numbers. Two numbers are factors of their LCM. The GCF of the two numbers is a factor of the LCM of the same numbers. That is, the LCM of two numbers always is divisible by the GCF of the two numbers. For example, the GCF of 4 and 6 is 2, and the LCM of 4 and 6 is 12. 12 is divisible by 2.

75. The LCM of 365 and 220 is 16,060. If both calendars begin on the same day, the number of years before the situation occurs again is:
For solar year: 16,060 ÷ 365 = 44 years
For ritual year: 16,060 ÷ 220 = 73 years.

SECTION 2.2

Objective A Exercises

1. Improper fraction

3. Proper fraction

5. $\dfrac{3}{4}$

7. $\dfrac{7}{8}$

9. $1\dfrac{1}{2}$

11. $2\dfrac{5}{8}$

13. $3\dfrac{3}{5}$

15. $\dfrac{5}{4}$

17. $\dfrac{8}{3}$

19. $\dfrac{28}{8}$

21.

23.

25.

27. $\begin{array}{r} 2 \\ 4\overline{)11} \\ -8 \\ \hline 3 \end{array}$ $\dfrac{11}{4}=2\dfrac{3}{4}$

29. $\begin{array}{r} 5 \\ 4\overline{)20} \\ -20 \\ \hline 0 \end{array}$ $\dfrac{20}{4}=5$

31. $\begin{array}{r} 1 \\ 8\overline{)9} \\ -8 \\ \hline 1 \end{array}$ $\dfrac{9}{8}=1\dfrac{1}{8}$

33. $\begin{array}{r} 2 \\ 10\overline{)23} \\ -20 \\ \hline 3 \end{array}$ $\dfrac{23}{10}=2\dfrac{3}{10}$

35. $\begin{array}{r} 3 \\ 16\overline{)48} \\ -48 \\ \hline 0 \end{array}$ $\dfrac{48}{16}=3$

37. $\begin{array}{r} 1 \\ 7\overline{)8} \\ -7 \\ \hline 1 \end{array}$ $\dfrac{8}{7}=1\dfrac{1}{7}$

39. $\begin{array}{r} 2 \\ 3\overline{)7} \\ -6 \\ \hline 1 \end{array}$ $\dfrac{7}{3}=2\dfrac{1}{3}$

41. $\begin{array}{r} 16 \\ 1\overline{)16} \\ -1 \\ \hline 06 \\ -6 \\ \hline 0 \end{array}$ $\dfrac{16}{1}=16$

43. $\begin{array}{r} 2 \\ 8\overline{)17} \\ -16 \\ \hline 1 \end{array}$ $\dfrac{17}{8}=2\dfrac{1}{8}$

45. $\begin{array}{r} 2 \\ 5\overline{)12} \\ -10 \\ \hline 2 \end{array}$ $\dfrac{12}{5}=2\dfrac{2}{5}$

47. $\begin{array}{r} 1 \\ 9\overline{)9} \\ -9 \\ \hline 0 \end{array}$ $\dfrac{9}{9}=1$

49. $\begin{array}{r} 9 \\ 8\overline{)72} \\ -72 \\ \hline 0 \end{array}$ $\dfrac{72}{8}=9$

51. $2\dfrac{1}{3}=\dfrac{6+1}{3}=\dfrac{7}{3}$

53. $6\dfrac{1}{2}=\dfrac{12+1}{2}=\dfrac{13}{2}$

55. $6\dfrac{5}{6}=\dfrac{36+5}{6}=\dfrac{41}{6}$

57. $9\dfrac{1}{4}=\dfrac{36+1}{4}=\dfrac{37}{4}$

59. $10\dfrac{1}{2}=\dfrac{20+1}{2}=\dfrac{21}{2}$

61. $8\dfrac{1}{9} = \dfrac{72+1}{9} = \dfrac{73}{9}$

63. $5\dfrac{3}{11} = \dfrac{55+3}{11} = \dfrac{58}{11}$

65. $2\dfrac{5}{8} = \dfrac{16+5}{8} = \dfrac{21}{8}$

67. $1\dfrac{5}{8} = \dfrac{8+5}{8} = \dfrac{13}{8}$

69. $11\dfrac{1}{9} = \dfrac{99+1}{9} = \dfrac{100}{9}$

71. $3\dfrac{3}{8} = \dfrac{24+3}{8} = \dfrac{27}{8}$

73. $6\dfrac{7}{13} = \dfrac{78+7}{13} = \dfrac{85}{13}$

Applying the Concepts

75. Students might mention any of the following: fractional parts of an hour, as in three-quarters of an hour; lengths of nails, as in three-quarter inch nail; lengths of fabric, as in one and five-eights yards of material; lengths of lumber, as in two and one-half feet of pine; ingredients in a recipe, as in one and one-half cups sugar; or innings pitched, as in four and two-thirds innings.

SECTION 2.3

Objective A Exercises

1. $\dfrac{1\cdot5}{2\cdot5} = \dfrac{5}{10}$

3. $\dfrac{3\cdot3}{16\cdot3} = \dfrac{9}{48}$

5. $\dfrac{3\cdot4}{8\cdot4} = \dfrac{12}{32}$

7. $\dfrac{3\cdot3}{17\cdot3} = \dfrac{9}{51}$

9. $\dfrac{3\cdot4}{4\cdot4} = \dfrac{12}{16}$

11. $\dfrac{3\cdot9}{1\cdot9} = \dfrac{27}{9}$

13. $\dfrac{1\cdot20}{3\cdot20} = \dfrac{20}{60}$

15. $\dfrac{11\cdot4}{15\cdot4} = \dfrac{44}{60}$

17. $\dfrac{2\cdot6}{3\cdot6} = \dfrac{12}{18}$

19. $\dfrac{5\cdot7}{7\cdot7} = \dfrac{35}{49}$

21. $\dfrac{5\cdot2}{9\cdot2} = \dfrac{10}{18}$

23. $\dfrac{7\cdot3}{1\cdot3} = \dfrac{21}{3}$

25. $\dfrac{7\cdot5}{9\cdot5} = \dfrac{35}{45}$

27. $\dfrac{15\cdot4}{16\cdot4} = \dfrac{60}{64}$

29. $\dfrac{3\cdot7}{14\cdot7} = \dfrac{21}{98}$

31. $\dfrac{5\cdot6}{8\cdot6} = \dfrac{30}{48}$

33. $\dfrac{5\cdot3}{14\cdot3} = \dfrac{15}{42}$

35. $\dfrac{17\cdot6}{24\cdot6} = \dfrac{102}{144}$

37. $\dfrac{3\cdot51}{8\cdot51} = \dfrac{153}{408}$

39. $\dfrac{17\cdot20}{40\cdot20} = \dfrac{340}{800}$

Objective B Exercises

41. $\dfrac{4}{12} = \dfrac{\overset{1}{\cancel{2}}\cdot\overset{1}{\cancel{2}}}{\underset{1}{\cancel{2}}\cdot\underset{1}{\cancel{2}}\cdot3} = \dfrac{1}{3}$

43. $\dfrac{22}{44} = \dfrac{\overset{1}{\cancel{2}}\cdot\overset{1}{\cancel{11}}}{2\cdot\underset{1}{\cancel{2}}\cdot\underset{1}{\cancel{11}}} = \dfrac{1}{2}$

45. $\dfrac{2}{12} = \dfrac{\overset{1}{\cancel{2}}}{\underset{1}{\cancel{2}}\cdot2\cdot3} = \dfrac{1}{6}$

47. $\dfrac{40}{36} = \dfrac{2\cdot2\cdot\overset{1}{\cancel{2}}\cdot5}{2\cdot2\cdot\underset{1}{\cancel{3}}\cdot3} = \dfrac{10}{9} = 1\dfrac{1}{9}$

49. $\dfrac{0}{30} = 0$

51. $\dfrac{9}{22} = \dfrac{3\cdot3}{2\cdot11} = \dfrac{9}{22}$

53. $\dfrac{75}{25} = \dfrac{3\cdot\overset{1}{\cancel{5}}\cdot\overset{1}{\cancel{5}}}{\underset{1}{\cancel{5}}\cdot\underset{1}{\cancel{5}}} = 3$

55. $\dfrac{16}{84} = \dfrac{\overset{1}{\cancel{2}}\cdot\overset{1}{\cancel{2}}\cdot2\cdot2}{\underset{1}{\cancel{2}}\cdot\underset{1}{\cancel{2}}\cdot3\cdot7} = \dfrac{4}{21}$

57. $\dfrac{12}{35} = \dfrac{2\cdot2\cdot3}{5\cdot7} = \dfrac{12}{35}$

59. $\dfrac{28}{44} = \dfrac{\overset{1}{2} \cdot \overset{1}{2} \cdot 7}{\underset{1}{2} \cdot \underset{1}{2} \cdot 11} = \dfrac{7}{11}$

61. $\dfrac{16}{12} = \dfrac{\overset{1}{2} \cdot \overset{1}{2} \cdot 2 \cdot 2}{\underset{1}{2} \cdot \underset{1}{2} \cdot 3} = \dfrac{4}{3} = 1\dfrac{1}{3}$

63. $\dfrac{24}{40} = \dfrac{\overset{1}{2} \cdot \overset{1}{2} \cdot \overset{1}{2} \cdot 3}{\underset{1}{2} \cdot \underset{1}{2} \cdot \underset{1}{2} \cdot 5} = \dfrac{3}{5}$

65. $\dfrac{8}{88} = \dfrac{\overset{1}{2} \cdot \overset{1}{2} \cdot \overset{1}{2}}{\underset{1}{2} \cdot \underset{1}{2} \cdot \underset{1}{2} \cdot 11} = \dfrac{1}{11}$

67. $\dfrac{144}{36} = \dfrac{\overset{1}{2} \cdot \overset{1}{2} \cdot 2 \cdot 2 \cdot \overset{1}{3} \cdot \overset{1}{3}}{\underset{1}{2} \cdot \underset{1}{2} \cdot \underset{1}{3} \cdot \underset{1}{3}} = 4$

69. $\dfrac{48}{144} = \dfrac{\overset{1}{2} \cdot \overset{1}{2} \cdot \overset{1}{2} \cdot \overset{1}{2} \cdot \overset{1}{3}}{\underset{1}{2} \cdot \underset{1}{2} \cdot \underset{1}{2} \cdot \underset{1}{2} \cdot 3 \cdot \underset{1}{3}} = \dfrac{1}{3}$

71. $\dfrac{60}{100} = \dfrac{\overset{1}{2} \cdot \overset{1}{2} \cdot 3 \cdot \overset{1}{5}}{\underset{1}{2} \cdot \underset{1}{2} \cdot 5 \cdot \underset{1}{5}} = \dfrac{3}{5}$

73. $\dfrac{36}{16} = \dfrac{\overset{1}{2} \cdot \overset{1}{2} \cdot 3 \cdot 3}{\underset{1}{2} \cdot \underset{1}{2} \cdot 2 \cdot 2} = \dfrac{9}{4} = 2\dfrac{1}{4}$

75. $\dfrac{32}{160} = \dfrac{\overset{1}{2} \cdot \overset{1}{2} \cdot \overset{1}{2} \cdot \overset{1}{2} \cdot \overset{1}{2}}{\underset{1}{2} \cdot \underset{1}{2} \cdot \underset{1}{2} \cdot \underset{1}{2} \cdot \underset{1}{2} \cdot 5} = \dfrac{1}{5}$

Applying the Concepts

77. $\dfrac{3}{1}, \dfrac{6}{2}, \dfrac{9}{3}, \dfrac{12}{4}, \dfrac{15}{5}$ are fractions that are equal to 3.

79a. $\dfrac{8}{50} = \dfrac{4}{25}$

Maine, Maryland, Massachusetts, Michigan, Minnesota, Mississippi, Missouri, Montana

79b. $\dfrac{8}{50} = \dfrac{4}{25}$

Alabama, Alaska, Arizona, Idaho, Indiana, Iowa, Ohio, Oklahoma

SECTION 2.4

Objective A Exercises

1. $\begin{array}{r} \dfrac{2}{7} \\ + \dfrac{1}{7} \\ \hline \dfrac{3}{7} \end{array}$

3. $\begin{array}{r} \dfrac{1}{2} \\ + \dfrac{1}{2} \\ \hline \dfrac{2}{2} = 1 \end{array}$

5. $\begin{array}{r} \dfrac{8}{11} \\ + \dfrac{7}{11} \\ \hline \dfrac{15}{11} = 1\dfrac{4}{11} \end{array}$

7. $\begin{array}{r} \dfrac{8}{5} \\ + \dfrac{9}{5} \\ \hline \dfrac{17}{5} = 3\dfrac{2}{5} \end{array}$

9. $\begin{array}{r} \dfrac{3}{5} \\ \dfrac{8}{5} \\ + \dfrac{3}{5} \\ \hline \dfrac{14}{5} = 2\dfrac{4}{5} \end{array}$

11. $\begin{array}{r} \dfrac{3}{4} \\ \dfrac{1}{4} \\ + \dfrac{5}{4} \\ \hline \dfrac{9}{4} = 2\dfrac{1}{4} \end{array}$

13. $\begin{array}{r} \dfrac{3}{8} \\ \dfrac{7}{8} \\ + \dfrac{1}{8} \\ \hline \dfrac{11}{8} = 1\dfrac{3}{8} \end{array}$

15. $\begin{array}{r} \dfrac{4}{15} \\ \dfrac{7}{15} \\ + \dfrac{11}{15} \\ \hline \dfrac{22}{15} = 1\dfrac{7}{15} \end{array}$

17.
$$\frac{3}{16}$$
$$\frac{5}{16}$$
$$+\frac{7}{16}$$
$$\frac{15}{16}$$

19.
$$\frac{3}{11}$$
$$\frac{5}{11}$$
$$+\frac{7}{11}$$
$$\frac{15}{11}=1\frac{4}{11}$$

21. $\frac{4}{9}+\frac{5}{9}=\frac{9}{9}=1$

23. $\frac{5}{8}+\frac{3}{8}+\frac{7}{8}=\frac{15}{8}=1\frac{7}{8}$

Objective B Exercises

25.
$$\frac{1}{2}=\frac{3}{6}$$
$$+\frac{2}{3}=\frac{4}{6}$$
$$\frac{7}{6}=1\frac{1}{6}$$

27.
$$\frac{3}{14}=\frac{3}{14}$$
$$+\frac{5}{7}=\frac{10}{14}$$
$$\frac{13}{14}$$

29.
$$\frac{8}{15}=\frac{32}{60}$$
$$+\frac{7}{20}=\frac{21}{60}$$
$$\frac{53}{60}$$

31.
$$\frac{3}{8}=\frac{21}{56}$$
$$+\frac{9}{14}=\frac{36}{56}$$
$$\frac{57}{56}=1\frac{1}{56}$$

33.
$$\frac{3}{20}=\frac{9}{60}$$
$$+\frac{7}{30}=\frac{14}{60}$$
$$\frac{23}{60}$$

35.
$$\frac{2}{3}=\frac{38}{57}$$
$$+\frac{6}{19}=\frac{18}{57}$$
$$\frac{56}{57}$$

37.
$$\frac{1}{3}=\frac{6}{18}$$
$$\frac{5}{6}=\frac{15}{18}$$
$$+\frac{7}{9}=\frac{14}{18}$$
$$\frac{35}{18}=1\frac{17}{18}$$

39.
$$\frac{5}{6}=\frac{40}{48}$$
$$\frac{1}{12}=\frac{4}{48}$$
$$+\frac{5}{16}=\frac{15}{48}$$
$$\frac{59}{48}=1\frac{11}{48}$$

41.
$$\frac{2}{3}=\frac{40}{60}$$
$$\frac{1}{5}=\frac{12}{60}$$
$$+\frac{7}{12}=\frac{35}{60}$$
$$\frac{87}{60}=1\frac{27}{60}=1\frac{9}{20}$$

43.
$$\frac{1}{4}=\frac{45}{180}$$
$$\frac{4}{5}=\frac{144}{180}$$
$$+\frac{5}{9}=\frac{100}{180}$$
$$\frac{289}{180}=1\frac{109}{180}$$

45.
$$\frac{5}{16}=\frac{45}{144}$$
$$\frac{11}{18}=\frac{88}{144}$$
$$+\frac{17}{24}=\frac{102}{144}$$
$$\frac{235}{144}=1\frac{91}{144}$$

47.
$$\frac{2}{3}=\frac{48}{72}$$
$$\frac{5}{8}=\frac{45}{72}$$
$$+\frac{7}{9}=\frac{56}{72}$$
$$\frac{149}{72}=2\frac{5}{72}$$

49.
$$\frac{3}{8}=\frac{15}{40}$$
$$+\ \frac{3}{5}=\frac{24}{40}$$
$$\frac{39}{40}$$

51.
$$\frac{3}{8}=\frac{9}{24}$$
$$\frac{5}{6}=\frac{20}{24}$$
$$+\ \frac{7}{12}=\frac{14}{24}$$
$$\frac{43}{24}=1\frac{19}{24}$$

53.
$$\frac{1}{2}=\frac{36}{72}$$
$$\frac{5}{8}=\frac{45}{72}$$
$$+\frac{7}{9}=\frac{56}{72}$$
$$\frac{137}{72}=1\frac{65}{72}$$

Objective C Exercises

55.
$$1\frac{1}{2}=1\frac{3}{6}$$
$$+\ 2\frac{1}{6}=2\frac{1}{6}$$
$$3\frac{4}{6}=3\frac{2}{3}$$

57.
$$4\frac{1}{2}=4\frac{6}{12}$$
$$+\ 5\frac{7}{12}=5\frac{7}{12}$$
$$9\frac{13}{12}=10\frac{1}{12}$$

59.
$$4$$
$$+\ 5\frac{2}{7}$$
$$9\frac{2}{7}$$

61.
$$3\ \frac{5}{8}=3\frac{25}{40}$$
$$+\ 2\frac{11}{20}=2\frac{22}{40}$$
$$5\frac{47}{40}=6\frac{7}{40}$$

63.
$$7\frac{5}{12}=7\frac{20}{48}$$
$$+\ 2\frac{9}{16}=2\frac{27}{48}$$
$$9\frac{47}{48}$$

65.
$$6\frac{1}{3}=6\frac{13}{39}$$
$$+\ 2\frac{3}{13}=2\frac{9}{39}$$
$$8\frac{22}{39}$$

67.
$$8\frac{29}{30}=8\frac{116}{120}$$
$$+\ 7\frac{11}{40}=7\frac{33}{120}$$
$$15\frac{149}{120}=16\frac{29}{120}$$

69.
$$17\frac{3}{8}=17\frac{15}{40}$$
$$+\ 7\frac{7}{20}=\ 7\frac{14}{40}$$
$$24\frac{29}{40}$$

71.
$$5\ \frac{7}{8}=\ 5\frac{21}{24}$$
$$+\ 27\frac{5}{12}=27\frac{10}{24}$$
$$32\frac{31}{24}=33\frac{7}{24}$$

73.
$$7\frac{5}{9}=7\frac{20}{36}$$
$$+\ 2\frac{7}{12}=2\frac{21}{36}$$
$$9\frac{41}{36}=10\frac{5}{36}$$

75.
$$2\frac{1}{2}=2\frac{6}{12}$$
$$3\frac{2}{3}=3\frac{8}{12}$$
$$+\ 4\frac{1}{4}=4\frac{3}{12}$$
$$9\frac{17}{12}=10\frac{5}{12}$$

77.
$$3\frac{1}{2}=3\frac{45}{90}$$
$$3\frac{1}{5}=3\frac{18}{90}$$
$$+\ 8\frac{1}{9}=8\frac{10}{90}$$
$$14\frac{73}{90}$$

79.
$$2\frac{3}{8}=2\frac{18}{48}$$
$$4\frac{7}{12}=4\frac{28}{48}$$
$$+3\frac{5}{16}=3\frac{15}{48}$$
$$9\frac{61}{48}=10\frac{13}{48}$$

81.
$$6\frac{5}{6} = 6\frac{45}{54}$$
$$17\frac{2}{9} = 17\frac{12}{54}$$
$$+\ 18\frac{5}{27} = 18\frac{10}{54}$$
$$41\frac{67}{54} = 42\frac{13}{54}$$

83.
$$2\frac{4}{9} = 2\frac{16}{36}$$
$$+\ 5\frac{7}{12} = 5\frac{21}{36}$$
$$7\frac{37}{36} = 8\frac{1}{36}$$

85.
$$4\frac{3}{4} = 4\frac{9}{12}$$
$$+\ 9\frac{1}{3} = 9\frac{4}{12}$$
$$13\frac{13}{12} = 14\frac{1}{12}$$

87.
$$2\frac{2}{3} = 2\frac{48}{72}$$
$$4\frac{5}{8} = 4\frac{45}{72}$$
$$+\ 2\frac{2}{9} = 2\frac{16}{72}$$
$$8\frac{109}{72} = 9\frac{37}{72}$$

Objective D Application Problems

89. Strategy To find the length of the shaft, add the three distances $\left(\frac{3}{8},\ \frac{11}{16},\ \text{and}\ \frac{1}{4}\ \text{inch}\right)$.

Solution
$$\frac{3}{8} = \frac{6}{16}$$
$$\frac{11}{16} = \frac{11}{16}$$
$$+\ \frac{1}{4} = \frac{4}{16}$$
$$\frac{21}{16} = 1\frac{5}{16}$$

The length of the shaft is $1\frac{5}{16}$ inches.

91. Strategy To find the total thickness of the table after the veneer has been applied, add the table-top thickness $\left(1\frac{1}{8}\ \text{inch}\right)$ to the veneer thickness $\left(\frac{3}{16}\ \text{inch}\right)$.

Solution
$$1\frac{1}{8} = 1\frac{2}{16}$$
$$+\ \frac{3}{16} = \frac{3}{16}$$
$$1\frac{5}{16}$$

The total thickness is $1\frac{5}{16}$ inches.

93a. Strategy To find the total number of hours worked, add the five amounts $\left(5,\ 3\frac{3}{4},\ 2\frac{1}{3},\ 1\frac{1}{4},\ \text{and}\ 7\frac{2}{3}\right)$.

Solution
$$5\ \ = 5$$
$$3\frac{3}{4} = 3\frac{9}{12}$$
$$2\frac{1}{3} = 2\frac{4}{12}$$
$$1\frac{1}{4} = 1\frac{3}{12}$$
$$+\ 7\frac{2}{3} = 7\frac{8}{12}$$
$$18\frac{24}{12} = 20$$

The total number of hours worked was 20.

93b. Strategy To find the total salary for the week, multiply the number of hours worked (20) by the pay for 1 hour ($9).

Solution
$$\begin{array}{r}\$9 \\ \times\ \ 20 \\ \hline \$180 \end{array}$$

Your total salary for the week is $180.

95. Strategy To find the total length of the wood beams, add the three lengths $\left(25\frac{3}{4},\ 12\frac{1}{2},\ \text{and}\ 17\frac{1}{2}\ \text{feet}\right)$.

Solution
$$25\frac{3}{4} = 25\frac{3}{4}$$
$$12\frac{1}{2} = 12\frac{2}{4}$$
$$+\ 17\frac{1}{2} = 17\frac{2}{4}$$
$$54\frac{7}{4} = 55\frac{3}{4}$$

The total length of wood needed is $55\frac{3}{4}$ feet.

97. Strategy   To find the fractional part of those who changed homes moved outside the county, add the fractional part of those who moved to a different state $\left(\dfrac{1}{7}\right)$ to the fractional part of those who moved to a different county in the same state $\left(\dfrac{4}{21}\right)$.

Solution

$$\begin{array}{r} \dfrac{1}{7}=\dfrac{3}{21} \\[6pt] +\dfrac{4}{21}=\dfrac{4}{21} \\[6pt] \hline \dfrac{7}{21}=\dfrac{1}{3} \end{array}$$

Those who changed homes outside the county are $\dfrac{1}{3}$ of the people.

Applying the Concepts

99. We can use 3 dimes and 1 quarter to model adding $\dfrac{3}{10}$ and $\dfrac{1}{4}$. Because the dime and the quarter are not the same unit, we change all of them to nickels and add.

$$\begin{array}{r} 3\text{ dimes} = 6\text{ nickels} \\ +\ 1\text{ quarter} = 5\text{ nickels} \\ \hline 11\text{ nickels} \end{array}$$

Now, consider the above problem in terms of dollars.

$$\begin{array}{r} \dfrac{3}{10}=\dfrac{6}{20} \\[6pt] +\dfrac{1}{4}=\dfrac{5}{20} \\[6pt] \hline \dfrac{11}{20} \end{array}$$

Note that

1 dime $=\dfrac{1}{10}$ of a dollar

1 quarter $=\dfrac{1}{4}$ of a dollar

1 nickel $=\dfrac{1}{20}$ of a dollar

SECTION 2.5

Objective A Exercises

1.
$$\begin{array}{r} \dfrac{9}{17} \\[6pt] -\dfrac{7}{17} \\[6pt] \hline \dfrac{2}{17} \end{array}$$

3.
$$\begin{array}{r} \dfrac{11}{12} \\[6pt] -\dfrac{7}{12} \\[6pt] \hline \dfrac{4}{12}=\dfrac{1}{3} \end{array}$$

5.
$$\begin{array}{r} \dfrac{9}{20} \\[6pt] -\dfrac{7}{20} \\[6pt] \hline \dfrac{2}{20}=\dfrac{1}{10} \end{array}$$

7.
$$\begin{array}{r} \dfrac{42}{65} \\[6pt] -\dfrac{17}{65} \\[6pt] \hline \dfrac{25}{65}=\dfrac{5}{13} \end{array}$$

9.
$$\begin{array}{r} \dfrac{23}{30} \\[6pt] -\dfrac{13}{30} \\[6pt] \hline \dfrac{10}{30}=\dfrac{1}{3} \end{array}$$

11.
$$\begin{array}{r} \dfrac{13}{14} \\[6pt] -\dfrac{5}{14} \\[6pt] \hline \dfrac{8}{14}=\dfrac{4}{7} \end{array}$$

13.
$$\begin{array}{r} \dfrac{7}{8} \\[6pt] -\dfrac{5}{8} \\[6pt] \hline \dfrac{2}{8}=\dfrac{1}{4} \end{array}$$

15.
$$\begin{array}{r} \dfrac{18}{23} \\[6pt] -\dfrac{9}{23} \\[6pt] \hline \dfrac{9}{23} \end{array}$$

17.
$$\begin{array}{r} \dfrac{17}{24} \\[6pt] -\dfrac{11}{24} \\[6pt] \hline \dfrac{6}{24}=\dfrac{1}{4} \end{array}$$

Objective B Exercises

19.
$$\frac{2}{3} = \frac{4}{6}$$
$$-\frac{1}{6} = \frac{1}{6}$$
$$\frac{3}{6} = \frac{1}{2}$$

21.
$$\frac{5}{8} = \frac{35}{56}$$
$$-\frac{2}{7} = \frac{16}{56}$$
$$\frac{19}{56}$$

23.
$$\frac{5}{7} = \frac{10}{14}$$
$$-\frac{3}{14} = \frac{3}{14}$$
$$\frac{7}{14} = \frac{1}{2}$$

25.
$$\frac{8}{15} = \frac{32}{60}$$
$$-\frac{7}{20} = \frac{21}{60}$$
$$\frac{11}{60}$$

27.
$$\frac{9}{14} = \frac{36}{56}$$
$$-\frac{3}{8} = \frac{21}{56}$$
$$\frac{15}{56}$$

29.
$$\frac{46}{51} = \frac{46}{51}$$
$$-\frac{3}{17} = \frac{9}{51}$$
$$\frac{37}{51}$$

31.
$$\frac{21}{35} = \frac{42}{70}$$
$$-\frac{5}{14} = \frac{25}{70}$$
$$\frac{17}{70}$$

33.
$$\frac{29}{60} = \frac{58}{120}$$
$$-\frac{3}{40} = \frac{9}{120}$$
$$\frac{49}{120}$$

35.
$$\frac{11}{15} = \frac{33}{45}$$
$$-\frac{5}{9} = \frac{25}{45}$$
$$\frac{8}{45}$$

37.
$$\frac{9}{14} = \frac{27}{42}$$
$$-\frac{5}{42} = \frac{5}{42}$$
$$\frac{22}{42} = \frac{11}{21}$$

39.
$$\frac{17}{20} = \frac{51}{60}$$
$$-\frac{7}{15} = \frac{28}{60}$$
$$\frac{23}{60}$$

41.
$$\frac{5}{6} = \frac{15}{18}$$
$$-\frac{7}{9} = \frac{14}{18}$$
$$\frac{1}{18}$$

Objective C Exercises

43.
$$16\frac{11}{15}$$
$$-11\frac{8}{15}$$
$$5\frac{3}{15} = 5\frac{1}{5}$$

45.
$$19\frac{16}{17}$$
$$-9\frac{7}{17}$$
$$10\frac{9}{17}$$

47.
$$5\frac{7}{8}$$
$$-1$$
$$4\frac{7}{8}$$

49.
$$3 = 2\frac{21}{21}$$
$$-2\frac{5}{21} = 2\frac{5}{21}$$
$$\frac{16}{21}$$

51.
$$16\frac{3}{8} = 15\frac{11}{8}$$
$$-\ 10\frac{7}{8} = 10\frac{7}{8}$$
$$5\frac{4}{8} = 5\frac{1}{2}$$

53.
$$8\frac{3}{7} = 7\frac{10}{7}$$
$$-\ 2\frac{6}{7} = 2\frac{6}{7}$$
$$5\frac{4}{7}$$

55.
$$23\frac{7}{8} = 23\frac{21}{24}$$
$$-\ 16\frac{2}{3} = 16\frac{16}{24}$$
$$7\frac{5}{24}$$

57.
$$65\frac{8}{35} = 65\frac{16}{70} = 64\frac{86}{70}$$
$$-\ 16\frac{11}{14} = 16\frac{55}{70} = 16\frac{55}{70}$$
$$48\frac{31}{70}$$

59.
$$101\frac{2}{9}$$
$$\underline{-16\phantom{\frac{2}{9}}}$$
$$85\frac{2}{9}$$

61.
$$17\phantom{\frac{13}{13}} = 16\frac{13}{13}$$
$$-7\frac{8}{13} = 7\frac{8}{13}$$
$$9\frac{5}{13}$$

63.
$$23\frac{3}{20} = 23\frac{3}{20} = 22\frac{23}{20}$$
$$-\ 7\frac{3}{5} = 7\frac{12}{20} = 7\frac{12}{20}$$
$$15\frac{11}{20}$$

65.
$$12\frac{3}{8} = 12\frac{9}{24} = 11\frac{33}{24}$$
$$-\ 7\frac{5}{12} = 7\frac{10}{24} = 7\frac{10}{24}$$
$$4\frac{23}{24}$$

67.
$$6\frac{1}{3} = 6\frac{5}{15} = 5\frac{20}{15}$$
$$-3\frac{3}{5} = 3\frac{9}{15} = 3\frac{9}{15}$$
$$2\frac{11}{15}$$

Objective D Application Problems

69.
$$12\frac{3}{8} = 11\frac{11}{8}$$
$$-\ 2\frac{7}{8} = \ 2\frac{7}{8}$$
$$9\frac{4}{8} = \ 9\frac{1}{2}$$

The missing dimension is $9\frac{1}{2}$ inches.

71. Strategy   To find the difference between Meyfarth's distance and Coachman's distance, subtract Coachman's distance $\left(66\frac{1}{8}\text{ inches}\right)$ from Meyfarth's distance $\left(75\frac{1}{2}\text{ inches}\right)$.

Solution
$$75\frac{1}{2} = 75\frac{4}{8}$$
$$-66\frac{1}{8} = 66\frac{1}{8}$$
$$9\frac{3}{8}$$

The difference between Meyfarth's distance and Coachman's distance was $9\frac{3}{8}$ inches.

Strategy   To find the difference between Kostadinova's distance and Meyfarth's distance, subtract Meyfarth's distance $\left(75\frac{1}{2}\text{ inches}\right)$ from Kostadinova's distance $\left(80\frac{3}{4}\text{ inches}\right)$.

Solution
$$80\frac{3}{4} = 80\frac{3}{4}$$
$$-75\frac{1}{2} = 75\frac{2}{4}$$
$$5\frac{1}{4}$$

The difference between Kostadinova's distance and Meyfarth's distance was $5\frac{1}{4}$ inches.

73a. Strategy To find the distance, add the distance from the starting point to the first checkpoint $\left(3\frac{3}{8} \text{ miles}\right)$ to the distance from the first checkpoint to the second checkpoint $\left(4\frac{1}{3} \text{ miles}\right)$.

Solution

$$3\frac{3}{8} = 3\frac{9}{24}$$
$$+4\frac{1}{3} = 4\frac{8}{24}$$
$$\overline{\phantom{+4\frac{1}{3}=}7\frac{17}{24}}$$

The distance from the starting point to the second checkpoint is $7\frac{17}{24}$ miles.

73b. Strategy To find the distance, subtract the distance from the starting point to the second checkpoint $\left(7\frac{17}{24} \text{ miles}\right)$ from the total distance (12 miles).

Solution

$$12 = 11\frac{24}{24}$$
$$-7\frac{17}{24} = 7\frac{17}{24}$$
$$\overline{\phantom{-7\frac{17}{24}=}4\frac{7}{24}}$$

The distance from the second checkpoint to the finish line is $4\frac{7}{24}$ miles.

75a. The wrestler has lost $5\frac{1}{4}$ pounds the first week and $4\frac{1}{4}$ pounds the second week. Thus the wrestler has lost more than 9 pounds the first two weeks. Since less than 13 pounds needs to be lost, the wrestler can attain the weight class by losing less than 4 pounds. This is less than the $4\frac{1}{4}$ pounds lost in the second week.

75b. Strategy To find how much weight must be lost to reach the desired weight:
• Add the amounts of weight lost during the first 2 weeks $\left(5\frac{1}{4} \text{ and } 4\frac{1}{4} \text{ pounds}\right)$.
• Subtract the total weight lost so far from the amount that is required $\left(12\frac{3}{4} \text{ pounds}\right)$.

Solution

$$5\frac{1}{4} \qquad\qquad 12\frac{3}{4} = 12\frac{3}{4}$$
$$+4\frac{1}{4} \qquad\qquad -9\frac{1}{2} = 9\frac{2}{4}$$
$$\overline{9\frac{2}{4} = 9\frac{1}{2}} \qquad\qquad \overline{\phantom{-9\frac{1}{2}=}3\frac{1}{4}}$$

The wrestler needs to lose $3\frac{1}{4}$ pounds to reach the desired weight.

Applying the Concepts

77. To find the missing number, subtract $2\frac{1}{2}$ from $5\frac{1}{3}$.

$$5\frac{1}{3} = 5\frac{2}{6} = 4\frac{8}{6}$$
$$-2\frac{1}{2} = 2\frac{3}{6} = 2\frac{3}{6}$$
$$\overline{\phantom{-2\frac{1}{2}=2\frac{3}{6}=}2\frac{5}{6}}$$

79. Right diagonal: $\dfrac{3}{4} + \dfrac{5}{8} + \dfrac{1}{2} = \dfrac{6}{8} + \dfrac{5}{8} + \dfrac{4}{8} = \dfrac{15}{8}$

Left diagonal: $\dfrac{5}{8} + \dfrac{7}{8} = \dfrac{12}{8}$; $\dfrac{15}{8} - \dfrac{12}{8} = \dfrac{3}{8}$

Top across: $\dfrac{3}{8} + \dfrac{3}{4} = \dfrac{3}{8} + \dfrac{6}{8} = \dfrac{9}{8}$; $\dfrac{15}{8} - \dfrac{9}{8} = \dfrac{6}{8} = \dfrac{3}{4}$

Left down: $\dfrac{3}{8} + \dfrac{1}{2} = \dfrac{3}{8} + \dfrac{4}{8} = \dfrac{7}{8}$; $\dfrac{15}{8} - \dfrac{7}{8} = \dfrac{8}{8} = 1$

Middle across: $1 + \dfrac{5}{8} = \dfrac{8}{8} + \dfrac{5}{8} = \dfrac{13}{8}$;
$\dfrac{15}{8} - \dfrac{13}{8} = \dfrac{2}{8} = \dfrac{1}{4}$

Bottom across: $\dfrac{1}{2} + \dfrac{7}{8} = \dfrac{4}{8} + \dfrac{7}{8} = \dfrac{11}{8}$;
$\dfrac{15}{8} - \dfrac{11}{8} = \dfrac{4}{8} = \dfrac{1}{2}$

$\frac{3}{8}$	$\frac{3}{4}$	$\frac{3}{4}$
1	$\frac{5}{8}$	$\frac{1}{4}$
$\frac{1}{2}$	$\frac{1}{2}$	$\frac{7}{8}$

SECTION 2.6

Objective A Exercises

1. $\dfrac{2\cdot 7}{3\cdot 8}=\dfrac{\overset{1}{2}\cdot 7}{3\cdot 2\cdot 2\cdot 2}=\dfrac{7}{12}$

3. $\dfrac{5\cdot 7}{16\cdot 15}=\dfrac{\overset{1}{5}\cdot 7}{2\cdot 2\cdot 2\cdot 2\cdot 3\cdot \underset{1}{5}}=\dfrac{7}{48}$

5. $\dfrac{1\cdot 1}{6\cdot 8}=\dfrac{1\cdot 1}{2\cdot 3\cdot 2\cdot 2\cdot 2}=\dfrac{1}{48}$

7. $\dfrac{11\cdot 6}{12\cdot 7}=\dfrac{11\cdot \overset{1}{2}\cdot \overset{1}{3}}{2\cdot 2\cdot \underset{1}{3}\cdot 7}=\dfrac{11}{14}$

9. $\dfrac{1\cdot 6}{6\cdot 7}=\dfrac{1\cdot \overset{1}{2}\cdot \overset{1}{3}}{2\cdot 3\cdot 7}=\dfrac{1}{7}$

11. $\dfrac{1\cdot 5}{5\cdot 8}=\dfrac{1\cdot \overset{1}{5}}{\underset{1}{5}\cdot 2\cdot 2\cdot 2}=\dfrac{1}{8}$

13. $\dfrac{8\cdot 27}{9\cdot 4}=\dfrac{\overset{1}{2}\cdot \overset{1}{2}\cdot 2\cdot \overset{1}{3}\cdot \overset{1}{3}\cdot 3}{\underset{1}{3}\cdot \underset{1}{3}\cdot \underset{1}{2}\cdot \underset{1}{2}}=6$

15. $\dfrac{5\cdot 1}{6\cdot 2}=\dfrac{5\cdot 1}{2\cdot 3\cdot 2}=\dfrac{5}{12}$

17. $\dfrac{16\cdot 27}{9\cdot 8}=\dfrac{\overset{1}{2}\cdot \overset{1}{2}\cdot \overset{1}{2}\cdot 2\cdot \overset{1}{3}\cdot \overset{1}{3}\cdot 3}{\underset{1}{3}\cdot \underset{1}{3}\cdot \underset{1}{2}\cdot \underset{1}{2}\cdot \underset{1}{2}}=6$

19. $\dfrac{3\cdot 4}{2\cdot 9}=\dfrac{\overset{1}{3}\cdot 2\cdot 2}{2\cdot \underset{1}{3}\cdot 3}=\dfrac{2}{3}$

21. $\dfrac{7\cdot 3}{8\cdot 14}=\dfrac{7\cdot \overset{1}{3}}{2\cdot 2\cdot 2\cdot 2\cdot \underset{1}{7}}=\dfrac{3}{16}$

23. $\dfrac{1\cdot 3}{10\cdot 8}=\dfrac{3}{2\cdot 5\cdot 2\cdot 2\cdot 2}=\dfrac{3}{80}$

25. $\dfrac{15\cdot 16}{8\cdot 3}=\dfrac{\overset{1}{3}\cdot 5\cdot \overset{1}{2}\cdot \overset{1}{2}\cdot \overset{1}{2}\cdot 2}{\underset{1}{2}\cdot \underset{1}{2}\cdot \underset{1}{2}\cdot \underset{1}{3}}=10$

27. $\dfrac{1\cdot 2}{2\cdot 15}=\dfrac{1\cdot \overset{1}{2}}{\underset{1}{2}\cdot 3\cdot 5}=\dfrac{1}{15}$

29. $\dfrac{5\cdot 14}{7\cdot 15}=\dfrac{\overset{1}{5}\cdot 2\cdot \overset{1}{7}}{\underset{1}{7}\cdot 3\cdot \underset{1}{5}}=\dfrac{2}{3}$

31. $\dfrac{5\cdot 42}{12\cdot 65}=\dfrac{\overset{1}{5}\cdot \overset{1}{2}\cdot \overset{1}{3}\cdot 7}{2\cdot \underset{1}{2}\cdot \underset{1}{3}\cdot \underset{1}{5}\cdot 13}=\dfrac{7}{26}$

33. $\dfrac{12\cdot 5}{5\cdot 3}=\dfrac{2\cdot 2\cdot \overset{1}{3}\cdot \overset{1}{5}}{\underset{1}{5}\cdot \underset{1}{3}}=4$

35. $\dfrac{16\cdot 125}{85\cdot 84}=\dfrac{\overset{1}{2}\cdot \overset{1}{2}\cdot 2\cdot 2\cdot \overset{1}{5}\cdot 5\cdot 5}{\underset{1}{5}\cdot 17\cdot \underset{1}{2}\cdot \underset{1}{2}\cdot 3\cdot 7}=\dfrac{100}{357}$

37. $\dfrac{7\cdot 15}{12\cdot 42}=\dfrac{7\cdot \overset{1}{3}\cdot 5}{2\cdot 2\cdot 3\cdot \underset{1}{3}\cdot 2\cdot 7}=\dfrac{5}{24}$

39. $\dfrac{5\cdot 3}{9\cdot 20}=\dfrac{\overset{1}{3}\cdot \overset{1}{5}}{2\cdot 2\cdot \underset{1}{3}\cdot 3\cdot \underset{1}{5}}=\dfrac{1}{12}$

41. $\dfrac{1\cdot 8}{2\cdot 15}=\dfrac{2\cdot 2\cdot 2}{2\cdot 3\cdot 5}=\dfrac{4}{15}$

Objective B Exercises

43. $\dfrac{4\cdot 3}{1\cdot 8}=\dfrac{\overset{1}{2}\cdot \overset{1}{2}\cdot 3}{\underset{1}{2}\cdot \underset{1}{2}\cdot 2}=\dfrac{3}{2}=1\dfrac{1}{2}$

45. $\dfrac{2\cdot 6}{3\cdot 1}=\dfrac{2\cdot 2\cdot \overset{1}{3}}{\underset{1}{3}\cdot 1}=4$

47. $\dfrac{1}{3}\times 1\dfrac{1}{3}=\dfrac{1}{3}\times \dfrac{4}{3}=\dfrac{1\cdot 2\cdot 2}{3\cdot 3}=\dfrac{4}{9}$

49. $1\dfrac{7}{8}\times \dfrac{4}{15}=\dfrac{15}{8}\times \dfrac{4}{15}=\dfrac{\overset{1}{3}\cdot \overset{1}{5}\cdot \overset{1}{2}\cdot \overset{1}{2}}{\underset{1}{2}\cdot \underset{1}{2}\cdot \underset{1}{2}\cdot \underset{1}{3}\cdot \underset{1}{5}}=\dfrac{1}{2}$

51. $55\times \dfrac{3}{10}=\dfrac{55\cdot 3}{1\cdot 10}=\dfrac{\overset{1}{5}\cdot 11\cdot 3}{2\cdot \underset{1}{5}}=\dfrac{33}{2}=16\dfrac{1}{2}$

53. $4\times 2\dfrac{1}{2}=\dfrac{4}{1}\times \dfrac{5}{2}=\dfrac{2\cdot \overset{1}{2}\cdot 5}{\underset{1}{2}}=10$

55. $2\dfrac{1}{7}\times 3=\dfrac{15}{7}\times \dfrac{3}{1}=\dfrac{3\cdot 5\cdot 3}{7}=\dfrac{45}{7}=6\dfrac{3}{7}$

57. $3\dfrac{2}{3}\times 5=\dfrac{11}{3}\times \dfrac{5}{1}=\dfrac{11\cdot 5}{3}=\dfrac{55}{3}=18\dfrac{1}{3}$

59. $\dfrac{1}{2}\times 3\dfrac{3}{7}=\dfrac{1}{2}\times \dfrac{24}{7}=\dfrac{2\cdot 2\cdot 2\cdot 3}{\underset{1}{2}\cdot 7}=\dfrac{12}{7}=1\dfrac{5}{7}$

61. $6\frac{1}{8} \times \frac{4}{7} = \frac{49}{8} \times \frac{4}{7} = \frac{\overset{1}{7}\cdot 7\cdot \overset{1}{2}\cdot \overset{1}{2}}{\underset{1}{2}\cdot \underset{1}{2}\cdot 2\cdot \underset{1}{7}} = \frac{7}{2} = 3\frac{1}{2}$

63. $5\frac{1}{8} \times 5 = \frac{41}{8} \times \frac{5}{1} = \frac{41\cdot 5}{2\cdot 2\cdot 2} = \frac{205}{8} = 25\frac{5}{8}$

65. $\frac{3}{8} \times 4\frac{1}{2} = \frac{3}{8} \times \frac{9}{2} = \frac{3\cdot 3\cdot 3}{2\cdot 2\cdot 2\cdot 2} = \frac{27}{16} = 1\frac{11}{16}$

67. $6 \times 2\frac{2}{3} = \frac{6}{1} \times \frac{8}{3} = \frac{2\cdot \overset{1}{3}\cdot 2\cdot 2\cdot 2}{\underset{1}{3}} = 16$

69. $1\frac{1}{3} \times 2\frac{1}{4} = \frac{4}{3} \times \frac{9}{4} = \frac{\overset{1}{2}\cdot \overset{1}{2}\cdot \overset{1}{3}\cdot 3}{\underset{1}{3}\cdot \underset{1}{2}\cdot \underset{1}{2}} = 3$

71. $2\frac{5}{8} \times 3\frac{2}{5} = \frac{21}{8} \times \frac{17}{5} = \frac{3\cdot 7\cdot 17}{2\cdot 2\cdot 2\cdot 5} = \frac{357}{40} = 8\frac{37}{40}$

73. $3\frac{1}{7} \times 2\frac{1}{8} = \frac{22}{7} \times \frac{17}{8} = \frac{\overset{1}{2}\cdot 11\cdot 17}{7\cdot 2\cdot 2\cdot 2} = \frac{187}{28} = 6\frac{19}{28}$

75. $2\frac{2}{5} \times 3\frac{1}{12} = \frac{12}{5} \times \frac{37}{12} = \frac{\overset{1}{2}\cdot \overset{1}{2}\cdot \overset{1}{3}\cdot 37}{5\cdot \underset{1}{2}\cdot \underset{1}{2}\cdot \underset{1}{3}} = \frac{37}{5} = 7\frac{2}{5}$

77. $5\frac{1}{5} \times 3\frac{1}{13} = \frac{26}{5} \times \frac{40}{13} = \frac{2\cdot \overset{1}{13}\cdot 2\cdot 2\cdot 2\cdot \overset{1}{5}}{\underset{1}{5}\cdot \underset{1}{13}} = 16$

79. $10\frac{1}{4} \times 3\frac{1}{5} = \frac{41}{4} \times \frac{16}{5} = \frac{41\cdot \overset{1}{2}\cdot \overset{1}{2}\cdot 2\cdot 2}{2\cdot 2\cdot 5} = \frac{164}{5} = 32\frac{4}{5}$

81. $5\frac{3}{7} \times 5\frac{1}{4} = \frac{38}{7} \times \frac{21}{4} = \frac{\overset{1}{2}\cdot 19\cdot 3\cdot \overset{1}{7}}{\underset{1}{7}\cdot 2\cdot 2} = \frac{57}{2} = 28\frac{1}{2}$

83. $2\frac{1}{2} \times 3\frac{3}{5} = \frac{5}{2} \times \frac{18}{5} = \frac{2\cdot 3\cdot 3\cdot \overset{1}{5}}{2\cdot \underset{1}{5}} = 9$

85. $2\frac{1}{8} \times \frac{5}{17} = \frac{17}{8} \times \frac{5}{17} = \frac{5\cdot \overset{1}{17}}{2\cdot 2\cdot 2\cdot \underset{1}{17}} = \frac{5}{8}$

87. $1\frac{3}{8} \times 2\frac{1}{5} = \frac{11}{8} \times \frac{11}{5} = \frac{121}{40} = 3\frac{1}{40}$

Objective C Application Problems

89. Strategy To find the cost of the salmon, multiply the amount of salmon $\left(2\frac{3}{4} \text{ pound}\right)$ by the cost per pound ($8).

Solution $2\frac{3}{4} \times 8 = \frac{11}{4} \times \frac{8}{1} = \frac{11\cdot 8}{4\cdot 1} = 22$
The salmon costs $22.

91a. No, $\frac{1}{3}$ of 9 feet is approximately 3 ft.

91b. Strategy To find the length cut, multiply the length of the board $\left(9\frac{1}{4} \text{ feet}\right)$ by $\frac{1}{3}$.

Solution $9\frac{1}{4} \times \frac{1}{3} = \frac{1}{3} \times \frac{37}{4} = \frac{1\cdot 37}{3\cdot 4} = \frac{37}{12} = 3\frac{1}{12}$

The length of the board cut off is $3\frac{1}{12}$ feet.

93. Strategy To find the area of the square, multiply the length of one side $\left(5\frac{1}{4} \text{ feet}\right)$ by itself $\left(5\frac{1}{4} \text{ feet}\right)$.

Solution

$5\frac{1}{4} \times 5\frac{1}{4} = \frac{21}{4} \times \frac{21}{4} = \frac{21\cdot 21}{4\cdot 4} = \frac{441}{16} = 27\frac{9}{16}$

The area of the square is $27\frac{9}{16}$ square feet.

95a. Strategy To find the amount budgeted for housing and utilities, multiply the total monthly income ($3200) by $\frac{2}{5}$.

Solution $\frac{2}{5} \times \$3200 = \frac{2\cdot 3200}{5} = \$1280$
The amount budgeted for housing and utilities is $1280.

95b. Strategy To find the amount remaining for other than housing and utilities, subtract the amount for housing and utilities ($1280) from the total monthly income ($3200).

Solution $3200
$\underline{-1280}$
$1920

The amount remaining for other than housing and utilities is $1920.

97. Strategy   To find the total cost of the robes, multiply the amount of material each robe requires $\left(2\dfrac{5}{8} \text{ yards}\right)$ by the cost of 1 yard (\$8) and by the number of robes needed (24).

Solution $2\dfrac{5}{8} \times \$8 \times 24 = \dfrac{21}{8} \times 8 \times 24$

$= \dfrac{21 \times 8 \times 24}{8} = \$504$

The total cost is \$504.

99. $12\dfrac{7}{12} \times 4\dfrac{1}{3} = \dfrac{151}{12} \times \dfrac{13}{3} = \dfrac{1963}{36} = 54\dfrac{19}{36}$

The weight of the $12\dfrac{7}{12}$ foot steel rod is $54\dfrac{19}{36}$ pounds.

101. $\dfrac{1}{2} \times \dfrac{3}{8} = \dfrac{3}{16}$

$\dfrac{3}{16}$ of the total portfolio is invested in corporate bonds.

Applying the Concepts

103. Student explanations should include the idea that every 4 years we must add one day to the usual 365-day year.

105. A. See problem 104.

SECTION 2.7

Objective A Exercises

1. $\dfrac{1}{3} \times \dfrac{5}{2} = \dfrac{1 \cdot 5}{3 \cdot 2} = \dfrac{5}{6}$

3. $\dfrac{3}{7} \times \dfrac{7}{3} = \dfrac{\overset{1}{\cancel{3}} \cdot \overset{1}{\cancel{7}}}{\underset{1}{\cancel{7}} \cdot \underset{1}{\cancel{3}}} = 1$

5. $0 \times \dfrac{4}{3} = 0$

7. $\dfrac{5}{24} \times \dfrac{36}{15} = \dfrac{\overset{1}{\cancel{5}} \cdot \overset{1}{\cancel{2}} \cdot \overset{1}{\cancel{2}} \cdot \overset{1}{\cancel{3}} \cdot \overset{1}{\cancel{3}}}{\underset{1}{\cancel{2}} \cdot \underset{1}{\cancel{2}} \cdot \underset{1}{\cancel{2}} \cdot \underset{1}{\cancel{3}} \cdot \underset{1}{\cancel{3}} \cdot \underset{}{\cancel{5}}} = \dfrac{1}{2}$

9. $\dfrac{15}{16} \times \dfrac{39}{16} = \dfrac{3 \cdot 5 \cdot 3 \cdot 13}{2 \cdot 2 \cdot 2 \cdot 2 \cdot 2 \cdot 2 \cdot 2 \cdot 2} = \dfrac{584}{256} = 2\dfrac{73}{256}$

11. $\dfrac{8}{9} \times \dfrac{5}{4} = \dfrac{2 \cdot \overset{1}{\cancel{2}} \cdot \overset{1}{\cancel{2}} \cdot 5}{3 \cdot 3 \cdot \underset{1}{\cancel{2}} \cdot \underset{1}{\cancel{2}}} = \dfrac{10}{9} = 1\dfrac{1}{9}$

13. $\dfrac{1}{9} \times \dfrac{3}{2} = \dfrac{\overset{1}{\cancel{3}}}{\underset{1}{\cancel{3}} \cdot 3 \cdot 2} = \dfrac{1}{6}$

15. $\dfrac{2}{5} \times \dfrac{7}{4} = \dfrac{\overset{1}{\cancel{2}} \cdot 7}{5 \cdot 2 \cdot \underset{1}{\cancel{2}}} = \dfrac{7}{10}$

17. $\dfrac{1}{2} \times \dfrac{4}{1} = \dfrac{2 \cdot 2}{\underset{1}{\cancel{2}}} = 2$

19. $\dfrac{1}{5} \times \dfrac{10}{1} = \dfrac{2 \cdot \overset{1}{\cancel{5}}}{\underset{1}{\cancel{5}}} = 2$

21. $\dfrac{7}{15} \times \dfrac{5}{14} = \dfrac{\overset{1}{\cancel{7}} \cdot \overset{1}{\cancel{5}}}{3 \cdot \underset{1}{\cancel{5}} \cdot 2 \cdot \underset{1}{\cancel{7}}} = \dfrac{1}{6}$

23. $\dfrac{14}{3} \times \dfrac{9}{7} = \dfrac{2 \cdot \overset{1}{\cancel{7}} \cdot \overset{1}{\cancel{3}} \cdot 3}{\underset{1}{\cancel{3}} \cdot \underset{1}{\cancel{7}}} = 6$

25. $\dfrac{5}{9} \times \dfrac{3}{25} = \dfrac{\overset{1}{\cancel{5}} \cdot \overset{1}{\cancel{3}}}{3 \cdot \underset{1}{\cancel{3}} \cdot \underset{1}{\cancel{5}} \cdot 5} = \dfrac{1}{15}$

27. $\dfrac{2}{3} \times \dfrac{3}{1} = \dfrac{2 \cdot \overset{1}{\cancel{3}}}{\underset{1}{\cancel{3}}} = 2$

29. $\dfrac{5}{7} \times \dfrac{7}{2} = \dfrac{5 \cdot \overset{1}{\cancel{7}}}{\underset{1}{\cancel{7}} \cdot 2} = \dfrac{5}{2} = 2\dfrac{1}{2}$

31. $\dfrac{2}{3} \times \dfrac{9}{2} = \dfrac{\overset{1}{\cancel{2}} \cdot \overset{1}{\cancel{3}} \cdot 3}{\underset{1}{\cancel{3}} \cdot \underset{1}{\cancel{2}}} = 3$

33. $\dfrac{4}{1} \times \dfrac{3}{2} = \dfrac{2 \cdot \overset{1}{\cancel{2}} \cdot 3}{\underset{1}{\cancel{2}}} = 6$

35. $\dfrac{3}{2} \times \dfrac{1}{3} = \dfrac{\overset{1}{\cancel{3}}}{2 \cdot \underset{1}{\cancel{3}}} = \dfrac{1}{2}$

37. $\dfrac{7}{8} \div \dfrac{3}{4} = \dfrac{7}{8} \times \dfrac{\overset{1}{\cancel{4}}}{3} = \dfrac{7}{6} = 1\dfrac{1}{6}$

39. $\dfrac{5}{7} \div \dfrac{3}{14} = \dfrac{5}{7} \times \dfrac{\overset{2}{\cancel{14}}}{3} = \dfrac{10}{3} = 3\dfrac{1}{3}$

Objective B Exercises

41. $\dfrac{5}{6} \times \dfrac{1}{25} = \dfrac{\overset{1}{\cancel{5}}}{2 \cdot 3 \cdot \underset{1}{\cancel{5}} \cdot 5} = \dfrac{1}{30}$

43. $\dfrac{6}{1} \div \dfrac{10}{3} = \dfrac{6}{1} \times \dfrac{3}{10} = \dfrac{2 \cdot 3 \cdot 3}{2 \cdot 5} = \dfrac{9}{5} = 1\dfrac{4}{5}$

45. $\dfrac{13}{2} \div \dfrac{1}{2} = \dfrac{13}{2} \times \dfrac{2}{1} = \dfrac{13 \cdot 2}{2} = 13$

47. $\dfrac{5}{12} \div \dfrac{24}{5} = \dfrac{5}{12} \times \dfrac{5}{24} = \dfrac{5 \cdot 5}{2 \cdot 2 \cdot 3 \cdot 2 \cdot 2 \cdot 3} = \dfrac{25}{288}$

49. $\dfrac{33}{4} \div \dfrac{11}{4} = \dfrac{33}{4} \times \dfrac{4}{11} = \dfrac{3 \cdot 11 \cdot 2 \cdot 2}{2 \cdot 2 \cdot 11} = 3$

51. $\dfrac{21}{5} \div \dfrac{21}{1} = \dfrac{21}{5} \times \dfrac{1}{21} = \dfrac{3 \cdot 7}{5 \cdot 3 \cdot 7} = \dfrac{1}{5}$

53. $\dfrac{11}{12} \div \dfrac{7}{3} = \dfrac{11}{12} \times \dfrac{3}{7} = \dfrac{11 \cdot 3}{2 \cdot 2 \cdot 3 \cdot 7} = \dfrac{11}{28}$

55. $\dfrac{5}{16} \div \dfrac{43}{8} = \dfrac{5}{16} \times \dfrac{8}{43} = \dfrac{5 \cdot 2 \cdot 2 \cdot 2}{2 \cdot 2 \cdot 2 \cdot 2 \cdot 43} = \dfrac{5}{86}$

57. $35 \div \dfrac{7}{24} = \dfrac{35}{1} \times \dfrac{24}{7} = \dfrac{5 \cdot 7 \cdot 2 \cdot 2 \cdot 2 \cdot 3}{7} = 120$

59. $\dfrac{11}{18} \div \dfrac{20}{9} = \dfrac{11}{18} \times \dfrac{9}{20} = \dfrac{11 \cdot 3 \cdot 3}{2 \cdot 3 \cdot 3 \cdot 2 \cdot 2 \cdot 5} = \dfrac{11}{40}$

61. $\dfrac{33}{16} \div \dfrac{5}{2} = \dfrac{33}{16} \times \dfrac{2}{5} = \dfrac{3 \cdot 11 \cdot 2}{2 \cdot 2 \cdot 2 \cdot 2 \cdot 5} = \dfrac{33}{40}$

63. $\dfrac{5}{3} \div \dfrac{3}{8} = \dfrac{5}{3} \times \dfrac{8}{3} = \dfrac{5 \cdot 2 \cdot 2 \cdot 2}{3 \cdot 3} = \dfrac{40}{9} = 4\dfrac{4}{9}$

65. $\dfrac{13}{8} \div \dfrac{4}{1} = \dfrac{13}{8} \times \dfrac{1}{4} = \dfrac{13}{2 \cdot 2 \cdot 2 \cdot 2 \cdot 2} = \dfrac{13}{32}$

67. $16 \div \dfrac{3}{2} = \dfrac{16}{1} \times \dfrac{2}{3} = \dfrac{2 \cdot 2 \cdot 2 \cdot 2 \cdot 2}{3} = \dfrac{32}{3} = 10\dfrac{2}{3}$

69. $\dfrac{133}{8} \div \dfrac{5}{3} = \dfrac{133}{8} \times \dfrac{3}{5} = \dfrac{133 \cdot 3}{2 \cdot 2 \cdot 2 \cdot 5} = \dfrac{399}{40} = 9\dfrac{39}{40}$

71. $\dfrac{4}{3} \div \dfrac{53}{9} = \dfrac{4}{3} \times \dfrac{9}{53} = \dfrac{2 \cdot 2 \cdot 3 \cdot 3}{3 \cdot 53} = \dfrac{12}{53}$

73. $\dfrac{413}{5} \div \dfrac{191}{10} = \dfrac{413}{5} \times \dfrac{10}{191} = \dfrac{7 \cdot 59 \cdot 2 \cdot 5}{5 \cdot 191}$
$= \dfrac{826}{191} = 4\dfrac{62}{191}$

75. $\dfrac{102}{1} \div \dfrac{3}{2} = \dfrac{102}{1} \times \dfrac{2}{3} = \dfrac{2 \cdot 3 \cdot 17 \cdot 2}{3} = 68$

77. $\dfrac{58}{7} \div 1 = \dfrac{58}{7} \times 1 = \dfrac{58}{7} = 8\dfrac{2}{7}$

79. $\dfrac{80}{9} \div \dfrac{49}{18} = \dfrac{80}{9} \times \dfrac{18}{49} = \dfrac{2 \cdot 2 \cdot 2 \cdot 2 \cdot 5 \cdot 2 \cdot 3 \cdot 3}{3 \cdot 3 \cdot 7 \cdot 7}$
$= \dfrac{160}{49} = 3\dfrac{13}{49}$

81. $\dfrac{59}{8} \div \dfrac{59}{32} = \dfrac{59}{8} \times \dfrac{32}{59} = \dfrac{59 \cdot 2 \cdot 2 \cdot 2 \cdot 2 \cdot 2}{2 \cdot 2 \cdot 2 \cdot 59} = 4$

83. $2\dfrac{3}{4} \div 1\dfrac{23}{32} = \dfrac{11}{4} \div \dfrac{55}{32} = \dfrac{11}{4} \times \dfrac{32}{55} = \dfrac{8}{5} = 1\dfrac{3}{5}$

85. $\dfrac{14}{77} \div 3\dfrac{1}{9} = \dfrac{14}{77} \div \dfrac{28}{9} = \dfrac{14}{17} \times \dfrac{9}{28} = \dfrac{9}{34}$

Objective C Application Problems

87. Strategy To find the number of servings in 16 ounces of cereal, divide 16 by the amount in each serving $\left(1\dfrac{1}{3} \text{ ounces}\right)$.

Solution $16 \div 1\dfrac{1}{3} = 16 \div \dfrac{4}{3}$
$= 16 \times \dfrac{3}{4} = \dfrac{16 \cdot 3}{4}$
$= 12$
There are 12 servings in 16 ounces of cereal.

89. Strategy To find the cost of each acre, divide the total cost ($200,000) by the number of acres $\left(8\dfrac{1}{3}\right)$.

Solution $\$200,000 \div 8\dfrac{1}{3} = 200,000 \div \dfrac{25}{3}$
$= 200,000 \times \dfrac{3}{25}$
$= \dfrac{200,000 \cdot 3}{25}$
$= \$24,000$
Each acre costs $24,000.

91. Strategy To find the number of turns, divide the distance for the nut to move $\left(1\frac{7}{8}\text{ inches}\right)$ by the distance the nut moves for each turn $\left(\frac{5}{32}\text{ inches}\right)$.

Solution $1\frac{7}{8} \div \frac{5}{32} = \frac{15}{8} \div \frac{5}{32}$

$= \frac{15}{8} \times \frac{32}{5}$

$= \frac{15 \cdot \overset{4}{\cancel{32}}}{\underset{1}{\cancel{8}} \cdot \underset{1}{\cancel{5}}}$

$= 12$

The nut will turn 12 times in moving $1\frac{7}{8}$ inches.

93a. Strategy To find the total weight of the fat and bone, subtract the weight after trimming $\left(9\frac{1}{3}\text{ pounds}\right)$ from the original weight $\left(10\frac{3}{4}\text{ pounds}\right)$.

Solution $10\frac{3}{4} = 10\frac{9}{12}$

$-9\frac{1}{3} = 9\frac{4}{12}$

$\rule{2cm}{0.4pt}$

$1\frac{5}{12}$

The total weight of the fat and bone was $1\frac{5}{12}$ pounds.

93. b. Strategy To find the number of servings, divide the weight after trimming $\left(9\frac{1}{3}\text{ pounds}\right)$ by the weight of one serving $\left(\frac{1}{3}\text{ pound}\right)$.

Solution $9\frac{1}{3} \div \frac{1}{3} = \frac{28}{3} \div \frac{1}{3} = \frac{28}{3} \times \frac{3}{1}$

$= \frac{28 \cdot \overset{1}{\cancel{3}}}{\underset{1}{\cancel{3}} \cdot 1} = 28$

The chef can cut 28 $\frac{1}{3}$-pound servings from the roast.

95. $6\frac{1}{4} \div \frac{1}{2} = \frac{25}{\underset{2}{\cancel{4}}} \times \frac{\overset{1}{\cancel{2}}}{1} = \frac{25}{2} = 12\frac{1}{2}$

The actual length of wall a is $12\frac{1}{2}$ feet.

$9 \div \frac{1}{2} = \frac{9}{1} \times \frac{2}{1} = 18$

The actual length of wall b is 18 feet.

$7\frac{7}{8} \div \frac{1}{2} = \frac{63}{\underset{4}{\cancel{8}}} \times \frac{\overset{1}{\cancel{2}}}{1} = \frac{63}{4} = 15\frac{3}{4}$

The actual length of wall c is $15\frac{3}{4}$ feet.

Applying the Concepts

97. No, for example, $4 \div \frac{1}{2} = 8$

99. First, find the spacing between the three columns.

$\frac{3}{8} \times 2 = \frac{3}{\underset{4}{\cancel{8}}} \times \frac{\overset{1}{\cancel{2}}}{1} = \frac{3}{4}$ inch

Second, find the remaining space for the columns.

$7\frac{1}{2} - \frac{3}{4} = 7\frac{2}{4} - \frac{3}{4} = 6\frac{6}{4} - \frac{3}{4} = 6\frac{3}{4}$ inches

Third, divide that space among the three columns.

$6\frac{3}{4} \div 3 = \frac{\overset{9}{\cancel{27}}}{4} \times \frac{1}{\underset{1}{\cancel{3}}} = \frac{9}{4} = 2\frac{1}{4}$ inches

101. a. $\frac{5}{6} \div \frac{1}{3} = \frac{5 \div 1}{6 \div 3} = \frac{5}{2} = 2\frac{1}{2}$

b. $\frac{4}{21} \div \frac{2}{7} = \frac{4 \div 2}{21 \div 7} = \frac{2}{3}$

c. $\frac{5}{9} \div \frac{5}{3} = \frac{5 \div 5}{9 \div 3} = \frac{1}{3}$

d. $\frac{15}{16} \div \frac{3}{8} = \frac{15 \div 3}{16 \div 8} = \frac{5}{2} = 2\frac{1}{2}$

SECTION 2.8

Objective A Exercises

1. $\frac{11}{40} < \frac{19}{40}$

3. $\frac{2}{3} = \frac{14}{21}, \frac{5}{7} = \frac{15}{21}, \frac{2}{3} < \frac{5}{7}$

5. $\frac{5}{8} = \frac{15}{24}, \frac{7}{12} = \frac{14}{24}, \frac{5}{8} > \frac{7}{12}$

7. $\frac{7}{9} = \frac{28}{36}, \frac{11}{12} = \frac{33}{36}, \frac{7}{9} < \frac{11}{12}$

9. $\frac{13}{14} = \frac{39}{42}, \frac{19}{21} = \frac{38}{42}, \frac{13}{14} > \frac{19}{21}$

11. $\dfrac{7}{24} = \dfrac{35}{120}, \ \dfrac{11}{30} = \dfrac{44}{120}, \ \dfrac{7}{24} < \dfrac{11}{30}$

Objective B Exercises

13. $\left(\dfrac{3}{8}\right)^2 = \dfrac{3}{8} \cdot \dfrac{3}{8} = \dfrac{9}{64}$

15. $\left(\dfrac{2}{9}\right)^3 = \dfrac{2}{9} \cdot \dfrac{2}{9} \cdot \dfrac{2}{9} = \dfrac{8}{729}$

17. $\left(\dfrac{2}{3}\right) \cdot \left(\dfrac{1}{2}\right)^4 = \left(\dfrac{2}{3}\right) \cdot \left(\dfrac{1}{2} \cdot \dfrac{1}{2} \cdot \dfrac{1}{2} \cdot \dfrac{1}{2}\right)$

$= \dfrac{\overset{1}{2} \cdot 1 \cdot 1 \cdot 1 \cdot 1}{3 \cdot \underset{1}{2} \cdot 2 \cdot 2 \cdot 2} = \dfrac{1}{24}$

19. $\left(\dfrac{2}{5}\right)^3 \cdot \left(\dfrac{5}{7}\right)^2 = \left(\dfrac{2}{5} \cdot \dfrac{2}{5} \cdot \dfrac{2}{5}\right) \cdot \left(\dfrac{5}{7} \cdot \dfrac{5}{7}\right)$

$= \dfrac{2 \cdot 2 \cdot 2 \cdot \overset{1}{5} \cdot \overset{1}{5}}{\underset{1}{5} \cdot \underset{1}{5} \cdot 5 \cdot 7 \cdot 7} = \dfrac{8}{245}$

21. $\left(\dfrac{1}{3}\right)^4 \cdot \left(\dfrac{9}{11}\right)^2 = \left(\dfrac{1}{3} \cdot \dfrac{1}{3} \cdot \dfrac{1}{3} \cdot \dfrac{1}{3}\right) \cdot \left(\dfrac{9}{11} \cdot \dfrac{9}{11}\right)$

$= \dfrac{1 \cdot 1 \cdot 1 \cdot 1 \cdot \overset{1}{3} \cdot \overset{1}{3} \cdot \overset{1}{3} \cdot \overset{1}{3}}{\underset{1}{3} \cdot \underset{1}{3} \cdot \underset{1}{3} \cdot \underset{1}{3} \cdot 11 \cdot 11} = \dfrac{1}{121}$

23. $\left(\dfrac{2}{3}\right)^4 \cdot \left(\dfrac{81}{100}\right)^2 = \left(\dfrac{2}{3} \cdot \dfrac{2}{3} \cdot \dfrac{2}{3} \cdot \dfrac{2}{3}\right)\left(\dfrac{81}{100} \cdot \dfrac{81}{100}\right)$

$= \dfrac{2 \cdot 2 \cdot 2 \cdot 2 \cdot \overset{1}{3} \cdot \overset{1}{3} \cdot \overset{1}{3} \cdot \overset{1}{3} \cdot \overset{1}{3} \cdot \overset{1}{3} \cdot 3 \cdot 3}{\underset{1}{3} \cdot \underset{1}{3} \cdot \underset{1}{3} \cdot \underset{1}{3} \cdot 2 \cdot \underset{1}{2} \cdot 5 \cdot 5 \cdot 2 \cdot \underset{1}{2} \cdot 5 \cdot 5} = \dfrac{81}{625}$

25. $\left(\dfrac{2}{7}\right) \cdot \left(\dfrac{7}{8}\right)^2 \cdot \left(\dfrac{8}{9}\right) = \left(\dfrac{2}{7}\right) \cdot \left(\dfrac{7}{8} \cdot \dfrac{7}{8}\right) \cdot \left(\dfrac{8}{9}\right)$

$= \dfrac{\overset{1}{2} \cdot \overset{1}{7} \cdot 7 \cdot \overset{1}{2} \cdot \overset{1}{2} \cdot \overset{1}{2}}{\underset{1}{7} \cdot \underset{1}{2} \cdot \underset{1}{2} \cdot \underset{1}{2} \cdot \underset{1}{2} \cdot 2 \cdot 2 \cdot 3 \cdot 3} = \dfrac{7}{36}$

27. $4 \cdot \left(\dfrac{3}{4}\right)^3 \cdot \left(\dfrac{4}{7}\right)^2 = \left(\dfrac{4}{1}\right) \cdot \left(\dfrac{3}{4} \cdot \dfrac{3}{4} \cdot \dfrac{3}{4}\right) \cdot \left(\dfrac{4}{7} \cdot \dfrac{4}{7}\right)$

$= \dfrac{\overset{1}{2} \cdot \overset{1}{2} \cdot 3 \cdot 3 \cdot 3 \cdot \overset{1}{2} \cdot \overset{1}{2} \cdot \overset{1}{2} \cdot \overset{1}{2}}{2 \cdot 2 \cdot \underset{1}{2} \cdot \underset{1}{2} \cdot \underset{1}{2} \cdot \underset{1}{2} \cdot 7 \cdot 7} = \dfrac{27}{49}$

Objective C Exercises

29. $\dfrac{1}{2} - \dfrac{1}{3} + \dfrac{2}{3} = \dfrac{3}{6} - \dfrac{2}{6} + \dfrac{2}{3}$

$= \dfrac{1}{6} + \dfrac{2}{3}$

$= \dfrac{1}{6} + \dfrac{4}{6}$

$= \dfrac{5}{6}$

31. $\dfrac{1}{3} \div \dfrac{1}{2} + \dfrac{3}{4} = \dfrac{1}{3} \cdot \dfrac{2}{1} + \dfrac{3}{4}$

$= \dfrac{2}{3} + \dfrac{3}{4}$

$= \dfrac{8}{12} + \dfrac{9}{12}$

$= \dfrac{17}{12} = 1\dfrac{5}{12}$

33. $\left(\dfrac{3}{4}\right)^2 - \dfrac{5}{12} = \dfrac{9}{16} - \dfrac{5}{12}$

$= \dfrac{27}{48} - \dfrac{20}{48}$

$= \dfrac{7}{48}$

35. $\dfrac{5}{6} \cdot \left(\dfrac{2}{3} - \dfrac{1}{6}\right) + \dfrac{7}{18} = \dfrac{5}{6} \cdot \left(\dfrac{4}{6} - \dfrac{1}{6}\right) + \dfrac{7}{18}$

$= \dfrac{5}{6} \cdot \dfrac{3}{6} + \dfrac{7}{18}$

$= \dfrac{5}{12} + \dfrac{7}{18}$

$= \dfrac{15}{36} + \dfrac{14}{36}$

$= \dfrac{29}{36}$

37. $\dfrac{7}{12} - \left(\dfrac{2}{3}\right)^2 + \dfrac{5}{8} = \dfrac{7}{12} - \dfrac{4}{9} + \dfrac{5}{8}$

$= \dfrac{21}{36} - \dfrac{16}{36} + \dfrac{5}{8}$

$= \dfrac{5}{36} + \dfrac{5}{8}$

$= \dfrac{10}{72} + \dfrac{45}{72}$

$= \dfrac{55}{72}$

39. $\dfrac{3}{4} \cdot \left(\dfrac{4}{9}\right)^2 + \dfrac{1}{2} = \dfrac{3}{4} \cdot \dfrac{16}{81} + \dfrac{1}{2}$

$= \dfrac{4}{27} + \dfrac{1}{2}$

$= \dfrac{8}{54} + \dfrac{27}{54}$

$= \dfrac{35}{54}$

41. $\left(\dfrac{1}{2} + \dfrac{3}{4}\right) \div \dfrac{5}{8} = \left(\dfrac{2}{4} + \dfrac{3}{4}\right) \div \dfrac{5}{8}$

$= \dfrac{5}{4} \cdot \dfrac{8}{5}$

$= 2$

43. $\dfrac{3}{8} \div \left(\dfrac{5}{12} + \dfrac{3}{8}\right) = \dfrac{3}{8} \div \left(\dfrac{10}{24} + \dfrac{9}{24}\right)$

$= \dfrac{3}{8} \div \dfrac{19}{24}$

$= \dfrac{3}{8} \cdot \dfrac{24}{19}$

$= \dfrac{9}{19}$

45. $\left(\dfrac{3}{8}\right)^2 \div \left(\dfrac{3}{7} + \dfrac{3}{14}\right) = \left(\dfrac{3}{8}\right)^2 \div \left(\dfrac{6}{14} + \dfrac{3}{14}\right)$

$= \left(\dfrac{3}{8}\right)^2 \div \dfrac{9}{14}$

$= \dfrac{9}{64} \cdot \dfrac{14}{9}$

$= \dfrac{7}{32}$

47. $\dfrac{2}{5} \div \dfrac{3}{8} \cdot \dfrac{4}{5} = \dfrac{2}{5} \cdot \dfrac{8}{3} \cdot \dfrac{4}{5}$

$= \dfrac{16}{15} \cdot \dfrac{4}{5}$

$= \dfrac{64}{75}$

Applying the Concepts

49. The "puzzle" works as it does because the sum of the fractions $\dfrac{1}{2}$, $\dfrac{1}{3}$, and $\dfrac{1}{9}$ is $\dfrac{17}{18}$, not 1. As a result, the first child actually received $\dfrac{9}{17}$ of the horses, not $\dfrac{1}{2}$; the second child received $\dfrac{6}{17}$ of the horses, not $\dfrac{1}{3}$; and the third child got $\dfrac{2}{17}$ of the horses, not $\dfrac{1}{9}$.

CHAPTER REVIEW

1. $\dfrac{30}{45} = \dfrac{2 \cdot \cancel{3} \cdot \cancel{5}}{3 \cdot \cancel{3} \cdot \cancel{5}} = \dfrac{2}{3}$

2. $\left(\dfrac{3}{4}\right)^3 \cdot \dfrac{20}{27} = \left(\dfrac{3}{4} \cdot \dfrac{3}{4} \cdot \dfrac{3}{4}\right)\left(\dfrac{20}{27}\right)$

$= \dfrac{\cancel{3} \cdot \cancel{3} \cdot \cancel{3} \cdot \cancel{2} \cdot \cancel{2} \cdot 5}{2 \cdot 2 \cdot 2 \cdot 2 \cdot \cancel{2} \cdot \cancel{2} \cdot \cancel{3} \cdot \cancel{3} \cdot \cancel{3}} = \dfrac{5}{16}$

3. $\dfrac{13}{4}$

4. $\begin{aligned} \dfrac{2}{3} &= \dfrac{12}{18} \\ \dfrac{5}{6} &= \dfrac{15}{18} \\ + \dfrac{2}{9} &= \dfrac{4}{18} \\ \hline \dfrac{31}{18} &= 1\dfrac{13}{18} \end{aligned}$

5. $\dfrac{11}{18} = \dfrac{44}{72}$, $\dfrac{17}{24} = \dfrac{51}{72}$, $\dfrac{11}{18} < \dfrac{17}{24}$

6. $\begin{aligned} 18\dfrac{1}{6} &= 18\dfrac{7}{42} = 17\dfrac{49}{42} \\ -3\dfrac{5}{7} &= 3\dfrac{30}{42} = 3\dfrac{30}{42} \\ \hline & 14\dfrac{19}{42} \end{aligned}$

7. $\dfrac{2}{7}\left[\dfrac{5}{8} - \dfrac{1}{3}\right] \div \dfrac{3}{5} = \dfrac{2}{7}\left[\dfrac{15}{24} - \dfrac{8}{24}\right] \div \dfrac{3}{5}$

$= \dfrac{2}{7}\left[\dfrac{7}{24}\right] \div \dfrac{3}{5} = \dfrac{2 \cdot \cancel{7}}{7 \cdot 24} \div \dfrac{3}{5}$

$= \dfrac{1}{12} \times \dfrac{5}{3} = \dfrac{5}{36}$

8. $2\dfrac{1}{3} \times 3\dfrac{7}{8} = \dfrac{7}{3} \times \dfrac{31}{8} = \dfrac{7 \cdot 31}{3 \cdot 8} = \dfrac{217}{24} = 9\dfrac{1}{24}$

9. $1\dfrac{1}{3} \div \dfrac{2}{3} = \dfrac{4}{3} \div \dfrac{2}{3} = \dfrac{4}{3} \times \dfrac{3}{2}$

$= \dfrac{4 \cdot 3}{3 \cdot 2} = \dfrac{\cancel{2} \cdot \cancel{2} \cdot \cancel{3}}{\cancel{3} \cdot \cancel{2}} = 2$

10. $\begin{aligned} \dfrac{17}{24} &= \dfrac{34}{48} \\ - \dfrac{3}{16} &= \dfrac{9}{48} \\ \hline & \dfrac{25}{48} \end{aligned}$

11. $8\dfrac{2}{3} \div 2\dfrac{3}{5} = \dfrac{26}{3} \div \dfrac{13}{5} = \dfrac{26}{3} \times \dfrac{5}{13}$

$= \dfrac{26 \cdot 5}{3 \cdot 13} = \dfrac{2 \cdot \cancel{13} \cdot 5}{3 \cdot \cancel{13}} = \dfrac{10}{3} = 3\dfrac{1}{3}$

12.

$20 =$	$\boxed{2 \cdot 2}$		5
$48 =$	$2 \cdot 2 \cdot 2 \cdot 2$	3	

GCF $= 2 \cdot 2 = 4$

13. $\dfrac{2 \cdot 12}{3 \cdot 12} = \dfrac{24}{36}$

14. $\dfrac{15}{28} \div \dfrac{5}{7} = \dfrac{15}{28} \times \dfrac{7}{5} = \dfrac{15 \cdot 7}{28 \cdot 5} = \dfrac{3 \cdot \cancel{5} \cdot \cancel{7}}{2 \cdot 2 \cdot \cancel{7} \cdot \cancel{5}} = \dfrac{3}{4}$

15. $\dfrac{8\cdot 4}{11\cdot 4}=\dfrac{32}{44}$

16. $2\dfrac{1}{4}\times 7\dfrac{1}{3}=\dfrac{9}{4}\times\dfrac{22}{3}=\dfrac{9\cdot 22}{4\cdot 3}=\dfrac{3\cdot\overset{1}{\cancel{3}}\cdot\overset{1}{\cancel{2}}\cdot 11}{\underset{1}{\cancel{2}}\cdot 2\cdot\underset{1}{\cancel{3}}}$

$=\dfrac{33}{2}=16\dfrac{1}{2}$

17.
$$\begin{array}{l}18=\begin{array}{|c|c|}\hline 2 & 3\\ \hline 2 & 3\cdot 3\\ \hline\end{array}\\ 12=\begin{array}{|c|c|}\hline (2\cdot 2) & (3\cdot 3)\\ \hline\end{array}\\ \text{LCM}=2\cdot 2\cdot 3\cdot 3=36\end{array}$$

18. $\dfrac{16}{44}=\dfrac{\overset{1}{\cancel{2}}\cdot\overset{1}{\cancel{2}}\cdot 2\cdot 2}{\underset{1}{\cancel{2}}\cdot\underset{1}{\cancel{2}}\cdot 11}=\dfrac{4}{11}$

19.
$$\begin{array}{r}\dfrac{3}{8}\\[4pt]\dfrac{5}{8}\\[4pt]+\dfrac{1}{8}\\[4pt]\hline\dfrac{9}{8}=1\dfrac{1}{8}\end{array}$$

20.
$$\begin{array}{r}16=15\dfrac{8}{8}\\[4pt]-5\dfrac{7}{8}=5\dfrac{7}{8}\\[4pt]\hline 10\dfrac{1}{8}\end{array}$$

21.
$$\begin{array}{r}4\dfrac{4}{9}=4\dfrac{24}{54}\\[4pt]2\dfrac{1}{6}=2\dfrac{9}{54}\\[4pt]+11\dfrac{17}{27}=11\dfrac{34}{54}\\[4pt]\hline 17\dfrac{67}{54}=18\dfrac{13}{54}\end{array}$$

22.
$$\begin{array}{l}15=\begin{array}{|c|c|}\hline 3 & 5\\ \hline 3 & (5)\\ \hline\end{array}\\ 25=\begin{array}{|c|c|}\hline & 5\cdot 5\\ \hline\end{array}\\ \text{GCF}=5\end{array}$$

23. $\begin{array}{r}3\\ 5\overline{)17}\\ \underline{-15}\\ 2\end{array}\qquad \dfrac{17}{5}=3\dfrac{2}{5}$

24. $\left[\dfrac{4}{5}-\dfrac{2}{3}\right]^2\div\dfrac{4}{15}=\left[\dfrac{12}{15}-\dfrac{10}{15}\right]^2\div\dfrac{4}{15}$

$=\left(\dfrac{2}{15}\right)^2\div\dfrac{4}{15}=\left(\dfrac{2}{15}\right)\left(\dfrac{2}{15}\right)\div\left(\dfrac{4}{15}\right)$

$=\dfrac{4}{225}\times\dfrac{15}{4}=\dfrac{4\cdot 15}{225\cdot 4}=\dfrac{1}{15}$

25.
$$\begin{array}{r}\dfrac{3}{8}=\dfrac{9}{24}\\[4pt]1\dfrac{2}{3}=1\dfrac{16}{24}\\[4pt]+3\dfrac{5}{6}=3\dfrac{20}{24}\\[4pt]\hline 4\dfrac{45}{24}=5\dfrac{21}{24}=5\dfrac{7}{8}\end{array}$$

26.
$$\begin{array}{l}18=\begin{array}{|c|c|}\hline 2 & 3\\ \hline (2) & 3\cdot 3\\ \hline\end{array}\\ 27=\begin{array}{|c|c|}\hline & (3\cdot 3\cdot 3)\\ \hline\end{array}\\ \text{LCM}=2\cdot 3\cdot 3\cdot 3=54\end{array}$$

27.
$$\begin{array}{r}\dfrac{11}{18}\\[4pt]-\dfrac{5}{18}\\[4pt]\hline\dfrac{6}{18}=\dfrac{1}{3}\end{array}$$

28. $2\dfrac{5}{7}=\dfrac{14+5}{7}=\dfrac{19}{7}$

29. $\dfrac{5}{6}\div\dfrac{5}{12}=\dfrac{5}{6}\cdot\dfrac{12}{5}=\dfrac{5\cdot 12}{6\cdot 5}=\dfrac{\overset{1}{\cancel{5}}\cdot 2\cdot 2\cdot\overset{1}{\cancel{3}}}{\underset{1}{\cancel{2}}\cdot\underset{1}{\cancel{3}}\cdot\underset{1}{\cancel{5}}}=2$

30. $\dfrac{5}{12}\times\dfrac{4}{25}=\dfrac{5\cdot 4}{12\cdot 25}=\dfrac{\overset{1}{\cancel{5}}\cdot 2\cdot 2}{2\cdot 2\cdot 3\cdot\underset{1}{\cancel{5}}\cdot 5}=\dfrac{1}{15}$

31. $\dfrac{11}{50}\times\dfrac{25}{44}=\dfrac{11\cdot 25}{50\cdot 44}=\dfrac{\overset{1}{\cancel{11}}\cdot\overset{1}{\cancel{5}}\cdot\overset{1}{\cancel{5}}}{2\cdot\underset{1}{\cancel{5}}\cdot\underset{1}{\cancel{5}}\cdot 2\cdot 2\cdot\underset{1}{\cancel{11}}}=\dfrac{1}{8}$

32. $1\dfrac{7}{8}$

33. Strategy   To find the total rainfall for the 3 months, add the amounts of rain from each month $\left(5\dfrac{7}{8},\ 6\dfrac{2}{3},\ \text{and}\ 8\dfrac{3}{4}\ \text{inches}\right).$

Solution
$$\begin{array}{r}5\dfrac{7}{8}=5\dfrac{21}{24}\\[4pt]6\dfrac{2}{3}=6\dfrac{16}{24}\\[4pt]+8\dfrac{3}{4}=8\dfrac{18}{24}\\[4pt]\hline 19\dfrac{55}{24}=21\dfrac{7}{24}\end{array}$$

The total rainfall for the 3 months was $21\dfrac{7}{24}$ inches.

34. Strategy   To find the cost of each acre, divide the total cost ($168,000) by the number of acres $\left(4\frac{2}{3}\right)$.

Solution $\$168,000 \div 4\frac{2}{3} = 168,000 \div \frac{14}{3}$
$$= 168,000 \times \frac{3}{14}$$
$$= \$36,000$$
The cost per acre was $36,000.

35. Strategy   To find how many miles the second checkpoint is from the finish line:

- Add the distance to the first checkpoint $\left(4\frac{1}{2} \text{ miles}\right)$ to the distance between the first checkpoint and the second checkpoint $\left(5\frac{3}{4} \text{ miles}\right)$.

- Subtract the total distance to the second checkpoint from the entire length of the race (15 miles).

Solution

$4\frac{1}{2} = 4\frac{2}{4}$ $\qquad$ $15 = 14\frac{4}{4}$

$+5\frac{3}{4} = 5\frac{3}{4}$ $\qquad$ $-10\frac{1}{4} = 10\frac{1}{4}$

$9\frac{5}{4} = 10\frac{1}{4}$ $\qquad$ $4\frac{3}{4}$

The second checkpoint is $4\frac{3}{4}$ miles from the finish line.

36. Strategy   To find how many miles the car can travel, multiply the number of miles the car can travel on 1 gallon (36) by the number of gallons used $\left(6\frac{3}{4}\right)$.

Solution $36 \times 6\frac{3}{4} = 36 \times \frac{27}{4} = \frac{36 \cdot 27}{4} = 243$
The car can travel 243 miles.

CHAPTER TEST

1. $\dfrac{9}{11} \times \dfrac{44}{81} = \dfrac{9 \cdot 44}{11 \cdot 81} = \dfrac{\overset{1}{\cancel{3}} \cdot \overset{1}{\cancel{3}} \cdot 2 \cdot 2 \cdot \overset{1}{\cancel{11}}}{\underset{1}{\cancel{11}} \cdot \underset{1}{\cancel{3}} \cdot \underset{1}{\cancel{3}} \cdot 3 \cdot 3} = \dfrac{4}{9}$

2.
$24 = \boxed{2 \cdot 2 \cdot 2} \;|\; 3 \;|\;$
$80 = \boxed{2 \cdot 2 \cdot 2} \cdot 2 \;|\; \;|\; 5$
$\text{GCF} = 2 \cdot 2 \cdot 2 = 8$

3. $\dfrac{5}{9} \div \dfrac{7}{18} = \dfrac{5}{9} \times \dfrac{18}{7} = \dfrac{5 \cdot 2 \cdot \overset{1}{\cancel{3}} \cdot \overset{1}{\cancel{3}}}{\underset{1}{\cancel{3}} \cdot \underset{1}{\cancel{3}} \cdot 7} = \dfrac{10}{7} = 1\dfrac{3}{7}$

4. $\left(\dfrac{3}{4}\right)^2 \div \left(\dfrac{2}{3} + \dfrac{5}{6}\right) - \dfrac{1}{12} = \left(\dfrac{3}{4} \cdot \dfrac{3}{4}\right) \div \left(\dfrac{4}{6} + \dfrac{5}{6}\right) - \dfrac{1}{12}$
$$= \dfrac{9}{16} \div \left(\dfrac{9}{6}\right) - \dfrac{1}{12}$$
$$= \dfrac{9}{16} \div \dfrac{3}{2} - \dfrac{1}{12}$$
$$= \dfrac{9}{16} \times \dfrac{2}{3} - \dfrac{1}{12}$$
$$\dfrac{\overset{1}{\cancel{3}} \cdot 3 \cdot \overset{1}{\cancel{2}}}{2 \cdot 2 \cdot 2 \cdot 2 \cdot \underset{1}{\cancel{3}}} - \dfrac{1}{12} = \dfrac{3}{8} - \dfrac{1}{12} = \dfrac{9}{24} - \dfrac{2}{24} = \dfrac{7}{24}$$

5. $9\dfrac{4}{5} = \dfrac{45+4}{5} = \dfrac{49}{5}$

6. $5\dfrac{2}{3} \times 1\dfrac{7}{17} = \dfrac{17}{3} \times \dfrac{24}{17} = \dfrac{17 \cdot 24}{3 \cdot 17} = \dfrac{\overset{1}{\cancel{17}} \cdot 2 \cdot 2 \cdot 2 \cdot \overset{1}{\cancel{3}}}{\underset{1}{\cancel{3}} \cdot \underset{1}{\cancel{17}}} = 8$

7. $\dfrac{40}{64} = \dfrac{\overset{1}{\cancel{2}} \cdot \overset{1}{\cancel{2}} \cdot \overset{1}{\cancel{2}} \cdot 5}{\underset{1}{\cancel{2}} \cdot \underset{1}{\cancel{2}} \cdot \underset{1}{\cancel{2}} \cdot 2 \cdot 2 \cdot 2} = \dfrac{5}{8}$

8. $\dfrac{3}{8} = \dfrac{9}{24}, \;\; \dfrac{5}{12} = \dfrac{10}{24}, \;\; \dfrac{3}{8} < \dfrac{5}{12}$

9. $\left(\dfrac{1}{4}\right)^3 \div \left(\dfrac{1}{8}\right)^2 - \dfrac{1}{6} = \left(\dfrac{1}{4} \cdot \dfrac{1}{4} \cdot \dfrac{1}{4}\right) \div \left(\dfrac{1}{8} \cdot \dfrac{1}{8}\right) - \dfrac{1}{6}$
$$= \dfrac{1}{64} \div \dfrac{1}{64} - \dfrac{1}{6}$$
$$= \dfrac{1}{64} \times \dfrac{64}{1} - \dfrac{1}{6}$$
$$= 1 - \dfrac{1}{6}$$
$$= \dfrac{6}{6} - \dfrac{1}{6} = \dfrac{5}{6}$$

10.
$24 = \boxed{2 \cdot 2 \cdot 2} \;|\; \boxed{3} \;|\;$
$40 = \boxed{2 \cdot 2 \cdot 2} \;|\; \;|\; \boxed{5}$
$\text{LCM} = 2 \cdot 2 \cdot 2 \cdot 3 \cdot 5 = 120$

11.
$\dfrac{17}{24}$
$-\dfrac{11}{24}$
$\overline{\dfrac{6}{24}} = \dfrac{1}{4}$

12.
$5\overline{)18}$ with quotient 3, -15, remainder 3 $\qquad$ $\dfrac{18}{5} = 3\dfrac{3}{5}$

13. $6\dfrac{2}{3} \div 3\dfrac{1}{6} = \dfrac{20}{3} \div \dfrac{19}{6} = \dfrac{20}{3} \times \dfrac{6}{19}$
$$= \dfrac{2 \cdot 2 \cdot 5 \cdot 2 \cdot \overset{1}{\cancel{3}}}{\underset{1}{\cancel{3}} \cdot 19} = \dfrac{40}{19} = 2\dfrac{2}{19}$$

14. $\dfrac{5 \cdot 9}{8 \cdot 9} = \dfrac{45}{72}$

15.
$$\dfrac{5}{6} = \dfrac{75}{90}$$
$$\dfrac{7}{9} = \dfrac{70}{90}$$
$$+\dfrac{1}{15} = \dfrac{6}{90}$$
$$\dfrac{151}{90} = 1\dfrac{61}{90}$$

16.
$$23\dfrac{1}{8} = 23\dfrac{11}{88} = 22\dfrac{99}{88}$$
$$-9\dfrac{9}{44} = 9\dfrac{18}{88} = 9\dfrac{18}{88}$$
$$13\dfrac{81}{88}$$

17.
$$\dfrac{9}{16} = \dfrac{27}{48}$$
$$-\dfrac{5}{12} = \dfrac{20}{48}$$
$$\dfrac{7}{48}$$

18. $\left(\dfrac{2}{3}\right)^4 \left(\dfrac{27}{23}\right) = \left(\dfrac{2}{3} \cdot \dfrac{2}{3} \cdot \dfrac{2}{3} \cdot \dfrac{2}{3}\right)\left(\dfrac{27}{32}\right)$

$$= \dfrac{\overset{1}{\cancel{2}} \cdot \overset{1}{\cancel{2}} \cdot \overset{1}{\cancel{2}} \cdot \overset{1}{\cancel{2}} \cdot \overset{1}{\cancel{3}} \cdot \overset{1}{\cancel{3}} \cdot \overset{1}{\cancel{3}}}{\underset{1}{\cancel{3}} \cdot \underset{1}{\cancel{3}} \cdot \underset{1}{\cancel{3}} \cdot \underset{1}{\cancel{3}} \cdot \underset{1}{\cancel{2}} \cdot \underset{1}{\cancel{2}} \cdot \underset{1}{\cancel{2}} \cdot \cancel{2} \cdot 2} = \dfrac{1}{6}$$

19.
$$\dfrac{7}{12}$$
$$\dfrac{11}{12}$$
$$+\dfrac{5}{12}$$
$$\dfrac{23}{12} = 1\dfrac{11}{12}$$

20.
$$12\dfrac{5}{12} = 12\dfrac{25}{60}$$
$$+9\dfrac{17}{20} = 9\dfrac{51}{60}$$
$$21\dfrac{76}{60} = 22\dfrac{16}{60} = 22\dfrac{4}{15}$$

21. $\dfrac{11}{4}$

22. Strategy To find the electrician's earnings, multiply daily earnings ($240) by the number of days worked $\left(3\dfrac{1}{2}\right)$.

Solution $\$240 \times 3\dfrac{1}{2} = 240 \times \dfrac{7}{2} = \dfrac{240 \cdot 7}{2} = \$840$
The electrician earns $840.

23. Strategy To find how many lots were available:

- Find how many acres were being developed by subtracting the amount set aside for the park $\left(1\dfrac{3}{4} \text{ acres}\right)$ from the total parcel $\left(7\dfrac{1}{4} \text{ acres}\right)$.

- Divide the amount being developed by the size of each lot $\left(\dfrac{1}{2} \text{ acre}\right)$.

Solution
$$7\dfrac{1}{4} = 6\dfrac{5}{4}$$
$$-1\dfrac{3}{4} = 1\dfrac{3}{4}$$
$$5\dfrac{2}{4} = 5\dfrac{1}{2}$$
$$5\dfrac{1}{2} \div \dfrac{1}{2} = \dfrac{11}{2} \times \dfrac{2}{1} = \dfrac{11 \cdot 2}{2} = 11$$

24. Strategy To find the value of a share of stock at the end of the second month:

- Find the value at the end of the first month by adding the gain $\left(\$5\dfrac{5}{8}\right)$ to the original price $\left(\$24\dfrac{1}{2}\right)$.

- Find the value at the end of the second month by subtracting the loss $\left(\$2\dfrac{1}{4}\right)$ from the price at the end of the first month.

Solution
$$\$24\dfrac{1}{2} = 24\dfrac{4}{8}$$
$$+ 5\dfrac{5}{8} = 5\dfrac{5}{8}$$
$$\$29\dfrac{9}{8} = \$30\dfrac{1}{8}$$
$$\$30\dfrac{1}{8} = 29\dfrac{9}{8}$$
$$- 2\dfrac{1}{4} = 2\dfrac{2}{8}$$
$$\$27\dfrac{7}{8}$$

The price at the end of the second month is $27\dfrac{7}{8}$.

25. Strategy To find the total rainfall for the 3-month period, add the rainfall amounts for each of the months $\left(11\frac{1}{2},\ 7\frac{5}{8},\ \text{and}\ 2\frac{1}{3}\ \text{inches}\right)$.

Solution
$$11\frac{1}{2} = 11\frac{12}{24}$$
$$7\frac{5}{8} = 7\frac{15}{24}$$
$$+2\frac{1}{3} = 2\frac{8}{24}$$
$$\overline{\qquad\qquad}$$
$$20\frac{35}{24} = 21\frac{11}{24}$$

The total rainfall for the 3-month period was $21\frac{11}{24}$ inches.

CUMULATIVE REVIEW

1. 290,000

2.
$$\overset{\overset{9}{8}\overset{9}{\cancel{4}}\overset{13}{\cancel{0}}\ \overset{13}{\cancel{0}}\overset{3}{\cancel{4}}\overset{17}{\cancel{7}}}{3\,9\,0\,,\,0\,4\,7}$$
$$-\ \ \ 9\,8\,,\,7\,6\,9$$
$$\overline{\quad 2\,9\,1\,,\,2\,7\,8\quad}$$

3.
$$\begin{array}{r} 926 \\ \times\ \ 79 \\ \hline 8334 \\ 6482 \\ \hline 73{,}154 \end{array}$$

4.
$$\begin{array}{r} 540\ \text{r}12 \\ 57\overline{)30{,}792} \\ \underline{-285} \\ 229 \\ \underline{-228} \\ 12 \\ \underline{-\ \ 0} \\ 12 \end{array}$$

5. $4 \cdot (6-3) \div 6 - 1 = 4 \cdot 3 \div 6 - 1$
$= 12 \div 6 - 1$
$= 2 - 1$
$= 1$

6. $44 = 2 \cdot 2 \cdot 11$
$$\begin{array}{r} 11 \\ 2\overline{)22} \\ 2\overline{)44} \end{array}$$

7.
	2	3	5	7
30 =	②	③	⑤	7
42 =	2	3		⑦

LCM $= 2 \cdot 3 \cdot 5 \cdot 7 = 210$

8.
	2	3	5
60 =	(2·2)	3	⑤
80 =	2·2·2·2		5

GCF $= 2 \cdot 2 \cdot 5 = 20$

9. $7\frac{2}{3} = \frac{21+2}{3} = \frac{23}{3}$

10.
$$\begin{array}{r} 6\ \text{r}1 \\ 4\overline{)25} \\ \underline{-24} \\ 1 \end{array} \qquad\qquad \frac{25}{4} = 6\frac{1}{4}$$

11. $\frac{5 \cdot 3}{16 \cdot 3} = \frac{15}{48}$

12. $\frac{24}{60} = \frac{\overset{1}{\cancel{2}} \cdot \overset{1}{\cancel{2}} \cdot 2 \cdot \overset{1}{\cancel{3}}}{\underset{1}{\cancel{2}} \cdot \underset{1}{\cancel{2}} \cdot \underset{1}{\cancel{3}} \cdot 5} = \frac{2}{5}$

13.
$$\frac{7}{12} = \frac{28}{48}$$
$$+\frac{9}{16} = \frac{27}{48}$$
$$\overline{\qquad\qquad}$$
$$\frac{55}{48} = 1\frac{7}{48}$$

14.
$$3\frac{7}{8} = 3\frac{42}{48}$$
$$7\frac{5}{12} = 7\frac{20}{48}$$
$$+2\frac{15}{16} = 2\frac{45}{48}$$
$$\overline{\qquad\qquad}$$
$$12\frac{107}{48} = 14\frac{11}{48}$$

15.
$$\frac{11}{12} = \frac{22}{24}$$
$$-\frac{3}{8} = \frac{9}{24}$$
$$\overline{\qquad\qquad}$$
$$\frac{13}{24}$$

16.
$$5\frac{1}{6} = 5\frac{3}{18} = 4\frac{21}{18}$$
$$-3\frac{7}{18} = 3\frac{7}{18} = 3\frac{7}{18}$$
$$\overline{\qquad\qquad}$$
$$1\frac{14}{18} = 1\frac{7}{9}$$

17. $\frac{3}{8} \times \frac{14}{15} = \frac{3 \cdot 14}{8 \cdot 15} = \frac{\overset{1}{\cancel{3}} \cdot \overset{1}{\cancel{2}} \cdot 7}{2 \cdot 2 \cdot 2 \cdot \underset{1}{\cancel{3}} \cdot 5} = \frac{7}{20}$

18. $3\frac{1}{8} \times 2\frac{2}{5} = \frac{25}{8} \times \frac{12}{5} = \frac{25 \cdot 12}{8 \cdot 5}$

$= \frac{5 \cdot \overset{1}{\cancel{5}} \cdot \overset{1}{\cancel{2}} \cdot \overset{1}{\cancel{2}} \cdot 3}{2 \cdot 2 \cdot 2 \cdot \underset{1}{\cancel{5}}} = \frac{15}{2} = 7\frac{1}{2}$

19. $\frac{7}{16} \div \frac{5}{12} = \frac{7}{16} \times \frac{12}{5} = \frac{7 \cdot 12}{16 \cdot 5}$

$= \frac{7 \cdot \overset{1}{\cancel{2}} \cdot \overset{1}{\cancel{2}} \cdot 3}{\underset{1}{\cancel{2}} \cdot 2 \cdot 2 \cdot \underset{1}{\cancel{2}} \cdot 5} = \frac{21}{20} = 1\frac{1}{20}$

20. $6\dfrac{1}{8} \div 2\dfrac{1}{3} = \dfrac{49}{8} \div \dfrac{7}{3} = \dfrac{49}{8} \times \dfrac{3}{7} = \dfrac{49 \cdot 3}{8 \cdot 7}$

$= \dfrac{7 \cdot \overset{1}{7} \cdot 3}{2 \cdot 2 \cdot 2 \cdot \underset{1}{7}} = \dfrac{21}{8} = 2\dfrac{5}{8}$

21. $\left(\dfrac{1}{2}\right)^3 \cdot \dfrac{8}{9} = \left(\dfrac{1}{2} \cdot \dfrac{1}{2} \cdot \dfrac{1}{2}\right) \cdot \dfrac{8}{9} = \dfrac{1}{8} \cdot \dfrac{\overset{1}{8}}{\underset{1}{9}} = \dfrac{1}{9}$

22. $\left(\dfrac{1}{2} + \dfrac{1}{3}\right) \div \left(\dfrac{2}{5}\right)^2 = \left(\dfrac{3}{6} + \dfrac{2}{6}\right) \div \left(\dfrac{2}{5} \cdot \dfrac{2}{5}\right)$

$= \dfrac{5}{6} \div \dfrac{4}{25} = \dfrac{5}{6} \times \dfrac{25}{4} = \dfrac{5 \cdot 25}{6 \cdot 4} = \dfrac{125}{24} = 5\dfrac{5}{24}$

23. Strategy   To find the amount in the checking account:
     Find the total of the checks written by adding the check amounts ($128, $54, and $315).
     Subtract the total of the checks written from the original balance in the checking account ($1359).

Solution
```
$128        $1359
  54       −  497
+ 315        $862
$497
```
The amount in the checking account at the end of the week was $862.

24. Strategy   To find the total income from the sale of the tickets:
     Find the income from the adult tickets by multiplying the ticket price ($5) by the number of tickets sold (87).
     Find the income from the student tickets by multiplying the ticket price ($2) by the number of tickets sold (135).
     Find the total income by adding the income from adult tickets to the income from the student tickets.

Solution
```
    87         135        $435
×  $5        ×  $2       +  270
$435        $270         $705
```
The total income from the tickets was $705.

25. Strategy   To find the total weight, add the three weights $\left(1\dfrac{1}{2},\ 7\dfrac{7}{8},\ \text{and } 2\dfrac{2}{3} \text{ pounds}\right)$.

Solution
$1\dfrac{1}{2} = 1\dfrac{12}{24}$
$7\dfrac{7}{8} = 7\dfrac{21}{24}$
$+2\dfrac{2}{3} = 2\dfrac{16}{24}$
$10\dfrac{49}{24} = 12\dfrac{1}{24}$

The total weight is $12\dfrac{1}{24}$ pounds.

26. Strategy   To find the length of the remaining piece, subtract the length of the cut piece $\left(2\dfrac{5}{8} \text{ feet}\right)$ from the original length of the board $\left(7\dfrac{1}{3} \text{ feet}\right)$.

Solution
$7\dfrac{1}{3} = 7\dfrac{8}{24} = 6\dfrac{32}{24}$
$-2\dfrac{5}{8} = 2\dfrac{15}{24} = 2\dfrac{15}{24}$
$4\dfrac{17}{24}$

The length of the remaining piece is $4\dfrac{17}{24}$ feet.

27. Strategy   To find how many miles the car can travel, multiply the number of gallons used $\left(8\dfrac{1}{3}\right)$ by the number of miles that the car travels on each gallon (27).

Solution $27 \times 8\dfrac{1}{3} = 27 \times \dfrac{25}{3} = 225$

The car travels 225 miles on $8\dfrac{1}{3}$ gallons of gas.

28. Strategy   To find how many parcels can be sold:
Find the amount of land that can be developed by subtracting the land donated for a park (2 acres) from the total purchase $\left(10\dfrac{1}{3}\text{ acres}\right)$.

Divide the amount of land that can be developed by the size of each parcel $\left(\dfrac{1}{3}\text{ acre}\right)$.

Solution
$$10\dfrac{1}{3}$$
$$\underline{-2}$$
$$8\dfrac{1}{3}$$

$$8\dfrac{1}{3} \div \dfrac{1}{3} = \dfrac{25}{3} \div \dfrac{1}{3} = \dfrac{25}{3} \times \dfrac{3}{1} = 25$$

The number of parcels that can be sold from the remaining land is 25.

Chapter 3: Decimals

PREP TEST and GO FIGURE

1. $\dfrac{3}{10}$

2. 36,900

3. four thousand seven hundred ninety-one

4. 6842

5. 9394

6. 1638

7.
```
    844
  × 91
    844
  7596
 76,804
```

8.
```
      278 r18
  23)6412
    −46
     181
    −161
     202
    −184
      18
```

Go Figure

There are 7 children in the family.
Maria has twice as many brothers as sisters. She has 4 brothers (including Pedro) and 2 sisters. Pedro has as many brothers as sisters. He has 3 sisters (including Maria) and 3 brothers.

SECTION 3.1

Objective A Exercises

1. hundredths

3. ten-thousandths

5. Twenty-seven hundredths

7. One and five thousandths

9. Thirty-six and four tenths

11. Thirty-five hundred-thousandths

13. Ten and seven thousandths

15. Fifty-two and ninety-five hundred-thousandths

17. Two hundred ninety-three ten-thousandths

19. Six and three hundred twenty-four thousandths

21. Two hundred seventy-six and three thousand two hundred ninety-seven ten-thousandths

23. Two hundred sixteen and seven hundred twenty-nine ten-thousandths

25. Four thousand six hundred twenty-five and three hundred seventy-nine ten-thousandths

27. One and one hundred-thousandth

29. 0.762

31. 0.000062

33. 8.0304

35. 304.07

37. 362.048

39. 3048.2002

Objective B Exercises

41.
```
        ┌── Given place value
  7.359
     └── 5 = 5
  7.4
```

43.
```
         ┌── Given place value
  23.009
      └── 0 < 5
  23.0
```

45.
```
           ┌── Given place value
  22.68259
        └── 2 < 5
  22.68
```

47.
```
          ┌── Given place value
  7.072854
       └── 8 > 5
  7.073
```

49.
```
            ┌── Given place value
  62.009435
         └── 4 < 5
  62.009
```

51.
```
           ┌── Given place value
  0.012346
        └── 4 < 5
  0.0123
```

53.
```
           ┌── Given place value
  2.079239
         └── 9 > 5
  2.07924
```

55.
```
            ┌── Given place value
  0.1009754
         └── 4 < 5
  0.100975
```

Applying the Concepts

57. A sales tax or meal tax is always rounded up to the nearest cent. Also, if a product is priced at "3 for $1.00," a customer is charged $.34 for one item.

Here is another situation in which a decimal is rounded up: 200 students are going on a field trip. Each bus holds 38 students. $200 \div 38 = 5.25$ Six buses are needed.

Generally, interest paid by an institution is rounded down – for example, interest paid by a bank on a saving account.

Here is another situation in which a decimal is rounded down: A bag of sour balls costs $3. You have $29. $29 \div 3 = 9.7$ You can buy 9 bags.

59. a. 5.5 and 6.4

    b. 10.15 and 10.24

SECTION 3.2

Objective A Exercises

1.
$$\begin{array}{r} {}^{1}\;{}^{11} \\ 16.008 \\ 2.0385 \\ +\;132.06 \\ \hline 150.1065 \end{array}$$

3.
$$\begin{array}{r} {}^{1}\;{}^{1} \\ 1.792 \\ 67. \\ +\;27.0526 \\ \hline 95.8446 \end{array}$$

5.
$$\begin{array}{r} {}^{1} \\ 3.02 \\ 62.7 \\ +\;3.924 \\ \hline 69.644 \end{array}$$

7.
$$\begin{array}{r} {}^{11}\;{}^{1} \\ 82.006 \\ 9.95 \\ +\;0.927 \\ \hline 92.883 \end{array}$$

9.
$$\begin{array}{r} {}^{21}\;{}^{11} \\ 4.307 \\ 99.82 \\ +\;9.078 \\ \hline 113.205 \end{array}$$

11.
$$\begin{array}{r} 0.29 \\ +\;0.4 \\ \hline 0.69 \end{array}$$

13.
$$\begin{array}{r} 7.3 \\ +\;9.005 \\ \hline 16.305 \end{array}$$

15.
$$\begin{array}{r} {}^{21}\;{}^{1} \\ 8.72 \\ 99.073 \\ +\;2.9736 \\ \hline 110.7666 \end{array}$$

17.
$$\begin{array}{r} {}^{2} \\ 8. \\ 89.43 \\ +\;7.0659 \\ \hline 104.4959 \end{array}$$

19.
$$\begin{array}{rcr} 219.9 & \approx & 220 \\ 0.872 & \approx & 1 \\ +\;13.42 & \approx & +\;13 \\ \hline \text{Cal.: } 234.192 & & \text{Est.: } 234 \end{array}$$

21.
$$\begin{array}{rcr} 678.92 & \approx & 679 \\ 97.6 & \approx & 98 \\ +\;5.423 & \approx & +\;5 \\ \hline \text{Cal.: } 781.943 & & \text{Est.: } 782 \end{array}$$

Objective B Application Problems

23. Strategy To find the length of the shaft, add the three measures on the shaft (1.87, 1.63, and 2.15 inches).

    Solution
$$\begin{array}{r} 1.87 \\ 1.63 \\ +\;2.15 \\ \hline 5.65 \end{array}$$

The total length of the shaft is 5.65 inches.

25. Strategy To find the amount in your checking account:
• Find the total amount of the four deposits ($210.98 + $45.32 + $1236.34 + $27.99).
• Add the total of the deposits to the previous balance ($2143.57).

    Solution
$$\begin{array}{rr} \$210.98 & \$1520.63 \\ 45.32 & +\;2143.57 \\ 1236.34 & \$3664.20 \\ +\;27.99 & \\ \hline \$1520.63 & \end{array}$$

The amount in the checking account is $3664.20.

27. Strategy To find total projected populations of Asia and African in 2050, add the expected population of Asia (5.3 billion) to the expected population of Africa (1.8 billion).

    Solution
$$\begin{array}{r} 5.3 \\ +\;1.8 \\ \hline 7.1 \end{array}$$

The combined populations of Asia and Africa in 2050 is expected to be 7.1 billion people.

29. Strategy To find how many self-employed people earn more than $5000, add the number of people from the chart who make more than $5000 (3.9, 1.6 and 1.0 million).

Solution
3.9 $5000 - $24,999
1.6 $25,000 - $49,999
+1.0 $50,000 or more
6.5

The number of self-employed people who earn more than $5000 is 6.5 million.

Applying the Concepts

31.
$3.45 Raisin bran
1.23 Bread
2.18 Milk
3.31 Lunch meat
+ 2.69 Butter
$12.86

No. $10 is not enough.

33.
1.4
× 4
5.6

No. A 4-foot rope cannot be wrapped around the box.

SECTION 3.3

Objective A Exercises

1.
13
1 3 10
24.037
−18.41
5.627

3.
12   9
1 2 10 6 10 10
12 3.0 7 0 0
−   9.4 2 7 3
113.6 4 2 7

5.
15 14 9 9
0 6 4 10 10 10
16.5 0 0 0
− 9.7 9 0 2
6.7 0 9 8

7.
18
6 8 10
235.7 9 0
− 20.0 9 3
215.6 9 7

9.
12 9 9 14
5 2 10 10 4 10
6 3.0 0 5 0
− 9.1 2 7 4
53.8 7 7 6

11.
11 9 9 9
8 1 10 10 10 10
9 2.0 0 0 0
− 19.2 9 0 9
7 2.7 0 9 1

13.
9
1 10 10
0.3 2 0 0
−0.0 0 5 8
0.3 1 4 2

15.
9
2 10 10
3.0 0 5
−1.9 8 2
1.0 2 3

17.
10 1
2 15 1 0 6 10
3 5 2.1 6 0
− 9 0.9 9 4
2 6 1.1 6 6

19.
11
6 1 14
7 2 4.32
− 69.
6 5 5.32

21.
11   13
5 1 13 8 3 10
3 6 2.3 9 4 0
− 19.4 6 7 2
3 4 2.9 2 6 8

23.
9 9
8 10 10 10
19.0 0 0
−10.3 7 2
8.6 2 8

25.
9 10
6 10 0 10
7.0 1 0
− 2.3 2 5
4.6 8 5

27.
12 14
8 2 4 10
19.3 3 0
− 8.9 6 7
10.3 8 3

29.
3.7529   ≈   4
−1.00784   ≈   −1
Cal.: 2.74506   Est.: 3

31.
9.07325   ≈   9
−1.924   ≈   −2
Cal.: 7.14925   Est.: 7

Objective B Application Problems

33. Strategy To find the missing dimension, subtract 6.79 from 14.34.

Solution
14.34
− 6.79
7.55

The missing dimension is 7.55 inches.

35a. Strategy To find the total amount of the checks written, add the three amounts ($67.92, $43.10, and $496.34).

Solution
$67.92
43.10
+ 496.34
$607.36

The total amount of the checks is $607.36.

35b. Strategy To find the new balance in your checking account, subtract the total amount of the checks written ($607.36) from the original balance ($1029.74).

Solution $1029.74
− 607.36
$422.38

Your new balance is $422.38.

37. Strategy To find how many inches above normal the rainfall was:
• Find the total rainfall (1.42 + 5.39 + 3.55 inches).
• Subtract the normal rainfall (13.25 inches) from the total rainfall (30.02 inches).

Solution
30.02
−13.25
16.77 inches

The rainfall was 16.77 inches above normal.

39. Strategy To find the growth in online shopping, subtract the number of households shopping online in 2000 (17.7 million) from the number of households shopping online in 2003 (40.3 million).

Solution
40.3
−17.7
22.6

The growth in online shopping from 2000 to 2003 is 22.6 million households.

Applying the Concepts

41a. Rounding to the tenths, the largest difference between the exact sum and the estimate is 0.05. Example: For numbers between 3.7 and 3.8, (1) Any number between 3.7 and 3.75 (not including 3.75) is rounded to 3.7, so the largest difference is *less than* 0.05. (2) Any number between 3.75 and 3.8 (including 3.75) is rounded to 3.8, so the largest difference is *equal to* 0.05. Therefore, rounding to the tenths, the largest amount by which the estimate of the sum of two decimals could differ from the exact sum is 0.05 + 0.05 = 0.1.

41b. For hundredths, 0.005 + 0.005 = 0.01.

41c. For thousandths, 0.0005 + 0.0005 = 0.001.

SECTION 3.4

Objective A Exercises

1.   0.9
× 0.4
0.36

3.   0.5
× 0.6
0.30

5.   0.5
× 0.5
0.25

7.   0.9
× 0.5
0.45

9.   7.7
× 0.9
6.93

11.   9.2
× 0.2
1.84

13.   7.2
× 0.6
4.32

15.   7.4
× 0.1
0.74

17.   7.9
× 5
39.5

19.   0.68
× 4
2.72

21.   0.67
× 0.9
0.603

23.   0.16
× 0.6
0.096

25.   2.5
× 5.4
13.50

27.   8.4
× 9.5
79.80

29.
```
     0.83
   ×  5.2
     166
     415
    4.316
```

31.
```
     0.46
   ×  3.9
     414
     138
    1.794
```

33.
```
     0.2
   × 0.3
    0.06
```

35.
```
     0.24
   ×  0.3
    0.072
```

37.
```
     1.47
   × 0.09
    0.1323
```

39.
```
     8.92
   × 0.004
    0.03568
```

41.
```
     0.49
   × 0.16
     294
      49
    0.0784
```

43.
```
     7.6
   × 0.01
    0.076
```

45.
```
    8.62
   ×   4
    34.48
```

47.
```
    64.5
   ×    9
    580.5
```

49.
```
     2.19
   ×  9.2
     438
    1971
    20.148
```

51.
```
      1.85
   ×  0.023
      555
      370
    0.04255
```

53.
```
     0.478
   ×  0.37
     3346
     1434
    0.17686
```

55.
```
     48.3
   × 0.0041
      483
     1932
    0.19803
```

57.
```
     2.437
   ×   6.1
     2437
    14622
    14.8657
```

59.
```
     0.413
   × 0.0016
     2478
      413
    0.0006608
```

61.
```
     94.73
   ×  0.57
     66311
    47365
    53.9961
```

63.
```
     8.005
   × 0.067
     56035
    48030
    0.536335
```

65.
```
    4.29
   × 0.1
    0.429
```

67.
```
    5.29
   × 0.4
    2.116
```

69.
```
    0.68
   × 0.7
    0.476
```

71.
```
     1.4
   × 0.73
      42
      98
    1.022
```

73.
```
     5.2
   × 7.3
     156
     364
    37.96
```

75.
$$\begin{array}{r} 3.8 \\ \times\ 0.61 \\ \hline 38 \\ 228 \\ \hline 2.318 \end{array}$$

77. $0.32 \times 10 = 3.2$

79. $0.065 \times 100 = 6.5$

81. $6.2856 \times 1000 = 6285.6$

83. $3.2 \times 1000 = 3200$

85. $3.57 \times 10{,}000 = 35{,}700$

87. $0.63 \times 10^1 = 6.3$

89. $0.039 \times 10^2 = 3.9$

91. $4.9 \times 10^4 = 49{,}000$

93. $0.067 \times 10^2 = 6.7$

95.
$$\begin{array}{r} 3.45 \\ \times\ 0.0035 \\ \hline 1725 \\ 1035 \\ \hline 0.012075 \end{array}$$

97.
$$\begin{array}{r} 0.00392 \\ \times\ 3.005 \\ \hline 1960 \\ 1176 \\ \hline 0.01177960 \\ \text{or } 0.0117796 \end{array}$$

99.
$$\begin{array}{r} 1.348 \\ \times\ 0.23 \\ \hline 4044 \\ 2696 \\ \hline 0.31004 \end{array}$$

101.
$$\begin{array}{r} 23.67 \\ \times\ 0.0035 \\ \hline 11835 \\ 7101 \\ \hline 0.082845 \end{array}$$

103.
$$\begin{array}{rr} 0.45 & 2.25 \\ \times\ 5 & \times\ 2.3 \\ \hline 2.25 & 675 \\ & 450 \\ \hline & 5.175 \end{array}$$

105.
$$\begin{array}{rcr} 28.5 & \approx & 30 \\ \times\ 3.2 & \approx & \times\ 3 \\ \hline \end{array}$$
Cal.: 91.2   Est.: 90

107.
$$\begin{array}{rcr} 2.38 & \approx & 2 \\ \times\ 0.44 & \approx & \times\ 0.4 \\ \hline \end{array}$$
Cal.: 1.0472   Est.: 0.8

109.
$$\begin{array}{rcr} 0.866 & \approx & 1 \\ \times\ 4.5 & \approx & \times\ 4.5 \\ \hline \end{array}$$
Cal.: 3.897   Est.: 4.5

111.
$$\begin{array}{rcr} 4.34 & \approx & 4 \\ \times\ 2.59 & \approx & \times\ 3 \\ \hline \end{array}$$
Cal.: 11.2406   Est.: 12

113.
$$\begin{array}{rcr} 8.434 & \approx & 8 \\ \times\ 0.044 & \approx & \times\ 0.04 \\ \hline \end{array}$$
Cal.: 0.371096   Est.: 0.32

115.
$$\begin{array}{rcr} 28.44 & \approx & 30 \\ \times\ 1.12 & \approx & \times\ 1 \\ \hline \end{array}$$
Cal.: 31.8528   Est.: 30

117.
$$\begin{array}{rcr} 49.6854 & \approx & 50 \\ \times\ 39.0672 & \approx & \times\ 40 \\ \hline \end{array}$$
Cal.: 1941.069459   Est.: 2000

119.
$$\begin{array}{rcr} 0.00456 & \approx & 0.005 \\ \times\ 0.009542 & \approx & \times\ 0.01 \\ \hline \end{array}$$
Cal.: 0.0000435112   Est.: 0.00005

Objective B Application Problems

121. Strategy To find the cost of operating the electric motor, multiply the hourly cost ($.027) by the number of hours it is used (56).

Solution
$$\begin{array}{r} 56 \\ \times\ .027 \\ \hline 392 \\ 112 \\ \hline 1.512 \approx 1.51 \end{array}$$
The motor costs $1.51 to operate.

123. Strategy To estimate the amount received, round each number so that all the digits are zero except the first digit, then multiply.

Solution
$$\begin{array}{rcr} 520 & \approx & 500 \\ \times\ \$.045 & \approx & .05 \\ \hline & & 25.00 \end{array}$$

The estimated amount received is $25.00.

Strategy To find the payment for recycling the newspapers, multiply the number of pounds (520) by the payment per pound (0.045).

Solution
$$\begin{array}{r} 520 \\ \times\ 0.045 \\ \hline 2600 \\ 2080 \\ \hline 23.400 \end{array}$$
The amount received was $23.40.

125. Strategy   To find the area of a square, multiply the length (6.75 feet) by the width (3.5 feet).

Solution

$$\begin{array}{r} 6.75 \\ \times\ 3.5 \\ \hline 3\,375 \\ 20\,25\ \\ \hline 23.625 \end{array}$$

The area is 23.625 square feet.

127a. Strategy   To find the amount of overtime pay, multiply the overtime rate ($28.35) by the number of hours worked (15).

Solution

$$\begin{array}{r} \$28.35 \\ \times\ 15 \\ \hline 141\,75 \\ 283\,5\ \\ \hline \$425.25 \end{array}$$

The amount of overtime pay is $425.25.

127b. Strategy   To find the nurse's total income for the week, add the overtime pay ($425.25) to the salary ($756).

Solution

$$\begin{array}{r} \$425.25 \\ +\ 756.00 \\ \hline \$1181.25 \end{array}$$

The nurse's total income is $1181.25.

129. Strategy   To find cost, multiply the rate for post office to addressee for up to $\frac{1}{2}$ pound ($12.25) by the number of packages (25).

Solution

$$\begin{array}{r} \$12.25 \\ \times\ 25 \\ \hline 61\,25 \\ 245\,0\ \\ \hline \$306.25 \end{array}$$

The cost is $306.25.

131. a. Grade 1

$$\begin{array}{r} 2.2 \\ \times\ 8 \\ \hline 17.6 \\ \times\ 1.2 \\ \hline 352 \\ 176\ \\ \hline 21.120 \end{array}$$

The total cost for grade 1 is $21.12.

Grade 2

$$\begin{array}{r} 3.4 \\ \times\ 6.5 \\ \hline 170 \\ 204\ \\ \hline 22.10 \\ \times\ 1.35 \\ \hline 11050 \\ 6630\ \\ 2210\ \ \\ \hline 29.8350 \end{array}$$

The total cost for grade 2 is $29.84.

Grade 3

$$\begin{array}{r} 6.75 \\ \times\ 15.4 \\ \hline 2700 \\ 3375\ \\ 675\ \ \\ \hline 103.950 \\ \times\ 1.94 \\ \hline 415800 \\ 935550\ \\ 103950\ \ \\ \hline 201.66300 \end{array}$$

The total cost for grade 3 is $201.66.

b. Total Cost:

Grade 1	$21.12
Grade 2	29.84
Grade 3	201.66
Total:	$252.62

The total cost is $252.62.

133. Car 1 $\dfrac{36}{367,921} = 0.0001$

Car 2 $\dfrac{42}{401,346} = 0.0001$

Car 3 $\dfrac{21}{298,773} = 0.00007$

Car 4 $\dfrac{32}{330,045} = 0.0001$

Car 5 $\dfrac{45}{432,989} = 0.0001$

All the cars would pass the test.

Applying the Concepts

135. When a number is multiplied by 10, 100, 1000, 10,000, etc., the decimal point is moved as many places to the right as there are zeros in the multiple of 10. For example, the decimal point in a number multiplied by 1000 would be moved three places to the right.

137. $1.3 = 1\dfrac{3}{10}$

$2.31 = 2\dfrac{31}{100}$

$1\dfrac{3}{10} \times 2\dfrac{31}{100} = \dfrac{13}{10} \times \dfrac{231}{100} = \dfrac{3003}{1000} = 3\dfrac{3}{1000} = 3.003$

SECTION 3.5

Objective A Exercises

1.
```
      0.82
  3 )2.46
     -24
      06
     - 06
       0
```

3.
```
        4.8
  0.8.)3.8.4
      -32
       64
      - 64
        0
```

5.
```
       89.
  0.7.)62.3.
      -56
       63
      - 63
        0
```

7.
```
       60.
  0.4.)24.0.
      -24
       00
      - 0
        0
```

9.
```
       84.3
  0.7.)59.0.1
      -56
       30
      - 28
        21
      - 21
        0
```

11.
```
       32.3
  0.5.)16.1.5
      -15
       11
      -10
       15
      -15
        0
```

13.
```
       5.06
  0.7.)3.5.42
      -35
        04
      -  0
        42
      - 42
         0
```

15.
```
       1.3
  6.3.)8.1.9
      -63
       189
      -189
         0
```

17.
```
        0.11
  3.6.)0.3.96
       - 36
         36
       - 36
          0
```

19.
```
        3.8
  6.9.)26.2.2
      -207
       552
      - 552
         0
```

21.
```
          6.32 ≈ 6.3
  8.8.)55.6.20
      -528
       282
      - 264
        180
      - 176
          4
```

23.
```
          0.57 ≈ 0.6
  9.5.)5.4.27
      -475
        677
      - 665
         12
```

25.
$$
\begin{array}{r}
2.52 \approx 2.5 \\
7.3.\overline{)18.4.00} \\
\underline{-146} \\
380 \\
\underline{-365} \\
150 \\
\underline{-146} \\
4
\end{array}
$$

27.
$$
\begin{array}{r}
1.07 \approx 1.1 \\
0.17.\overline{)0.18.30} \\
\underline{-17} \\
13 \\
\underline{-0} \\
130 \\
\underline{-119} \\
11
\end{array}
$$

29.
$$
\begin{array}{r}
130.64 \approx 130.6 \\
0.053.\overline{)6.924.00} \\
\underline{-53} \\
162 \\
\underline{-159} \\
34 \\
\underline{-0} \\
340 \\
\underline{-318} \\
220 \\
\underline{-212} \\
8
\end{array}
$$

31.
$$
\begin{array}{r}
0.808 \approx 0.81 \\
8\overline{)6.467} \\
\underline{-64} \\
06 \\
\underline{-0} \\
67 \\
\underline{-64} \\
3
\end{array}
$$

33.
$$
\begin{array}{r}
0.089 \\
0.72.\overline{)0.06.47} \approx 0.09 \\
\underline{-576} \\
71
\end{array}
$$

35.
$$
\begin{array}{r}
40.70 \\
0.95.\overline{)38.66.5} \\
\underline{-380} \\
66 \\
\underline{-0} \\
665 \\
\underline{-665} \\
0
\end{array}
$$

37.
$$
\begin{array}{r}
0.46 \\
60.3.\overline{)27.7.38} \\
\underline{-2412} \\
3618 \\
\underline{-3618} \\
0
\end{array}
$$

39.
$$
\begin{array}{r}
0.0190 \approx 0.019 \\
54\overline{)1.0280} \\
\underline{-54} \\
488 \\
\underline{-486} \\
20 \\
\underline{-0} \\
0
\end{array}
$$

41.
$$
\begin{array}{r}
0.0874 \approx 0.087 \\
0.5.\overline{)0.0.4370} \\
\underline{-40} \\
37 \\
\underline{-35} \\
20 \\
\underline{-20} \\
0
\end{array}
$$

43.
$$
\begin{array}{r}
0.3600 \approx 0.360 \\
95.3.\overline{)34.3.1000} \\
\underline{-2859} \\
5720 \\
\underline{-5718} \\
20 \\
\underline{-0} \\
200 \\
\underline{-0} \\
200
\end{array}
$$

45.
$$
\begin{array}{r}
0.1031 \approx 0.103 \\
4.72.\overline{)0.48.7100} \\
\underline{-472} \\
151 \\
\underline{-0} \\
1510 \\
\underline{-1416} \\
940 \\
\underline{-472} \\
468
\end{array}
$$

47.
$$
\begin{array}{r}
0.0086 \approx 0.009 \\
26.7.\overline{)0.2.3070} \\
\underline{-2136} \\
1710 \\
\underline{-1602} \\
108
\end{array}
$$

49.
$$
\begin{array}{r}
0.9 \approx 1 \\
90\overline{)89.76} \\
\underline{-810} \\
87
\end{array}
$$

51.
$$
\begin{array}{r}
2.5 \approx 3 \\
0.413.\overline{)1.047.8} \\
\underline{-826} \\
2218 \\
\underline{-2065} \\
153
\end{array}
$$

53. $$0.778.\overline{)0.790.0}\quad \frac{1.0}{}\approx 1$$
$$-\underline{778}$$
$$120$$

55. $$6.9.\overline{)392.0.0}\quad \frac{56.8}{}\approx 57$$
$$-\underline{345}$$
$$470$$
$$-\underline{414}$$
$$560$$
$$-\underline{552}$$
$$8$$

57. $4.07 \div 10 = 0.407$

59. $42.67 \div 10 = 4.267$

61. $1.037 \div 100 = 0.01037$

63. $8.295 \div 1000 = 0.008295$

65. $825.37 \div 1000 = 0.82537$

67. $0.32 \div 10^1 = 0.032$

69. $23.627 \div 10^2 = 0.23627$

71. $0.0053 \div 10^2 = 0.000053$

73. $1.8932 \div 10^3 = 0.0018932$

75. $$2.4.\overline{)44.2.08}\quad \frac{18.42}{}$$
$$-\underline{24}$$
$$202$$
$$-\underline{192}$$
$$100$$
$$-\underline{96}$$
$$48$$
$$-\underline{48}$$
$$0$$

77. $$45\overline{)723.15}\quad \frac{16.07}{}$$
$$-\underline{45}$$
$$273$$
$$-\underline{270}$$
$$31$$
$$-\underline{0}$$
$$315$$
$$-\underline{315}$$
$$0$$

79. $13.5 \div 10^3 = 0.0135$

81. $23.678 \div 1000 = 0.023678$

83. $$0.05.\overline{)0.00.560}\quad \frac{0.112}{}$$
$$-\underline{5}$$
$$06$$
$$-\underline{5}$$
$$10$$
$$-\underline{10}$$
$$0$$

85. Cal.: $42.42 \div 3.8 = 11.1632$
Est.: $40 \div 4 = 10$

87. Cal.: $389 \div 0.44 = 884.0909$
Est.: $400 \div 0.4 = 1000$

89. Cal.: $6.394 \div 3.5 = 1.8269$
Est.: $6 \div 4 = 1.5$

91. Cal.: $1.235 \div 0.021 = 58.8095$
Est.: $1 \div 0.02 = 50$

93. Cal.: $95.443 \div 1.32 = 72.3053$
Est.: $100 \div 1 = 100$

95. Cal.: $1.000523 \div 429.07 = 0.0023$
Est.: $1 \div 400 = 0.0025$

Objective B Application Problems

97. Strategy To find the number of yards per carry, divide the total number of yards (162) by the number of carries (26).

Solution $$26\overline{)162.000}\quad \frac{6.230}{}\approx 6.23$$
$$-\underline{156}$$
$$60$$
$$-\underline{52}$$
$$80$$
$$-\underline{78}$$
$$20$$
$$-\underline{0}$$
$$20$$

The number of yards gained per carry is 6.23.

99. Strategy To find the cost per can, divide the cost of a case ($6.79) by the number of cans in a case (24).

Solution $$24\overline{)6.79}\quad \frac{0.282}{}\approx 0.28$$
$$-\underline{48}$$
$$199$$
$$-\underline{192}$$
$$70$$
$$-\underline{48}$$
$$22$$

The cost per can is $.28.

101. Strategy To find the monthly payment, divide the yearly premium ($703.80) by 12.

Solution

$$
\begin{array}{r}
58.65 \\
12\overline{)703.80} \\
-60 \\
\hline
103 \\
-96 \\
\hline
78 \\
-72 \\
\hline
60 \\
-60 \\
\hline
0
\end{array}
$$

The monthly payment is $58.65.

103. Strategy To find the cost per mile, divide the toll ($5.60) by the number of miles (136 miles).

Solution $5.60 \div 136 = 0.041$
$0.041 \approx 0.04$
The cost per mile is $.04.

Applying the Concepts

105. When a number is divided by 10, 100, 1000, 10,000, etc., the decimal point is moved as many places to the left as there are zeros in the multiple of 10. For example, the decimal point in a number divided by 1000 would be moved three places to the left.

107. To calculate a batting average, divide the number of hits by the number of times at bat. Round to the nearest thousandth. Nomar Garciaparra's batting average $= \dfrac{190}{532} = 0.357$.

109. $3.45 \div 0.5 = 6.9$

111. $6.009 - 4.68 = 1.329$

113. $9.876 + 23.12 = 32.996$

115. 5.217

117. 0.025

SECTION 3.6

Objective A Exercises

1. $8\overline{)5.000}$ (0.625)

3. $3\overline{)2.0000}$ $0.6666 \approx 0.667$

5. $6\overline{)1.0000}$ $0.1666 \approx 0.167$

7. $12\overline{)5.0000}$ $0.4166 \approx 0.417$

9. $4\overline{)7.000}$ (1.750)

11. $1\frac{1}{2} = \frac{3}{2};\quad 2\overline{)3.000}$ (1.500)

13. $4\overline{)16.000}$ (4.000)

15. $1000\overline{)3.000}$ (0.003)

17. $7\frac{2}{25} = \frac{177}{25};\quad 25\overline{)177.000}$ (7.080)

19. $37\frac{1}{2} = \frac{75}{2};\quad 2\overline{)75.000}$ (37.500)

21. $25\overline{)4.000}$ (0.160)

23. $8\frac{2}{5} = \frac{42}{5};\quad 5\overline{)42.000}$ (8.400)

Objective B Exercises

25. $0.8 = \dfrac{8}{10} = \dfrac{4}{5}$

27. $0.32 = \dfrac{32}{100} = \dfrac{8}{25}$

29. $0.125 = \dfrac{125}{1000} = \dfrac{1}{8}$

31. $1.25 = 1\dfrac{25}{100} = 1\dfrac{1}{4}$

33. $16.9 = 16\dfrac{9}{10}$

35. $8.4 = 8\dfrac{4}{10} = 8\dfrac{2}{5}$

37. $8.437 = 8\dfrac{437}{1000}$

39. $2.25 = 2\dfrac{25}{100} = 2\dfrac{1}{4}$

41. $0.15\dfrac{1}{3} = \dfrac{15\frac{1}{3}}{100} = 15\dfrac{1}{3} \div 100 = \dfrac{46}{3} \times \dfrac{1}{100}$
$= \dfrac{46}{300} = \dfrac{23}{150}$

43. $0.87\dfrac{7}{8} = \dfrac{87\frac{7}{8}}{100} = 87\dfrac{7}{8} \div 100 = \dfrac{703}{8} \times \dfrac{1}{100} = \dfrac{703}{800}$

45. $7.38 = 7\dfrac{38}{100} = 7\dfrac{19}{50}$

47. $0.57 = \dfrac{57}{100}$

49. $0.66\frac{2}{3} = \frac{66\frac{2}{3}}{100} = 66\frac{2}{3} \div 100 = \frac{200}{3} \times \frac{1}{100} = \frac{2}{3}$

Objective C Exercises

51. $0.6 > 0.45$

53. $3.89 < 3.98$

55. $0.025 < 0.105$

57. $\frac{4}{5} = 0.8$

$0.8 < 0.802$

$\frac{4}{5} < 0.802$

59. $\frac{7}{8} = 0.875$

$0.85 < 0.875$

$0.85 < \frac{7}{8}$

61. $\frac{7}{12} \approx 0.583$

$0.583 > 0.58$

$\frac{7}{12} > 0.58$

63. $\frac{11}{12} \approx 0.9167$

$0.9167 < 0.92$

$\frac{11}{12} < 0.92$

65. $0.623 > 0.6023$

67. $0.87 > 0.087$

69. $0.033 < 0.3$

Applying the Concepts

71. Strategy To determine if the number is more than or less than:

- Multiply the number of hearing-impaired individuals aged 0 – 17 (1.37) by 4

- Compare to the number of hearing-impaired individuals aged 65 – 74.

Solution
$$\begin{array}{r} 1.37 \\ \times \quad 4 \\ \hline 5.48 \end{array}$$
$5.41 < 5.48$
The number of hearing-impaired individuals aged 65 – 74 is less than four times the number of hearing-impaired individuals aged 0 – 17.

73. No. 0.0402 rounded to hundredths is 0.04; to thousandths it is 0.040.

CHAPTER REVIEW

1.
$$\begin{array}{r} 54.5 \\ 0.067.\overline{)3.651.5} \\ -335 \\ \hline 301 \\ -268 \\ \hline 335 \\ -335 \\ \hline 0 \end{array}$$

2.
$$\begin{array}{r} {\scriptstyle 2\,3\,1\ \ 1} \\ 369.41 \\ 88.3 \\ 9.774 \\ + 366.474 \\ \hline 833.958 \end{array}$$

3. $0.055 < 0.1$

4. Twenty-two and ninety-two ten-thousandths

5.
┌─── Given place value
0.05678235
└─── 2 < 5
0.05678

6. $2\frac{1}{3} = \frac{7}{3}$; $\begin{array}{r} 2.333 \approx 2.33 \\ 3\overline{)7.000} \end{array}$

7. $0.375 = \frac{375}{1000} = \frac{3}{8}$

8.
$$\begin{array}{r} {\scriptstyle 1\ 1} \\ 3.42 \\ 0.794 \\ + 32.5 \\ \hline 36.714 \end{array}$$

9. 34.025

10. $\frac{5}{8} = 0.625$, $\frac{5}{8} > 0.62$

11.
$$\begin{array}{r} 0.7777 \approx 0.778 \\ 9\overline{)7.0000} \end{array}$$

12. $0.66\frac{2}{3} = \frac{66\frac{2}{3}}{100} = 66\frac{2}{3} \div 100 = \frac{200}{3} \div 100$

$= \frac{200}{3} \times \frac{1}{100} = \frac{2}{3}$

13.
$$\begin{array}{r} {\scriptstyle 12\,10\ 9} \\ {\scriptstyle 6\ \ 2\ \ 0\,10\,10} \\ 27.3100 \\ - 4.4465 \\ \hline 22.8635 \end{array}$$

14.
┌─── Given place value
7.93704
└─── 7 > 5
7.94

15.
```
    3.08
  ×  2.9
   2772
    616
  8.932
```

16. Three hundred forty-two and thirty-seven hundredths

17. 3.06753

18.
```
    34.79
  ×  0.74
   13916
   24353
  25.7446
```

19.
```
            6.594
  0.053.)0.349.482
          −318
           314
          −265
           498
          −477
           212
          −212
             0
```

20.
```
        15
    6 17 8 5 10
    7.7 9 6 0
  − 2.9 1 7 5
    4.8 7 8 5
```

21. Strategy   To find the new balance in your checking account:

   • Find the total amount of the checks by adding the check amounts ($145.72 and $88.45).

   • Subtract the total check amounts from the original balance ($895.68).

   Solution
```
    $145.72        $895.68
  +   88.45      −  234.17
    $234.17        $661.51
```
   The new balance in your account is $661.51.

22. Strategy   To find the total number of children, add the number of children in public school (46.353 million), in private school (5.863 million) and in home-schooling (1.23 million).

   Solution
```
       46.353
        5.863
      + 1.23
       53.446
```
   There are 53.446 million children in grades K − 12.

23. Strategy   To find the difference in the number of children, subtract the number of children in private school (5.863 million) from the number of children in public school (46.353 million).

   Solution
```
      46.353
    − 5.863
      40.490
```
   There are 40.49 million more children in public school than private school.

24. Strategy   To find the amount of milk served during a 5-day school week, multiply the amount of milk served daily (1.9 million gallons) by 5 days.

   Solution
```
      1.9
    × 5
      9.5
```
   During a 5-day school week, 9.5 million gallons of milk is served.

25. Strategy   To find how many times greater the number of acres planted in Texas is than the number planted in Mississippi, divide the number of acres planted in Texas (6.2 million) by the number of acres planted in Mississippi (1.2 million).

   Solution
```
           5.16
    1.2)6.200
        −60
         20
        −12
         80
        −72
          8
```
   $5.16 \approx 5.2$
   The number of acres planted in Texas is 5.2 times greater than the number of acres planted in Mississippi.

CHAPTER TEST

1. $0.66 < 0.666$

2.
```
      9 10
    6 10 010
    7.0 1 0
  − 2.3 2 5
    4.6 8 5
```

3. Forty-five and three hundred two ten-thousandths

4. $\dfrac{9}{13} = 13\overline{)9.0000}$ with $0.6923 \approx 0.692$

5. $0.825 = \dfrac{825}{1000} = \dfrac{33}{40}$

6.
```
          ┌──Given place value
  0.07395
          └──5 = 5
  0.0740
```

7.
$$
\begin{array}{r}
1.5378 \approx 1.538 \\
0.037\,\overline{)0.056.9000} \\
-\underline{37} \\
199 \\
-\underline{185} \\
140 \\
-\underline{111} \\
290 \\
-\underline{259} \\
310 \\
-\underline{296} \\
14
\end{array}
$$

8.
$$
\begin{array}{r}
{}^{16}\;\;{}^{9}\;{}^{9}\;{}^{12}{}^{9} \\
{}^{2}{}^{6}\;{}^{10}{}^{10}\;{}^{2}{}^{10}{}^{10} \\
37.00300 \\
-\;\;9.23674 \\
\hline
27.76626
\end{array}
$$

9.
$$
\begin{array}{l}
\overline{}\text{Given place value} \\
7.0954625 \\
\underline{}\;4<5 \\
7.095
\end{array}
$$

10.
$$
\begin{array}{r}
232. \\
0.006\,\overline{)1.392.} \\
-\underline{12} \\
19 \\
-\underline{18} \\
12 \\
-\underline{12} \\
0
\end{array}
$$

11.
$$
\begin{array}{r}
{}^{2}\;{}^{12}\;{}^{11} \\
270.93 \\
97. \\
1.976 \\
+\;\;88.675 \\
\hline
458.581
\end{array}
$$

12. **Strategy**   To find the missing dimension, subtract the given length (4.86 inches) from the total length (6.23 inches).

   Solution
$$
\begin{array}{r}
6.23 \\
-\;4.86 \\
\hline
1.37
\end{array}
$$
   The missing dimension is 1.37 inches.

13.
$$
\begin{array}{r}
1.37 \\
\times\;0.004 \\
\hline
0.00548
\end{array}
$$

14.
$$
\begin{array}{r}
{}^{1\,1} \\
62.3 \\
4.007 \\
+\;189.65 \\
\hline
255.957
\end{array}
$$

15. 209.07086

16. **Strategy**   To find the amount of each payment:

   - Find the total amount to be paid by subtracting the down payment ($2500) from the cost of the car ($16,734.40).

   - Divide the amount remaining to be paid by the number of payments (36).

   Solution
$$
\begin{array}{r}
\$16,734.40 \\
-\;2,500.00 \\
\hline
\$14,234.40
\end{array}
$$

$$
\begin{array}{r}
395.40 \\
36\,\overline{)14234.40} \\
-\underline{108} \\
343 \\
-\;\underline{324} \\
194 \\
-\;\underline{180} \\
144 \\
-\underline{144} \\
00 \\
-\underline{0} \\
0
\end{array}
$$

   Each payment is $395.40.

17. **Strategy**   To find your total income, add the salary ($727.50), commission ($1909.64), and bonus ($450).

   Solution
$$
\begin{array}{r}
\$727.50 \\
1909.64 \\
+\;450.00 \\
\hline
\$3087.14
\end{array}
$$
   Your total income is $3087.14.

18. **Strategy**   To find the cost of the 12-minute call:

   - Find the number of additional minutes charged above the 3-minute base by subtracting the base (3 minutes) from the total length (12 minutes).

   - Multiply the number of additional minutes by the rate ($.42).

   - Add the charge for additional minutes to the base rate ($.85).

   Solution $12 - 3 = 9$
$$
\begin{array}{cc}
\$.42 & \$3.78 \\
\times\;\;9 & +\;.85 \\
\hline
\$3.78 & \$4.63
\end{array}
$$
   The cost of the call is $4.63.

19. Strategy To find average hours per year, multiply the weekly computer use by the 10th-grade student (6.7 hours) by 52 weeks.

Solution

$$\begin{array}{r} 6.7 \\ \times\ 52 \\ \hline 134 \\ 335\ \\ \hline 348.4 \end{array}$$

The yearly average computer use by a 10th-grade student is 348.4 hours.

20. Strategy To find how many more hours a 2nd-grade student uses a computer than a 5th-grade student:

- Subtract the number of hours the 5th-grade student uses the computer (4.2) from the number of hours the 2nd-grade student uses the computer (4.9).
- Multiply the difference by 52 weeks.

Solution

$$\begin{array}{r} 4.9 \\ -4.2 \\ \hline 0.7 \end{array} \qquad \begin{array}{r} 52 \\ \times\ 0.7 \\ \hline 36.4 \end{array}$$

On average a 2nd-grade student uses the computer 36.4 hours more per year than a 5th-grade student.

CUMULATIVE REVIEW

1.
$$\begin{array}{r} 235\text{ r}17 \\ 89\overline{)20932} \\ -178 \\ \hline 313 \\ -267 \\ \hline 462 \\ -445 \\ \hline 17 \end{array}$$

2. $2^3 \cdot 4^2 = 8 \cdot 16 = 128$

3. $2^2 - (7-3) \div 2 + 1$

$4 - 4 \div 2 + 1$

$4 - 2 + 1$

3

4.

	2	3
9 =		(3·3)
12 =	2·2	3
24 =	(2·2·2)	3

LCM = 2·2·2·3·3 = 72

5. $\dfrac{22}{5} = \begin{array}{r} 4\text{ r}2 \\ 5\overline{)\ 22} \\ -20 \\ \hline 2 \end{array} = 4\dfrac{2}{5}$

6. $4\dfrac{5}{8} = \dfrac{32+5}{8} = \dfrac{37}{8}$

7. $\dfrac{5\cdot 5}{12\cdot 5} = \dfrac{25}{60}$

8.
$$\begin{array}{rl} \dfrac{3}{8} = & \dfrac{18}{48} \\ \dfrac{5}{12} = & \dfrac{20}{48} \\ +\dfrac{9}{16} = & \dfrac{27}{48} \\ \hline & \dfrac{65}{48} = 1\dfrac{17}{48} \end{array}$$

9.
$$\begin{array}{rl} 5\dfrac{7}{12} = & 5\dfrac{21}{36} \\ +\ 3\dfrac{7}{18} = & 3\dfrac{14}{36} \\ \hline & 8\dfrac{35}{36} \end{array}$$

10.
$$\begin{array}{rl} 9\dfrac{5}{9} = 9\dfrac{20}{36} = & 8\dfrac{56}{36} \\ -\ 3\dfrac{11}{12} = 3\dfrac{33}{36} = & 3\dfrac{33}{36} \\ \hline & 5\dfrac{23}{36} \end{array}$$

11. $\dfrac{9}{16} \times \dfrac{4}{27} = \dfrac{9 \times 4}{16 \times 27} = \dfrac{\overset{1}{\cancel{3}} \cdot \overset{1}{\cancel{3}} \cdot \overset{1}{\cancel{2}} \cdot \overset{1}{\cancel{2}}}{\underset{1}{\cancel{2}} \cdot \underset{1}{\cancel{2}} \cdot 2 \cdot 2 \cdot \underset{1}{\cancel{3}} \cdot \underset{1}{\cancel{3}} \cdot 3} = \dfrac{1}{12}$

12. $2\dfrac{1}{8} \times 4\dfrac{5}{17} = \dfrac{17}{8} \times \dfrac{73}{17} = \dfrac{17 \cdot 73}{8 \cdot 17} = \dfrac{73}{8} = 9\dfrac{1}{8}$

13. $\dfrac{11}{12} \div \dfrac{3}{4} = \dfrac{11}{12} \times \dfrac{4}{3} = \dfrac{11 \cdot 4}{12 \cdot 3} = \dfrac{11 \cdot \overset{1}{\cancel{2}} \cdot \overset{1}{\cancel{2}}}{\underset{1}{\cancel{2}} \cdot \underset{1}{\cancel{2}} \cdot 3 \cdot 3} = \dfrac{11}{9} = 1\dfrac{2}{9}$

14. $2\dfrac{3}{8} \div 2\dfrac{1}{2} = \dfrac{19}{8} \div \dfrac{5}{2} = \dfrac{19}{8} \times \dfrac{2}{5} = \dfrac{19 \cdot 2}{8 \cdot 5}$

$= \dfrac{19 \cdot \overset{1}{\cancel{2}}}{2 \cdot 2 \cdot \underset{1}{\cancel{2}} \cdot 5} = \dfrac{19}{20}$

15. $\left(\dfrac{2}{3}\right)^2 \left(\dfrac{3}{4}\right)^3 = \left(\dfrac{2}{3} \cdot \dfrac{2}{3}\right)\left(\dfrac{3}{4} \cdot \dfrac{3}{4} \cdot \dfrac{3}{4}\right)$

$= \dfrac{\overset{1}{\cancel{2}} \cdot \overset{1}{\cancel{2}} \cdot \overset{1}{\cancel{3}} \cdot \overset{1}{\cancel{3}} \cdot 3}{\underset{1}{\cancel{3}} \cdot \underset{1}{\cancel{3}} \cdot \underset{1}{\cancel{2}} \cdot \underset{1}{\cancel{2}} \cdot 2 \cdot 2 \cdot 2 \cdot 2} = \dfrac{3}{16}$

16. $\left(\dfrac{2}{3}\right)^2 - \left(\dfrac{2}{3} - \dfrac{1}{2}\right) + 2$

$= \left(\dfrac{2}{3} \cdot \dfrac{2}{3}\right) - \left(\dfrac{4}{6} - \dfrac{3}{6}\right) + 2$

$= \dfrac{4}{9} - \dfrac{1}{6} + 2$

$= \dfrac{8}{18} - \dfrac{3}{18} + \dfrac{36}{18}$

$= \dfrac{41}{18} = 2\dfrac{5}{18}$

17. Sixty-five and three hundred nine ten-thousandths

18.
$$\begin{array}{r}
{}^{2\ 3\ \ \ 1\ 1\ 1}\\
379.006\\
27.523\\
9.8707\\
+\ 8\ 8.2994\\
\hline
504.6991
\end{array}$$

19.
$$\begin{array}{r}
{}^{8\ \overset{9}{10}\ \overset{9}{10}\ \overset{14}{4}\ 10}\\
29.0\ 0\ 5\ 0\\
-7.9\ 2\ 8\ 6\\
\hline
21.0\ 7\ 6\ 4
\end{array}$$

20.
$$\begin{array}{r}
9.074\\
\times\ \ 6.09\\
\hline
81666\\
544440\\
\hline
55.26066
\end{array}$$

21.
$$
\begin{array}{r}
2.1544 \approx 2.154
\end{array}
$$
$$8.09.\overline{)17.42.9630}$$
$$\begin{array}{r}
-16\,18\\
\hline
1249\\
-809\\
\hline
4406\\
-4045\\
\hline
3613\\
-3236\\
\hline
3770\\
-3236\\
\hline
534
\end{array}$$

22. $\dfrac{11}{15} = 15\overline{)11.000}\ \ \dfrac{0.7333 \approx 0.733}{}$

23. $0.16\dfrac{2}{3} = \dfrac{16\frac{2}{3}}{100} = \dfrac{\frac{50}{3}}{100} = \dfrac{50}{3} \div 100 = \dfrac{50}{3} \times \dfrac{1}{100}$
$$= \dfrac{50}{300} = \dfrac{1}{6}$$

24. $\dfrac{8}{9} \approx 0.89, \dfrac{8}{9} < 0.98$

25. Strategy   To find how many more vacation days are mandated in Sweden than in Germany, subtract the number of days mandated in Germany (18) from the number of days mandated in Sweden (30).

     Solution
$$\begin{array}{r}
32\\
-18\\
\hline
14
\end{array}$$
There are 14 more vacation days mandated in Sweden than in Germany.

26. Strategy   To find the value of each share at the end of the 2 months:

- Find the price after the stock lost by subtracting the loss $\left(\$\dfrac{3}{4}\right)$ from the original price $\left(\$32\dfrac{1}{8}\right)$.

- Add the amount of the gain $\left(\$1\dfrac{1}{2}\right)$ to the price after the loss.

     Solution
$$\begin{array}{r}
\$32\dfrac{1}{8} = 32\dfrac{1}{8} = 31\dfrac{9}{8}\\
-\dfrac{3}{4} = \dfrac{6}{8} = \dfrac{6}{8}\\
\hline
\$31\dfrac{3}{8}
\end{array}$$

$$\begin{array}{r}
\$31\dfrac{3}{8} = 31\dfrac{3}{8}\\
+1\dfrac{1}{2} = 1\dfrac{4}{8}\\
\hline
\$32\dfrac{7}{8}
\end{array}$$

The price at the end of 2 months was $\$32\dfrac{7}{8}$.

27. Strategy   To find your balance after you write the checks:

- Find the total of the checks written by adding the amount of the checks ($42.98, $16.43, and $137.56).

- Subtract the total of the checks written from the original balance ($814.35).

     Solution
$$\begin{array}{rr}
\$42.98 & \$814.35\\
16.43 & -196.97\\
+137.56 & \overline{\$617.38}\\
\hline
\$196.97 &
\end{array}$$

Your checking account balance is $617.38.

28. Strategy   To find the resulting thickness, subtract the amount removed (0.017 inch) from the original thickness (1.412 inches).

     Solution
$$\begin{array}{r}
1.412\\
-0.017\\
\hline
1.395
\end{array}$$

The resulting thickness is 1.395 inches.

29. Strategy   To find the amount of income tax you paid:

- Find the amount of tax paid on profit by multiplying the profit ($64,860) by the rate (0.08).

- Add the amount of tax paid on profit to the base tax ($820).

Solution

$$\begin{array}{r} \$64{,}860 \\ \times\ \ 0.08 \\ \hline \$5188.80 \end{array} \qquad \begin{array}{r} \$5188.80 \\ +820.00 \\ \hline \$6008.80 \end{array}$$

You paid $6008.80 in income tax last year.

30. Strategy   To find the amount of the monthly payment:

- Find the amount to be paid in payments by subtracting the down payment ($20) from the cost ($210.96).

- Divide the amount to be paid in payments by the number of payments (8).

Solution

$$\begin{array}{r} \$210.96 \\ -\ 20.00 \\ \hline \$190.96 \end{array}$$

$$\begin{array}{r} 23.87 \\ 8)\overline{190.96} \\ -16 \\ \hline 30 \\ -24 \\ \hline 69 \\ -64 \\ \hline 56 \\ -56 \\ \hline 0 \end{array}$$

The amount of each payment is $23.87.

Chapter 4: Ratio and Proportion

PREP TEST and GO FIGURE

1. $\dfrac{8}{10} = \dfrac{\overset{1}{2} \cdot 2 \cdot 2}{2 \cdot 5} = \dfrac{4}{5}$

2. $\dfrac{450}{650 + 250} = \dfrac{450}{900} = \dfrac{\cancel{450}}{2 \cdot \cancel{450}} = \dfrac{1}{2}$

3. $15\overline{)372.0}$ quotient 24.8

4. $4 \times 33 = 132$
$62 \times 2 = 124$
$132 > 124$
4×33 is greater.

5. $5\overline{)20}$ quotient 4

Go Figure

From the third statement, we know that the order of three of the four men is either Luis, Kim, and Reggie or Reggie, Kim and Luis. From the fourth statement, Luis is standing between Dave and Kim. So building from what we already know, then either the order is Dave, Luis, Kim, and Reggie or Reggie, Kim, Luis, and Dave. From the second statement, we know that Dave is not first. Therefore, the order is Reggie, Kim, Luis, and Dave.

SECTION 4.1

Ratio and Proportion

1. $\dfrac{3 \text{ pints}}{15 \text{ pints}} = \dfrac{3}{15} = \dfrac{1}{5}$
3 pints:15 pints = 3:15 = 1:5
3 pints to 15 pints = 3 to 15 = 1 to 5

3. $\dfrac{\$40}{\$20} = \dfrac{40}{20} = \dfrac{2}{1}$
$40:$20 = 40:20 = 2:1
$40 to $20 = 40 to 20 = 2 to 1

5. $\dfrac{3 \text{ miles}}{8 \text{ miles}} = \dfrac{3}{8}$
3 miles:8 miles = 3:8
3 miles to 8 miles = 3 to 8

7. $\dfrac{37 \text{ hours}}{24 \text{ hours}} = \dfrac{37}{24}$
37 hours:24 hours = 37:24
37 hours to 24 hours = 37 to 24

9. $\dfrac{6 \text{ minutes}}{6 \text{ minutes}} = \dfrac{6}{6} = \dfrac{1}{1}$
6 minutes:6 minutes = 6:6 = 1:1
6 minutes to 6 minutes = 6 to 6 = 1 to 1

11. $\dfrac{35 \text{ cents}}{50 \text{ cents}} = \dfrac{35}{50} = \dfrac{7}{10}$
35 cents:50 cents = 35:50 = 7:10
35 cents to 50 cents = 35 to 50 = 7 to 10

13. $\dfrac{30 \text{ minutes}}{60 \text{ minutes}} = \dfrac{30}{60} = \dfrac{1}{2}$
30 minutes:60 minutes = 30:60 = 1:2
30 minutes to 60 minutes = 30 to 60 = 1 to 2

15. $\dfrac{32 \text{ ounces}}{16 \text{ ounces}} = \dfrac{32}{16} = \dfrac{2}{1}$
32 ounces:16 ounces = 32:16 = 2:1
32 ounces to 16 ounces = 32 to 16 = 2 to 1

17. $\dfrac{3 \text{ cups}}{4 \text{ cups}} = \dfrac{3}{4}$
3 cups:4 cups = 3:4
3 cups to 4 cups = 3 to 4

19. $\dfrac{\$5}{\$3} = \dfrac{5}{3}$
$5:$3 = 5:3
$5 to $3 = 5 to 3

21. $\dfrac{12 \text{ quarts}}{18 \text{ quarts}} = \dfrac{12}{18} = \dfrac{2}{3}$
12 quarts:18 quarts = 12:18 = 2:3
12 quarts to 18 quarts = 12 to 18 = 2 to 3

23. $\dfrac{14 \text{ days}}{7 \text{ days}} = \dfrac{14}{7} = \dfrac{2}{1}$
14 days:7 days = 14:7 = 2:1
14 days to 7 days = 14 to 7 = 2 to 1

Objective B Application Problems

25. Strategy To find the ratio, write the ratio of housing ($800) to total expenses ($2400) in simplest form.

Solution $\dfrac{\$800}{\$2400} = \dfrac{800}{2400} = \dfrac{1}{3}$
The ratio is $\dfrac{1}{3}$.

27. Strategy To find the ratio, write the ratio of utilities ($150) to food ($400) in simplest form.

Solution $\dfrac{\$150}{\$400} = \dfrac{150}{400} = \dfrac{3}{8}$
The ratio is $\dfrac{3}{8}$.

29. Strategy   To find the ratio, write in simplest form the number of college freshmen playing basketball over the number of high school seniors playing basketball.

Solution $\dfrac{4000}{154,000} = \dfrac{2}{77}$

The ratio is $\dfrac{2}{77}$.

31. Strategy   To find the ratio, write the ratio of turns in the primary coil (40) to the number of turns in the secondary coil (480) in simplest form.

Solution $\dfrac{40}{480} = \dfrac{1}{12}$

The ratio is $\dfrac{1}{12}$.

33a. Strategy   To find the amount of the increase, subtract the original value ($90,000) from the increased value ($110,000).

Solution $\begin{array}{r} \$110,000 \\ -\ \ 90,000 \\ \hline \$20,000 \end{array}$

The amount of the increase is $20,000.

33b. Strategy   To find the ratio, write the ratio of the increase ($20,000) to the original value ($90,000) in simplest form.

Solution $\dfrac{\$20,000}{\$90,000} = \dfrac{20,000}{90,000} = \dfrac{2}{9}$

The ratio is $\dfrac{2}{9}$.

Applying the Concepts

35. Income = $3500 + $250 + $140 = $3890
Debts = $900 + $170 + $160 + $95 = $1325
The ratio = $\dfrac{\$1325}{\$3890} = \dfrac{265}{778}$

37. No. Income = $3400 + $83 + $650 + $34 = $4167
Debts = $1800 + $104 + $35 + $120 + $234
     + $197 = $2490
The ratio = $\dfrac{\$2490}{\$4167} = \dfrac{830}{1389} = 0.5976$

The ratio is greater than $\dfrac{2}{5}$ (0.4).

SECTION 4.2

Objective A   Exercises

1. $\dfrac{3 \text{ pounds}}{4 \text{ people}}$

3. $\dfrac{\$80}{12 \text{ boards}} = \dfrac{\$20}{3 \text{ boards}}$

5. $\dfrac{300 \text{ miles}}{15 \text{ gallons}} = \dfrac{20 \text{ miles}}{1 \text{ gallon}}$

7. $\dfrac{20 \text{ children}}{8 \text{ families}} = \dfrac{5 \text{ children}}{2 \text{ families}}$

9. $\dfrac{16 \text{ gallons}}{2 \text{ hours}} = \dfrac{8 \text{ gallons}}{1 \text{ hour}}$

Objective B   Exercises

11. $\dfrac{10 \text{ feet}}{4 \text{ seconds}} = 2.5$ feet/second

13. $\dfrac{\$1300}{4 \text{ weeks}} = \$325$/week

15. $\dfrac{1100 \text{ trees}}{10 \text{ acres}} = 110$ trees/acre

17. $\dfrac{\$32.97}{7 \text{ hours}} = \$4.71$/hour

19. $\dfrac{628.8 \text{ miles}}{12 \text{ hours}} = 52.4$ miles/hour

21. $\dfrac{344.4 \text{ miles}}{12.3 \text{ gallons}} = 28$ miles/gallon

23. $\dfrac{\$349.80}{212 \text{ pounds}} = \$1.65$/pound

Objective C   Application Problems

25. Strategy   To find the number of miles driven per gallon of gas, divide the total number of miles (326.6) by the total number of gallons (11.5).

Solution $11.5\overline{)326.6}^{\,28.4}$

The number of miles driven per gallon of gas is 28.4.

27. Strategy   To find how much fuel the rocket uses in 1 minute, divide the total fuel (534,000 gallons) by the number of minutes (2.5).

Solution $2.5\overline{)534,000}^{\,213,600}$

The number of gallons used in 1 minute is 213,600.

29a. Strategy   To find how many pounds of beef were packaged, subtract the waste (75 pounds) from the original weight (250 pounds).

Solution $\begin{array}{r} 250 \\ -\ 75 \\ \hline 175 \end{array}$

The number of pounds of beef packaged was 175.

29b. Strategy To find the cost per pound of the packaged beef, divide the total cost ($365.75) by the weight of the packaged beef (175 pounds).

Solution $175\overline{)\$365.75}$ with quotient $\$2.09$

The cost per pound was $2.09.

31a. Strategy To find the rate per minute, multiply the rate (5.6 feet per second) by 60 seconds per minute.

Solution

$$\frac{5.6 \text{ ft}}{1 \text{ sec}} \times \frac{60 \text{ sec}}{1 \text{ min}} = 336 \text{ ft/min}$$

The camera goes through film at the rate of 336 feet per minute.

31b. Strategy To find how fast the camera uses a 500-foot roll, divide the length of the roll (500 feet) by the rate that it is used (5.6 feet per second).

Solution $5.6\overline{)500}$ with quotient 89.2

$89.2 \approx 89$

The camera uses the 500-foot roll in 89 seconds.

33a. Strategy To find the price of the computer hardware in euros, multiply the price ($38,000) by the euro exchange rate (1.1118 euros per U.S. dollar).

Solution

$$\frac{\$38,000}{1} \times \frac{1.118 \text{ euros}}{\$1} = 42,484 \text{ euros}$$

The price of the computer hardware would be 42,484 euros.

33b. Strategy To find the price of a car in yen, multiply the price ($24,000) by the Japanese yen exchange rate (106.500 yen per U.S. dollar).

Solution

$$\frac{\$24,000}{1} \times \frac{106.500 \text{ yen}}{\$1} = 2,556,000 \text{ yen}$$

The price of the car would be 2,556,000 yen.

Applying the Concepts

35. The price-earnings ratio of a company's stock is computed by dividing the current price by per-share of the stock by the annual earnings per share. For example, if the price-earnings ratio of a company's stock is 8.5, the price of the stock is 8.5 times the earnings per share of the stock.

SECTION 4.3

Objective A Exercises

1. $\dfrac{4}{8} \times \dfrac{10}{20} \rightarrow 8 \times 10 = 80$
$\rightarrow 4 \times 20 = 80$

The proportion is true.

3. $\dfrac{7}{8} \times \dfrac{11}{12} \rightarrow 8 \times 11 = 88$
$\rightarrow 7 \times 12 = 84$

The proportion is not true.

5. $\dfrac{27}{8} \times \dfrac{9}{4} \rightarrow 8 \times 9 = 72$
$\rightarrow 27 \times 4 = 108$

The proportion is not true.

7. $\dfrac{45}{135} \times \dfrac{3}{9} \rightarrow 135 \times 3 = 405$
$\rightarrow 45 \times 9 = 405$

The proportion is true.

9. $\dfrac{16}{3} \times \dfrac{48}{9} \rightarrow 3 \times 48 = 144$
$\rightarrow 16 \times 9 = 144$

The proportion is true.

11. $\dfrac{7}{40} \times \dfrac{7}{8} \rightarrow 40 \times 7 = 280$
$\rightarrow 7 \times 8 = 56$

The proportion is not true.

13. $\dfrac{50}{2} \times \dfrac{25}{1} \rightarrow 2 \times 25 = 50$
$\rightarrow 50 \times 1 = 50$

The proportion is true.

15. $\dfrac{6}{5} \times \dfrac{30}{25} \rightarrow 5 \times 30 = 150$
$\rightarrow 6 \times 25 = 150$

The proportion is true.

17. $\dfrac{15}{4} \times \dfrac{45}{12} \rightarrow 4 \times 45 = 180$
$\rightarrow 15 \times 12 = 180$

The proportion is true.

19. $\dfrac{300}{4} \times \dfrac{450}{7} \rightarrow 4 \times 450 = 1800$
$\rightarrow 300 \times 7 = 2100$

The proportion is not true.

21. $\dfrac{65}{5} \times \dfrac{26}{2} \rightarrow 5 \times 26 = 130$
$\rightarrow 65 \times 2 = 130$

The proportion is true.

23.
$$\frac{7}{4} \diagdown \frac{42}{20} \rightarrow 4 \times 42 = 168 \\ \rightarrow 7 \times 20 = 140$$

The proportion is not true.

Objective B Exercises

25. $n \times 8 = 4 \times 6$
$n \times 8 = 24$
$n = 24 \div 8$
$n = 3$

27. $12 \times 9 = 18 \times n$
$108 = 18 \times n$
$108 \div 18 = n$
$6 = n$

29. $6 \times 36 = n \times 24$
$216 = n \times 24$
$216 \div 24 = n$
$9 = n$

31. $n \times 135 = 45 \times 17$
$n \times 135 = 765$
$n = 765 \div 135$
$n \approx 5.67$

33. $n \times 3 = 6 \times 2$
$n \times 3 = 12$
$n = 12 \div 3$
$n = 4$

35. $n \times 8 = 5 \times 7$
$n \times 8 = 35$
$n = 35 \div 8$
$n \approx 4.38$

37. $n \times 4 = 11 \times 32$
$n \times 4 = 352$
$n = 352 \div 4$
$n = 88$

39. $5 \times 8 = 12 \times n$
$40 = 12 \times n$
$40 \div 12 = n$
$3.33 \approx n$

41. $n \times 12 = 15 \times 21$
$n \times 12 = 315$
$n = 315 \div 12$
$n = 26.25$

43. $32 \times 3 = n \times 1$
$96 = n \times 1$
$96 \div 1 = n$
$96 = n$

45. $18 \times n = 11 \times 16$
$18 \times n = 176$
$n = 176 \div 18$
$n \approx 9.78$

47. $28 \times n = 8 \times 12$
$28 \times n = 96$
$n = 96 \div 28$
$n \approx 3.43$

49. $0.3 \times 25 = 5.6 \times n$
$7.5 = 5.6 \times n$
$7.5 \div 5.6 = n$
$1.34 \approx n$

51. $0.7 \times n = 9.8 \times 3.6$
$0.7 \times n = 35.28$
$n = 35.28 \div 0.7$
$n = 50.4$

Objective C Application Problems

53. Strategy To find out how many calories are in a 0.5-ounce serving of cereal, write and solve a proportion using n to represent the calories.

Solution $\dfrac{6 \text{ ounces}}{600 \text{ calories}} = \dfrac{0.5 \text{ ounces}}{n \text{ calories}}$
$6 \times n = 600 \times 0.5$
$6 \times n = 300$
$n = 300 \div 6$
$n = 50$
A 0.5-ounce serving contains 50 calories.

55. Strategy To find out how many pounds of fertilizer are used, write and solve a proportion using n to represent the pounds of fertilizer.

Solution $\dfrac{2 \text{ pounds}}{100 \text{ square feet}} = \dfrac{n \text{ pounds}}{3500 \text{ square feet}}$
$2 \times 3500 = 100 \times n$
$7000 = 100 \times n$
$7000 \div 100 = n$
$70 = n$
The homeowner used 70 pounds of fertilizer.

57. Strategy To find the number of wooden bats produced, write and solve a proportion using n to represent the number of wooden bats.

Solution $\dfrac{4 \text{ alum. bats}}{15 \text{ wooden bats}} = \dfrac{100 \text{ alum. bats}}{n \text{ wooden bats}}$
$4 \times n = 15 \times 100$
$4 \times n = 1500$
$n = 1500 \div 4$
$n = 375$
There were 375 wooden bats produced.

59. Strategy   To find the distance between two cities that are 2 inches apart on the map, write and solve a proportion using n to represent the number of miles.

Solution
$$\frac{1.25 \text{ inches}}{10 \text{ miles}} = \frac{2 \text{ inches}}{n \text{ miles}}$$
$$1.25 \times n = 10 \times 2$$
$$1.25 \times n = 20$$
$$n = 20 \div 1.25$$
$$n = 16$$
The distance is 16 miles.

61. Strategy   To find the dosage for a person who weighs 150 pounds, write and solve a proportion using n to represent the number of ounces.

Solution
$$\frac{n}{150 \text{ pounds}} = \frac{\frac{1}{3} \text{ ounce}}{40 \text{ pounds}}$$
$$40 \times n = \frac{1}{3} \times 150$$
$$40 \times n = 50$$
$$n = 50 \div 40$$
$$n = 1.25$$
The number of ounces required is 1.25.

63. Strategy   To find how many people in a county of 240,000 eligible voters would vote in the election, write and solve a proportion using n to represent the number of voters.

Solution
$$\frac{n}{240,000} = \frac{2}{3}$$
$$2 \times 240,000 = 3 \times n$$
$$480,000 = 3 \times n$$
$$480,000 \div 3 = n$$
$$160,000 = n$$
The number of voters would be 160,000.

65. Strategy   To find the annual payment, write and solve a proportion using n to represent the annual payment.

Solution
$$\frac{\$35.35}{\$10,000} = \frac{n}{\$50,000}$$
$$35.35 \times 50,000 = 10,000 \times n$$
$$1,767,500 = 10,000 \times n$$
$$1,767,500 \div 10,000 = n$$
$$176.75 = n$$
The annual payment is $176.75.

67. Strategy   To find how many shares of stock you own after a split, write and solve a proportion using n to represent the number of shares.

Solution
$$\frac{5}{3} = \frac{n}{240}$$
$$5 \times 240 = n \times 3$$
$$1200 = n \times 3$$
$$1200 \div 3 = n$$
$$400 = n$$
You own 400 shares.

69. Strategy   To find how much a bowling ball weighs on the moon, write and solve a proportion using n to represent the weight on the moon.

Solution
$$\frac{1}{6} = \frac{n}{16}$$
$$1 \times 16 = n \times 6$$
$$16 = n \times 6$$
$$16 \div 6 = n$$
$$2.67 = n$$
The bowling ball would weigh 2.67 pounds on the moon.

71. Strategy   To find what dividend an investor would receive after purchasing additional shares:
• Find the total number of shares owned by adding the original number (50) to the amount purchased (300).
• Find the dividend by writing and solving a proportion using n to represent the dividend.

Solution
$$\begin{array}{r} 300 \\ + \ \ 50 \\ \hline 350 \text{ shares} \end{array}$$
$$\frac{n}{350 \text{ shares}} = \frac{\$153}{50 \text{ shares}}$$
$$153 \times 350 = n \times 50$$
$$53,550 = n \times 50$$
$$53,550 \div 50 = n$$
$$\$1071 = n$$
The dividend would be $1071.

Applying the Concepts

73. No, it is not possible. The sum of the fractions is $\frac{2}{5} + \frac{3}{4} = \frac{23}{20} = 1\frac{3}{20}$, which is greater than 1. In order for the responses to be possible, the sum of the fractions must be 1.

CHAPTER REVIEW

1. $\dfrac{2}{9} \times \dfrac{10}{45} \rightarrow \begin{array}{l} 9 \times 10 = 90 \\ 2 \times 45 = 90 \end{array}$

   The proportion is true.

2. $\dfrac{\$32}{\$80} = \dfrac{32}{80} = \dfrac{2}{5}$

   $\$32{:}\$80 = 32{:}80 = 2{:}5$

   $\$32 \text{ to } \$80 = 32 \text{ to } 80 = 2 \text{ to } 5$

3. $\dfrac{250 \text{ miles}}{4 \text{ hours}} = 62.5 \text{ miles/hour}$

4. $\dfrac{8}{15} \times \dfrac{32}{60} \rightarrow \begin{array}{l} 15 \times 32 = 480 \\ 8 \times 60 = 480 \end{array}$

   The proportion is true.

5. $\dfrac{16}{n} = \dfrac{4}{17}$

   $16 \times 17 = n \times 4$

   $272 = n \times 4$

   $272 \div 4 = n$

   $68 = n$

6. $\dfrac{\$300}{40 \text{ hours}} = \$7.50/\text{hour}$

7. $\dfrac{\$8.75}{5 \text{ pounds}} = \$1.75/\text{pound}$

8. $\dfrac{8 \text{ feet}}{28 \text{ feet}} = \dfrac{8}{28} = \dfrac{2}{7}$

   $8 \text{ feet}{:}28 \text{ feet} = 8{:}28 = 2{:}7$

   $8 \text{ feet to } 28 \text{ feet} = 8 \text{ to } 28 = 2 \text{ to } 7$

9. $\dfrac{n}{8} = \dfrac{9}{2}$

   $n \times 2 = 8 \times 9$

   $n \times 2 = 72$

   $n = 72 \div 2$

   $n = 36$

10. $\dfrac{18}{35} = \dfrac{10}{n}$

    $n \times 18 = 35 \times 10$

    $n \times 18 = 350$

    $n = 350 \div 18$

    $n = 19.44$

11. $\dfrac{6 \text{ inches}}{15 \text{ inches}} = \dfrac{6}{15} = \dfrac{2}{5}$

    $6 \text{ inches}{:}15 \text{ inches} = 6{:}15 = 2{:}5$

    $6 \text{ inches to } 15 \text{ inches} = 6 \text{ to } 15 = 2 \text{ to } 5$

12. $\dfrac{3}{8} \times \dfrac{10}{24} \rightarrow \begin{array}{l} 8 \times 10 = 80 \\ 3 \times 24 = 72 \end{array}$

    The proportion is not true.

13. $\dfrac{\$15}{4 \text{ hours}}$

14. $\dfrac{326.4 \text{ miles}}{12 \text{ gallons}} = 27.2 \text{ miles/gallon}$

15. $\dfrac{12 \text{ days}}{12 \text{ days}} = \dfrac{12}{12} = \dfrac{1}{1}$

    $12 \text{ days}{:}12 \text{ days} = 12{:}12 = 1{:}1$

    $12 \text{ days to } 12 \text{ days} = 12 \text{ to } 12 = 1 \text{ to } 1$

16. $\dfrac{5}{7} \times \dfrac{25}{35} \rightarrow \begin{array}{l} 7 \times 25 = 175 \\ 5 \times 35 = 175 \end{array}$

    The proportion is true.

17. $\dfrac{24}{11} = \dfrac{n}{30}$

    $24 \times 30 = n \times 11$

    $720 = n \times 11$

    $720 \div 11 = n$

    $65.45 = n$

18. $\dfrac{100 \text{ miles}}{3 \text{ hours}}$

19. Strategy   To find the ratio:
    - Find the amount of the decrease by subtracting the current price ($24) from the original price ($40).
    - Write the ratio between the decrease and the original price.

    Solution $\begin{array}{r} \$40 \\ -\ 24 \\ \hline 16 \end{array}$

    $\dfrac{\$16}{\$40} = \dfrac{16}{40} = \dfrac{2}{5}$

    The ratio is $\dfrac{2}{5}$.

20. Strategy   To find the property tax on a home valued at $120,000, write and solve a proportion using n to represent the property tax.

    Solution $\dfrac{n}{\$120,000} = \dfrac{\$900}{\$45,000}$

    $900 \times 120,000 = n \times 45,000$

    $108,000,000 = n \times 45,000$

    $108,000,000 \div 45,000 = n$

    $2400 = n$

    The property tax is $2400.

21. Strategy   To find the ratio, write the ratio of the high temperature (84 degrees) to the low temperature (42 degrees).

    Solution $\dfrac{84 \text{ degrees}}{42 \text{ degrees}} = \dfrac{84}{42} = \dfrac{2}{1}$

    The ratio is $\dfrac{2}{1}$.

22. Strategy To find the cost per radio of the radios that did pass inspection:
• Find the number of radios that did pass inspection by subtracting the number that did not pass inspection (24) from the total (1000).
• Divide the total manufacturing cost ($36,600) by the number of radios that did pass inspection.

Solution
$$1000$$
$$\underline{-\;\;24}$$
$$976$$

$$976\overline{)\underset{\$37.50}{\$36,600}}$$

The cost per radio was $37.50.

23. Strategy To find how many concrete blocks would be needed to build a wall 120 feet long, write and solve a proportion using n to represent the number of concrete blocks.

Solution
$$\frac{n}{120 \text{ feet}} = \frac{448 \text{ concrete blocks}}{40 \text{ feet}}$$
$$n \times 40 = 120 \times 448$$
$$n \times 40 = 53,760$$
$$n = 53,760 \div 40$$
$$n = 1344$$

The number of concrete blocks that would be needed is 1344.

24. Strategy To find the ratio, write a ratio of TV advertising ($30,000) to newspaper advertising ($12,000).

Solution
$$\frac{\$30,000}{\$12,000} = \frac{30,000}{12,000} = \frac{5}{2}$$

The ratio is $\frac{5}{2}$.

25. Strategy To find the cost per pound, divide the total cost ($10.20) by the number of pounds (15).

Solution
$$15\overline{)\underset{\$.68}{\$10.20}}$$

The cost per pound is $.68.

26. Strategy To find the average number of miles driven per hour, divide the total number of miles driven (198.8) by the number of hours (3.5).

Solution
$$3.5\overline{)\underset{56.8}{198.8}}$$

The average number of miles driven per hour was 56.8.

27. Strategy To find the cost of $50,000 of insurance, write and solve a proportion using n to represent the cost.

Solution
$$\frac{n}{\$50,000} = \frac{\$3.87}{\$1000}$$
$$n \times 1000 = 3.87 \times 50,000$$
$$n \times 1000 = 193,500$$
$$n = 193,500 \div 1000$$
$$n = 193.50$$

The cost is $193.50.

28. Strategy To find the cost per share, divide the total cost ($3580) by the number of shares (80).

Solution
$$80\overline{)\underset{\$44.75}{\$3580}}$$

The cost per share is $44.75.

29. Strategy To find how many pounds of fertilizer are used on a lawn that measures 3000 square feet, write and solve a proportion using n to represent the number of pounds of fertilizer.

Solution
$$\frac{n}{3000 \text{ square feet}} = \frac{1.5 \text{ pounds}}{200 \text{ square feet}}$$
$$n \times 200 = 1.5 \times 3000$$
$$n \times 200 = 4500$$
$$n = 4500 \div 200$$
$$n = 22.5$$

The number of pounds of fertilizer used is 22.5.

30. Strategy To find the ratio:
• Find the amount of the increase by subtracting the original value ($80,000) from the increased value ($120,000).
• Write the ratio of the amount of the increase to the original value ($80,000).

Solution
$$\$120,000$$
$$\underline{-\;\;80,000}$$
$$\$40,000$$

$$\frac{\$40,000}{\$80,000} = \frac{40,000}{80,000} = \frac{1}{2}$$

The ratio is $\frac{1}{2}$.

CHAPTER TEST

1. $\frac{22,036.80}{12 \text{ months}} = \$1836.40/\text{month}$

2. $\frac{40 \text{ miles}}{240 \text{ miles}} = \frac{40}{240} = \frac{1}{6}$
40 miles:240 miles = 40:240 = 1:6
40 miles to 240 miles = 40 to 240 = 1 to 6

3. $\dfrac{18 \text{ supports}}{8 \text{ feet}} = \dfrac{9 \text{ supports}}{4 \text{ feet}}$

4. $\dfrac{40}{125} \times \dfrac{5}{25} \begin{array}{l} \rightarrow 125 \times 5 = 625 \\ \rightarrow 40 \times 25 = 1000 \end{array}$

   The proportion is not true.

5. $\dfrac{12 \text{ days}}{8 \text{ days}} = \dfrac{12}{8} = \dfrac{3}{2}$

   12 days:8 days = 12:8 = 3:2

   12 days to 8 days = 12 to 8 = 3 to 2

6. $\dfrac{5}{12} = \dfrac{60}{n}$

   $n \times 5 = 12 \times 60$

   $n \times 5 = 720$

   $n = 720 \div 5$

   $n = 144$

7. $\dfrac{256.2 \text{ miles}}{8.4 \text{ gallons}} = 30.5 \text{ miles/gallon}$

8. $\dfrac{\$27}{\$81} = \dfrac{27}{81} = \dfrac{1}{3}$

   \$27:\$81 = 27:81 = 1:3

   \$27 to \$81 = 27 to 81 = 1 to 3

9. $\dfrac{5}{14} \times \dfrac{25}{70} \begin{array}{l} \rightarrow 14 \times 25 = 350 \\ \rightarrow 5 \times 70 = 350 \end{array}$

   The proportion is true.

10. $\dfrac{n}{18} = \dfrac{9}{4}$

    $n \times 4 = 9 \times 18$

    $n \times 4 = 162$

    $n = 162 \div 4$

    $n = 40.5$

11. $\dfrac{\$81}{12 \text{ boards}} = \dfrac{\$27}{4 \text{ boards}}$

12. $\dfrac{18 \text{ feet}}{30 \text{ feet}} = \dfrac{18}{30} = \dfrac{3}{5}$

    18 feet:30 feet = 18:30 = 3:5

    18 feet to 30 feet = 18 to 30 = 3 to 5

13. Strategy To find the dividend on 500 shares of the utility stock, write and solve a proportion using n to represent the dividend.

    Solution $\dfrac{n}{500 \text{ shares}} = \dfrac{\$62.50}{50 \text{ shares}}$

    $n \times 50 = 500 \times \$62.50$

    $n \times 50 = 31,250$

    $n = 625$

    The dividend is \$625.

14. Strategy To find the ratio, write the ratio of the city temperature (86°) to the desert temperature (112°).

    Solution $\dfrac{86 \text{ degrees}}{112 \text{ degrees}} = \dfrac{86}{112} = \dfrac{43}{56}$

    The ratio is $\dfrac{43}{56}$.

15. $\dfrac{2421 \text{ miles}}{4.5 \text{ hours}} = 538 \text{ miles/hour}$

16. Strategy To estimate the number of pounds of water in a college student weighing 150 pounds, write and solve a proportion using n to represent the number of pounds of water.

    Solution $\dfrac{88 \text{ pounds water}}{100 \text{ pounds body weight}}$

    $= \dfrac{n}{150 \text{ pounds body weight}}$

    $88 \times 150 = n \times 100$

    $13,200 = n \times 100$

    $13,200 \div 100 = n$

    $132 = n$

    The college student's body contains 132 pounds of water.

17. $\dfrac{\$69.20}{40 \text{ feet}} = \$1.73/\text{foot}$

18. Strategy To find how many ounces of medication are required for a person who weighs 175 pounds, write and solve a proportion using n to represent the ounces of medication.

    Solution $\dfrac{\frac{1}{4} \text{ ounce}}{50 \text{ pounds}} = \dfrac{n}{175 \text{ pounds}}$

    $\dfrac{1}{4} \times 175 = n \times 50$

    $43.75 = n \times 50$

    $43.75 \div 50 = n$

    $0.875 = n$

    The amount of medication required is 0.875 ounces.

19. Strategy To find the ratio:
• Find the total cost of advertising by adding the cost of television advertising ($25,000) to the cost of radio advertising ($40,000).
• Write the ratio of radio advertising ($40,000) to the total cost of advertising.

Solution
$$\begin{array}{r} \$25,000 \\ +\ \ 40,000 \\ \hline \$65,000 \end{array}$$

$$\frac{\$40,000}{\$65,000} = \frac{40,000}{65,000} = \frac{8}{13}$$

The ratio is $\frac{8}{13}$.

20. Strategy To find the property tax on a house valued at $140,000, write and solve a proportion using n to represent the amount of the property tax.

Solution
$$\frac{n}{\$140,000} = \frac{\$1500}{\$175,000}$$
$$n \times 175,000 = 140,000 \times 1500$$
$$n \times 175,000 = 210,000,000$$
$$n = 210,000,000 \div 175,000$$
$$n = 1200$$
The property tax is $1200.

CUMULATIVE REVIEW

1.
$$\begin{array}{r} \overset{9}{\cancel{1}} \overset{10}{\cancel{10}} \overset{10}{\cancel{10}} \overset{8}{\cancel{8}} \overset{15}{\cancel{15}} \\ 2\,0,\,0\,9\,5 \\ -\ 10,\,9\,3\,7 \\ \hline 9,\,1\,5\,8 \end{array}$$

2. $2 \cdot 2 \cdot 2 \cdot 2 \cdot 3 \cdot 3 \cdot 3 = 2^4 \cdot 3^3$

3. $4 - (5-2)^2 \div 3 + 2 = 4 - (-3)^2 \div 3 + 2$
$$= 4 - 9 \div 3 + 2$$
$$= 4 - 3 + 2$$
$$= 1 + 2 = 3$$

4. $160 = 2 \cdot 2 \cdot 2 \cdot 2 \cdot 2 \cdot 5$
$$\begin{array}{r} 5 \\ 2\overline{)10} \\ 2\overline{)20} \\ 2\overline{)40} \\ 2\overline{)80} \\ 2\overline{)160} \end{array}$$

5.

	2	3
9 =		3·3
12 =	②·2	3
18 =	2	③·③

LCM = $2 \cdot 2 \cdot 3 \cdot 3 = 36$

6.

	2	3	7
28 =	2·2		⑦
42 =	②	3	7

GCF = $2 \cdot 7 = 14$

7. $\frac{40}{64} = \frac{8 \cdot 5}{8 \cdot 8} = \frac{5}{8}$

8.
$$\begin{array}{r} 3\frac{5}{6} = 3\frac{25}{30} \\ +\ 4\frac{7}{15} = 4\frac{14}{30} \\ \hline 7\frac{39}{30} = 8\frac{9}{30} = 8\frac{3}{10} \end{array}$$

9.
$$\begin{array}{r} 10\frac{1}{6} = 10\frac{3}{18} = 9\frac{21}{18} \\ -\ 4\frac{5}{9} = 4\frac{10}{18} = 4\frac{10}{18} \\ \hline 5\frac{11}{18} \end{array}$$

10. $\frac{11}{12} \times 3\frac{1}{11} = \frac{11}{12} \times \frac{34}{11}$
$$= \frac{11 \times 34}{12 \times 11}$$
$$= \frac{\overset{1}{\cancel{11}} \cdot 2 \cdot 17}{2 \cdot 2 \cdot 3 \cdot \underset{1}{\cancel{11}}} = \frac{17}{6} = 2\frac{5}{6}$$

11. $3\frac{1}{3} \div \frac{5}{7} = \frac{10}{3} \div \frac{5}{7}$
$$= \frac{10}{3} \times \frac{7}{5}$$
$$= \frac{10 \cdot 7}{3 \cdot 5} = \frac{2 \cdot \overset{1}{\cancel{5}} \cdot 7}{3 \cdot \underset{1}{\cancel{5}}} = \frac{14}{3} = 4\frac{2}{3}$$

12. $\left(\frac{2}{5} + \frac{3}{4}\right) \div \frac{3}{2} = \left(\frac{8}{20} + \frac{15}{20}\right) \div \frac{3}{2}$
$$= \frac{23}{20} \times \frac{2}{3}$$
$$= \frac{23 \times 2}{20 \times 3} = \frac{23 \cdot \overset{1}{\cancel{2}}}{2 \cdot 2 \cdot 5 \cdot 3} = \frac{23}{30}$$

13. Four and seven hundred nine ten-thousandths

14.   ┌─Given place value
2.09762
     └─7 > 5
2.10

15.
$$8.09)\overline{16.097600} \quad \frac{1.9898 \approx 1.990}{}$$
$$\underline{-809}$$
$$8007$$
$$\underline{-7281}$$
$$7266$$
$$\underline{-6472}$$
$$7940$$
$$\underline{-7281}$$
$$6590$$
$$\underline{-6472}$$
$$118$$

16. $0.06\frac{2}{3} = \frac{6\frac{2}{3}}{100} = 6\frac{2}{3} \div 100 = \frac{20}{3} \div 100$
$$= \frac{20}{3} \times \frac{1}{100}$$
$$= \frac{20 \cdot 1}{3 \cdot 100} = \frac{1}{15}$$

17. $\frac{25 \text{ miles}}{200 \text{ miles}} = \frac{25}{200} = \frac{1}{8}$

18. $\frac{87\cancel{c}}{6 \text{ bars of soap}} = \frac{29\cancel{c}}{2 \text{ bars of soap}}$

19. $\frac{250.5 \text{ miles}}{7.5 \text{ gallons of gas}} = 33.4 \text{ miles/gallon}$

20. $\frac{40}{n} = \frac{160}{17}$
$$40 \times 17 = n \times 160$$
$$680 = n \times 160$$
$$680 \div 160 = n$$
$$4.25 = n$$

21. $\frac{457.6 \text{ miles}}{8 \text{ hours}} = 57.2 \text{ miles/hour}$

22. $\frac{12}{5} = \frac{n}{15}$
$$12 \times 15 = n \times 5$$
$$180 = n \times 5$$
$$180 \div 5 = n$$
$$36 = n$$

23. Strategy To find your new checking account balance:
• Find the total of the checks written by adding the two checks ($192 and $88).
• Subtract the total of the checks written from the original balance ($1024).

Solution
$$\begin{array}{ll} \$192 & \$1024 \\ \underline{+\ 88} & \underline{-\ 280} \\ \$280 & \$744 \end{array}$$
Your new balance is $744.

24. Strategy To find the monthly payments:
• Find the amount to be paid by subtracting the down payment ($5000) from the original cost ($22,760).
• Divide the amount remaining to be paid by the number of payments (48).

Solution
$$\begin{array}{ll} \$22,760 & 48)\overline{\$17,760} \quad \frac{\$370}{} \\ \underline{-\ 5,000} & \\ \$17,760 & \end{array}$$
The monthly payments are $370 each.

25. Strategy To find how many pages remain to be read:
• Find the amount read during vacation by multiplying the total (175 pages) by $\frac{2}{5}$.
• Subtract the number of pages read during vacation from the total (175 pages).

Solution $\frac{2}{5} \times 175 = \frac{2}{5} \times \frac{175}{1} = 70$
$$175 - 70 = 105$$
The number of pages that remain to be read is 105.

26. Strategy To find the cost per acre, divide the total cost ($84,000) by the number of acres $\left(2\frac{1}{3}\right)$.

Solution $\$84,000 \div 2\frac{1}{3} = 84,000 \div \frac{7}{3}$
$$= 84,000 \times \frac{3}{7} = \$36,000$$
The cost per acre was $36,000.

27. Strategy To find the amount of change:
• Find the total amount of the purchases by adding the two purchases ($22.79 and $9.59).
• Subtract the total amount of the purchases from $50.

Solution
$$\begin{array}{ll} \$22.79 & \$50.00 \\ \underline{+\ 9.59} & \underline{-\ 32.38} \\ \$32.38 & \$17.62 \end{array}$$
The change was $17.62.

28. Strategy   To find monthly salary, divide the annual salary ($41,691) by 12 months.

Solution

$$\begin{array}{r}
3468.25 \\
12\overline{)41,619.00} \\
\underline{-36} \\
56 \\
\underline{-48} \\
81 \\
\underline{-72} \\
99 \\
\underline{-96} \\
30 \\
\underline{-24} \\
60 \\
\underline{-60} \\
0
\end{array}$$

Your monthly salary is $3468.25.

29. Strategy   To find how many inches will be eroded in 50 months, write and solve a proportion using n to represent the number of inches.

Solution $\dfrac{3 \text{ inches}}{6 \text{ months}} = \dfrac{n}{50 \text{ months}}$

$3 \times 50 = n \times 6$

$150 = n \times 6$

$150 \div 6 = n$

$25 = n$

The number of inches will be 25.

30. Strategy   To find how many ounces of medication are required for a person who weighs 160 pounds, write and solve a proportion using n to represent the number of ounces.

Solution $\dfrac{n}{160} = \dfrac{\frac{1}{2} \text{ ounce}}{50 \text{ pounds}}$

$n \times 50 = \dfrac{1}{2} \times 160$

$n \times 50 = 80$

$n = 80 \div 50$

$n = 1.6$

The number of ounces required is 1.6.

Chapter 5: Percents

PREP TEST and GO FIGURE

1. $\dfrac{19}{100}$

2. 0.23

3. 47

4. 2850

5.
$$0.015\overline{)60.000.}\quad\begin{array}{r}4000.\\[2pt]\underline{-60}\\00\\\underline{-0}\\00\\\underline{-0}\\0\end{array}$$

6. $8 \div \dfrac{1}{4} = \dfrac{8}{1} \times \dfrac{4}{1} = 32$

7. $\dfrac{5}{8} \times \dfrac{100}{1} = \dfrac{5 \cdot \overset{1}{\cancel{2}} \cdot \overset{1}{\cancel{2}} \cdot 5 \cdot 5}{\underset{1}{\cancel{2}} \cdot \underset{1}{\cancel{2}} \cdot 2} = \dfrac{125}{2} = 62\dfrac{1}{2} = 62.5$

8. $66\dfrac{2}{3}$

9. $16\overline{)28.00}\quad\begin{array}{r}1.75\end{array}$

Go Figure

a. The smallest three-digit number is 101. However, any number that ends in "1" is not divisible by 2, and hence not by 6. So the numbers 111, 121, … 191 are also eliminated. The next smallest three-digit palindrome is 202, which is not divisible by 3, and neither is 212. However, 222 is divisible by 2 and 3, which means that it is a multiple of 6. So 222 is the smallest three-digit multiple of 6 that is a palindrome.

b. To use the process, add 874 and 478.

$$\begin{array}{r}478\\+874\\\hline 1352\end{array}$$

Then add 1352 and 2531.

$$\begin{array}{r}1352\\+2531\\\hline 3883\end{array}$$

The number 3883 is a palindrome.

SECTION 5.1

Objective A Exercises

1. $25\% = 25 \times \dfrac{1}{100} = \dfrac{25}{100} = \dfrac{1}{4}$

   $25\% = 25 \times 0.01 = 0.25$

3. $130\% = 130 \times \dfrac{1}{100} = \dfrac{130}{100} = 1\dfrac{3}{10}$

   $130\% = 130 \times 0.01 = 1.30$

5. $100\% = 100 \times \dfrac{1}{100} = \dfrac{100}{100} = 1$

   $100\% = 100 \times 0.01 = 1.00$

7. $73\% = 73 \times \dfrac{1}{100} = \dfrac{73}{100}$

   $73\% = 73 \times 0.01 = 0.73$

9. $383\% = 383 \times \dfrac{1}{100} = \dfrac{383}{100} = 3\dfrac{83}{100}$

   $383\% = 383 \times 0.01 = 3.83$

11. $70\% = 70 \times \dfrac{1}{100} = \dfrac{70}{100} = \dfrac{7}{10}$

    $70\% = 70 \times 0.01 = 0.70$

13. $88\% = 88 \times \dfrac{1}{100} = \dfrac{88}{100} = \dfrac{22}{25}$

    $88\% = 88 \times 0.01 = 0.88$

15. $32\% = 32 \times \dfrac{1}{100} = \dfrac{32}{100} = \dfrac{8}{25}$

    $32\% = 32 \times 0.01 = 0.32$

17. $66\dfrac{2}{3}\% = 66\dfrac{2}{3} \times \dfrac{1}{100} = \dfrac{200}{3} \times \dfrac{1}{100}$

    $= \dfrac{200}{300} = \dfrac{2}{3}$

19. $83\dfrac{1}{3}\% = 83\dfrac{1}{3} \times \dfrac{1}{100} = \dfrac{250}{3} \times \dfrac{1}{100}$

    $= \dfrac{250}{300} = \dfrac{5}{6}$

21. $11\dfrac{1}{9}\% = 11\dfrac{1}{9} \times \dfrac{1}{100} = \dfrac{100}{9} \times \dfrac{1}{100}$

    $= \dfrac{100}{900} = \dfrac{1}{9}$

23. $45\dfrac{5}{11}\% = 45\dfrac{5}{11} \times \dfrac{1}{100} = \dfrac{500}{11} \times \dfrac{1}{100}$

    $= \dfrac{500}{1100} = \dfrac{5}{11}$

25. $4\dfrac{2}{7}\% = 4\dfrac{2}{7} \times \dfrac{1}{100} = \dfrac{30}{7} \times \dfrac{1}{100}$

    $= \dfrac{30}{700} = \dfrac{3}{70}$

27. $6\dfrac{2}{3}\% = 6\dfrac{2}{3} \times \dfrac{1}{100} = \dfrac{20}{3} \times \dfrac{1}{100} = \dfrac{20}{300} = \dfrac{1}{15}$

29. $6.5\% = 6.5 \times 0.01 = 0.065$

31. $12.3\% = 12.3 \times 0.01 = 0.123$

33. $0.55\% = 0.55 \times 0.01 = 0.0055$

35. $8.25\% = 8.25 \times 0.01 = 0.0825$

37. $5.05\% = 5.05 \times 0.01 = 0.0505$

39. $2\% = 2 \times 0.01 = 0.02$

41. $80.4\% = 80.4 \times 0.01 = 0.804$

43. $4.9\% = 4.9 \times 0.01 = 0.049$

Objective B Exercises

45. $0.73 = 0.73 \times 100\% = 73\%$

47. $0.01 = 0.01 \times 100\% = 1\%$

49. $2.94 = 2.94 \times 100\% = 294\%$

51. $0.006 = 0.006 \times 100\% = 0.6\%$

53. $3.106 = 3.106 \times 100\% = 310.6\%$

55. $0.70 = 0.70 \times 100\% = 70\%$

57. $\dfrac{37}{100} = \dfrac{37}{100} \times 100\% = \dfrac{3700}{100}\% = 37\%$

59. $\dfrac{2}{5} = \dfrac{2}{5} \times 100\% = \dfrac{200}{5}\% = 40\%$

61. $\dfrac{1}{8} = \dfrac{1}{8} \times 100\% = \dfrac{100}{8}\% = 12.5\%$

63. $1\dfrac{1}{2} = 1\dfrac{1}{2} \times 100\% = \dfrac{3}{2} \times 100\% = \dfrac{300}{2}\% = 150\%$

65. $1\dfrac{2}{3} = 1\dfrac{2}{3} \times 100\% = \dfrac{5}{3} \times 100\% = \dfrac{500}{3}\% \approx 166.7\%$

67. $\dfrac{7}{8} = \dfrac{7}{8} \times 100\% = \dfrac{700}{8}\% = 87.5\%$

69. $\dfrac{12}{25} = \dfrac{12}{25} \times 100\% = \dfrac{1200}{25}\% = 48\%$

71. $\dfrac{1}{3} = \dfrac{1}{3} \times 100\% = \dfrac{100}{3}\% = 33\dfrac{1}{3}\%$

73. $1\dfrac{2}{3} = 1\dfrac{2}{3} \times 100\% = \dfrac{5}{3} \times 100\% = \dfrac{500}{3}\% = 166\dfrac{2}{3}\%$

75. $\dfrac{7}{8} = \dfrac{7}{8} \times 100\% = \dfrac{700}{8}\% = 87\dfrac{1}{2}\%$

Applying the Concepts

77. **a.** False. $4 \times 200\% = 8$

    b. False. $4 \div 200\% = 2$

    c. True.

79. $\dfrac{1}{3} \times 100\% = 33\dfrac{1}{3}\%$; this represents $33\dfrac{1}{3}\%$ off the regular price.

81. $1 - \dfrac{2}{5} = \dfrac{5}{5} - \dfrac{2}{5} = \dfrac{3}{5}$

$\dfrac{3}{5} \times 100\% = 60\%$;

60% of the population did not vote.

SECTION 5.2

Objective A Exercises

1. $0.08 \times 100 = n$
        $8 = n$

3. $0.27 \times 40 = n$
       $10.8 = n$

5. $0.0005 \times 150 = n$
      $0.075 = n$

7. $1.25 \times 64 = n$
      $80 = n$

9. $n = 0.107 \times 485$
   $n = 51.895$

11. $n = 0.0025 \times 3000$
   $n = 7.5$

13. $0.80 \times 16.25 = n$
       $13 = n$

15. $n = 0.015 \times 250$
   $n = 3.75$

17. $\dfrac{1}{6} \times 120 = n$
      $20 = n$

19. $n = \dfrac{1}{3} \times 630$
   $n = 210$

21. $0.05 \times 95 = n$ or $0.75 \times 6 = n$
      $4.75 = n$ $4.5 = n$
Because $4.75 > 4.5$, 5% of 95 is larger.

23. $0.79 \times 16 = n$ or $0.20 \times 65 = n$
     $12.64 = n$ $13 = n$
Because $12.64 < 13$, 79% of 16 is smaller.

25. $0.02 \times 1500 = n$ or $0.72 \times 40 = n$
      $30 = n$ $28.8 = n$
Because $20.8 < 30$, 72% of 40 is smaller.

27. $n = 0.31294 \times 82,460$
   $n = 25,805.0324$

Objective B Application Problems

29. Strategy To the number of people who do not have health insurance, write and solve the basic percent equation using n to the number of people between ages 18 and 24 who do not have health insurance. The percent is 30% and the base is 44.

Solution
$$30\% \times 44 = n$$
$$0.30 \times 44 = n$$
$$13.2 = n$$

About 13.2 million people aged 18 to 24 do not have health insurance.

31. Strategy To find the number of computers sold by Dell, write and solve the basic percent equation using n to represent the number of computers sold by Dell. The percent is 17% and the base is 11 million.

Solution
$$17\% \times 11 = n$$
$$0.17 \times 11 = n$$
$$1.87 = n$$

The number of computers sold by Dell was 1.87 million.

33a. Strategy To find the sales tax, write and solve the basic percent equation using n as the sales tax. The percent is 6% and the base is $19,500.

Solution
$$6\% \times \$19,500 = n$$
$$0.06 \times 19,500 = n$$
$$1170 = n$$

The sales tax is $1170.

33b. Strategy To find the total cost of the car, add the sales tax ($1170) to the purchase price of the car ($19,500).

Solution
$$\$19,500$$
$$\underline{+ 1,170}$$
$$\$20,670$$

The total cost of the car is $20,670.

35a. Strategy To find the increase in the number of eggs eaten in 2000, write and solve the basic percent equation using n to represent the number of eggs eaten in 2000. The percent is 9% and the base is 236.

Solution
$$9\% \times 236 = n$$
$$0.09 \times 236 = n$$
$$21.24 = n$$

Rounded to the nearest whole number, 21 more eggs were eaten in 2000 than in 1995.

35b. Strategy To find the number of eggs eaten in 2000, add the increase in the number of eggs eaten in 2000 (21) to the number of eggs eaten in 1995 (236).

Solution
$$236$$
$$\underline{+ 21}$$
$$257$$

Americans ate an average of 257 eggs during 2000.

Applying the Concepts

37. $43\% \times 112 = 48.16$
Employees spent 48.2 hours with family and friends.

39. Actual time: $112 \times 20\% = 22.4$
Preferred time: $112 \times 23\% = 25.76$
$25.76 - 22.4 = 3.36$
There are approximately 3.4 hours difference between the actual and preferred amount of time the employees spent on self.

SECTION 5.3

Objective A Exercises

1. $n \times 75 = 24$
$\quad n = 24 \div 75$
$\quad n = 0.32$
$\quad n = 32\%$

3. $n \times 90 = 15$
$\quad n = 15 \div 90$
$\quad n = 0.16\frac{2}{3}$
$\quad n = 16\frac{2}{3}\%$

5. $n \times 12 = 24$
$\quad n = 24 \div 12$
$\quad n = 2$
$\quad n = 200\%$

7. $n \times 16 = 6$
$\quad n = 6 \div 16$
$\quad n = 0.375$
$\quad n = 37.5\%$

9. $\quad 18 = n \times 100$
$\quad 18 \div 100 = n$
$\quad 0.18 = n$
$\quad 18\% = n$

11. $\quad 5 = n \times 2000$
$\quad 5 \div 2000 = n$
$\quad 0.0025 = n$
$\quad 0.25\% = n$

13. $n \times 6 = 1.2$
$\quad n = 1.2 \div 6$
$\quad n = 0.2$
$\quad n = 20\%$

15.
$$16.4 = n \times 4.1$$
$$16.4 \div 4.1 = n$$
$$4 = n$$
$$400\% = n$$

17.
$$1 = n \times 40$$
$$1 \div 40 = n$$
$$0.025 = n$$
$$2.5\% = n$$

19.
$$n \times 48 = 18$$
$$n = 18 \div 48$$
$$n = 0.375$$
$$n = 37.5\%$$

21.
$$n \times 2800 = 7$$
$$n = 7 \div 2800$$
$$n = 0.0025$$
$$n = 0.25\%$$

23.
$$4.2 = n \times 175$$
$$4.2 \div 175 = n$$
$$0.024 = n$$
$$2.4\% = n$$

25.
$$n \times 86.5 = 8.304$$
$$n = 8.304 \div 86.5$$
$$n = 0.096$$
$$n = 9.6\%$$

Objective B Application Problems

27. Strategy To find what percent of couples disagree about financial matters, write and solve the basic percent equation using n to represent the unknown percent. The base is 10 and the amount is 7.

Solution
$$n \times 10 = 7$$
$$n = 7 \div 10$$
$$n = 0.70$$
The percent of couples who disagree about financial matters is 70%.

29. Strategy To find what percent of the vegetables were wasted, write and solve the basic percent equation using n to represent the unknown percent. The base is 63 billion and the amount is 16 billion.

Solution
$$n \times 63 \text{ billion} = 16 \text{ billion}$$
$$n = 16 \text{ billion} \div 63 \text{ billion}$$
$$n \approx 0.2539$$
Approximately 25.4% of the vegetables were wasted.

31. Strategy To find what percent of the total amount spent on energy utilities is spent on lighting, write and solve the basic percent equation using n to represent the unknown percent. The base is $1355 and the amount is $81.30.

Solution
$$n \times \$1355 = \$81.30$$
$$n = 81.30 \div 1355$$
$$n = 0.06$$
The typical American household spends 6% of their total energy utilities is spent on lighting.

33. Strategy To find what percent of the workday is not spent at a computer:
• Find how many hours are not spent at a computer by subtracting the number of hours that are spent in front of the computer (3.1) from the total (8).
• Find the percent by writing and solving the basic percent equation using n to represent the unknown percent. The number of hours not spent at a computer (4.9) is the amount and the total (8) is the base.

Solution
$$\begin{array}{r} 8.0 \\ -3.1 \\ \hline 4.9 \end{array}$$
$$n \times 8 = 4.9$$
$$n = 4.9 \div 8$$
$$n = 0.6125 = 61.25\%$$
The percent of the workday that is not spent at a computer is 61.25%

Applying the Concepts

35.
$$\begin{array}{r} \$1,200 \\ 1,400 \\ 4,000 \\ 3,900 \\ 3,000 \\ +\ 1,100 \\ \hline 14,600 \end{array}$$

$14,600 is the total amount spent.
$4,000 is spent for food.
$$\frac{\$4,000}{\$14,600} \approx 0.2739$$
Approximately 27.4% of the total expenses is spent for food.

37.
$$\frac{\$76}{\$134} \times 100\% \approx 56.7\%$$
The rebate was approximately 56.7% of the final cost.

SECTION 5.4

Objective A Exercises

1. $0.12 \times n = 9$
   $n = 9 \div 0.12$
   $n = 75$

3. $8 = 0.16 \times n$
   $8 \div 0.16 = n$
   $50 = n$

5. $10 = 0.10 \times n$
   $10 \div 0.10 = n$
   $100 = n$

7. $0.30 \times n = 25.5$
   $n = 25.5 \div 0.30$
   $n = 85$

9. $0.025 \times n = 30$
   $n = 30 \div 0.025$
   $n = 1200$

11. $1.25 \times n = 24$
    $n = 24 \div 1.25$
    $n = 19.2$

13. $18 = 2.4 \times n$
    $18 \div 2.4 = n$
    $7.5 = n$

15. $4.8 = 0.15 \times n$
    $4.8 \div 0.15 = n$
    $32 = n$

17. $25.6 = 0.128 \times n$
    $25.6 \div 0.128 = n$
    $200 = n$

19. $0.007 \times n = 0.56$
    $n = 0.56 \div 0.007$
    $n = 80$

21. $0.30 \times n = 2.7$
    $n = 2.7 \div 0.30$
    $n = 9$

23. $84 = \frac{1}{6} \times n$
    $84 \div \frac{1}{6} = n$
    $504 = n$

25. $\frac{2}{3} \times n = 72$
    $n = 72 \div \frac{2}{3}$
    $n = 108$

Objective B Application Problems

27. **Strategy** To find the number of travelers who allowed their children to miss school, write and solve the basic percent equation using n to represent of travelers. The percent is 12% and the amount is 1.738 million.

    Solution $11\% \times n = 1.738$
    $0.11 \times n = 1.738$
    $n = 1.738 \div 0.11$
    $n = 15.8$
    There were 15.8 million travelers who allowed their children to miss school to go along on a trip.

29. **Strategy** To find the average cost of attending college in 1998-1999, write and solve the basic percent equation using n to represent the average cost of attending college in 1998-1999. The percent is 103.4% and the amount is $3356.

    Solution $103.4\% \times n = 3356$
    $1.034 \times n = 3356$
    $n = 3356 \div 1.034$
    $n \approx 3246$
    The average cost of attending a 4-year public college in 1998-1999 was approximately $3246.

31a. **Strategy** To find the number of computer boards tested, write and solve the basic percent equation using n to represent the number of computer boards tested. The percent is 0.8% and the amount is 24.

    Solution $n \times 0.8\% = 24$
    $n \times 0.008 = 24$
    $n = 24 \div 0.008$
    $n = 3000$
    The number of boards tested was 3000.

31b. **Strategy** To find the number of boards that were tested as not defective, subtract the number of defective boards (24) from the total tested (3000).

    Solution $\begin{array}{r} 3000 \\ - 24 \\ \hline 2976 \end{array}$
    The number of boards that were tested as not defective was 2976.

33. Strategy   To find the number of Democratic votes that were not for Al Gore:
• Find the total number of Democratic votes cast, write and solve the basic percent equation using n to represent the unknown total number of Democratic votes. The percent is 52% and the known amount is 76,897.
• Find the number of Democratic votes that were not for Al Gore by subtracting the number of votes for Al Gore (76,897) from the total number of votes (147,879).

Solution

$$52\% \times n = 76{,}897$$
$$0.52 \times n = 76{,}897$$
$$n = 76{,}897 \div 0.52$$
$$n \approx 147{,}879$$

$$\begin{array}{r} 147{,}879 \\ -\ 76{,}897 \\ \hline 70{,}982 \end{array}$$

The number of Democratic votes not for Al Gore was 70,982

Applying the Concepts

35. $0.08 \div 0.04 = 2$ milligrams of copper.

37.   The buyer is paying 25% of the owed price.

SECTION 5.5

Objective A Exercises

1.
$$\frac{26}{100} = \frac{n}{250}$$
$$26 \times 250 = n \times 100$$
$$6500 = n \times 100$$
$$6500 \div 100 = n$$
$$65 = n$$

3.
$$\frac{37}{148} = \frac{n}{100}$$
$$37 \times 100 = 148 \times n$$
$$3700 = 148 \times n$$
$$3700 \div 148 = n$$
$$25\% = n$$

5.
$$\frac{68}{100} = \frac{51}{n}$$
$$68 \times n = 100 \times 51$$
$$68 \times n = 5100$$
$$n = 5100 \div 68$$
$$n = 75$$

7.
$$\frac{n}{100} = \frac{43}{344}$$
$$n \times 344 = 100 \times 43$$
$$n \times 344 = 4300$$
$$n = 4300 \div 344$$
$$n = 12.5\%$$

9.
$$\frac{82}{n} = \frac{20.5}{100}$$
$$82 \times 100 = n \times 20.5$$
$$8200 = n \times 20.5$$
$$8200 \div 20.5 = n$$
$$400 = n$$

11.
$$\frac{n}{300} = \frac{6.5}{100}$$
$$n \times 100 = 300 \times 6.5$$
$$n \times 100 = 1950$$
$$n = 1950 \div 100$$
$$n = 19.5$$

13.
$$\frac{7.4}{50} = \frac{n}{100}$$
$$7.4 \times 100 = 50 \times n$$
$$740 = 50 \times n$$
$$740 \div 50 = n$$
$$14.8\% = n$$

15.
$$\frac{50.5}{100} = \frac{n}{124}$$
$$50.5 \times 124 = n \times 100$$
$$6262 = n \times 100$$
$$6262 \div 100 = n$$
$$62.62 = n$$

17.
$$\frac{120}{100} = \frac{6}{n}$$
$$120 \times n = 6 \times 100$$
$$120 \times n = 600$$
$$n = 600 \div 120$$
$$n = 5$$

19.
$$\frac{n}{18} = \frac{250}{100}$$
$$n \times 100 = 250 \times 18$$
$$n \times 100 = 4500$$
$$n = 4500 \div 100$$
$$n = 45$$

21.
$$\frac{33}{n} = \frac{220}{100}$$
$$33 \times 100 = n \times 220$$
$$3300 = n \times 220$$
$$3300 \div 220 = n$$
$$15 = n$$

Objective B Application Problems

23. Strategy To find the total amount the charity organization collected, write and solve a proportion using n to represent the total collected (base). The percent is 12% and the amount is $2940.

Solution
$$\frac{12}{100} = \frac{2940}{n}$$
$$12 \times n = 2940 \times 100$$
$$12 \times n = 294,000$$
$$n = 294,000 \div 12$$
$$n = 24,500$$

The total amount collected was $24,500.

25. Strategy To find the total land area, write and solve a proportion using n to represent the total land area. The percent is 16% and the amount is 9,400,000 square miles.

Solution
$$\frac{16}{100} = \frac{9,400,000}{n}$$
$$16 \times n = 9,400,000 \times 100$$
$$16 \times n = 9,400,000,000$$
$$n = 9,400,000,000 \div 16$$
$$n = 58,750,000$$

The world's total land area is 58,750,000 square miles.

27. Strategy To find the amount that came from the federal government, write and solve a proportion using n to represent the amount. The percent is 7% and the base is $335 billion.

Solution
$$\frac{7}{100} = \frac{n}{335}$$
$$100 \times n = 335 \times 7$$
$$100 \times n = 2345$$
$$n = 2345 \div 100$$
$$n = 23.45$$

The amount that came from the federal government was $23.45 billion.

29. Strategy To find what percent of the retail price of Corel WordPerfect the mail order price is, write and solve a proportion using n to represent the percent. The base is $249 and the amount is $169.95.

Solution
$$\frac{n}{100} = \frac{\$169.95}{\$249}$$
$$n \times \$249 = 100 \times \$169.95$$
$$n \times \$249 = \$16,995$$
$$n = \$16,995 \div 249$$
$$n \approx 68.25$$

The mail order price of Corel WordPerfect is approximately 68.3% of the retail price.

Applying the Concept

31. Strategy To find the percent of the deaths due to traffic accidents:
• Find the total number of deaths.
• To find the percent of deaths due to traffic accidents, write and solve a proportion using n to represent the percent. The base is the total deaths (156) and the amount is 73.

Solution
$$
\begin{array}{r}
19 \\
6 \\
58 \\
+\ 73 \\
\hline
156
\end{array}
$$
$$\frac{n}{100} = \frac{73}{156}$$
$$n \times 156 = 100 \times 73$$
$$n \times 156 = 7300$$
$$n = 7300 \div 156$$
$$n \approx 46.8\%$$

The percent of deaths due to traffic accidents was approximately 46.8%

33. After 15 hours, the potency is $100\% - 50\% = 50\%$.
After 30 hours, the potency is $50\% - 50\%(50\%) = 25\%$
After 45 hours, the potency is $25\% - 50\%(25\%) = 25\% - 12.5\% = 12.5\%$
After 60 hours, the potency is $12.5\% - 50\%(12.5\%) = 12.5\% - 6.25\% = 6.25\%$
The potency has decreased by $100\% - 6.25\% = 93.75\%$.

CHAPTER REVIEW

1. $0.30 \times 200 = n$
$60 = n$

2. $n \times 80 = 16$
$n = 16 \div 80$
$n = 0.2$
$n = 20\%$

3. $1\frac{3}{4} \times 100\% = 1.75 \times 100\% = 175\%$

4. $0.20 \times n = 15$
$n = 15 \div 0.20$
$n = 75$

5. $12\% = 12 \times \frac{1}{100} = \frac{12}{100} = \frac{3}{25}$

6. $0.22 \times 88 = n$
$19.36 = n$

7. $n \times 20 = 30$
$n = 30 \div 20$
$n = 1.5$
$n = 150\%$

8. $0.16\frac{2}{3} \times n = 84$

   $\frac{1}{6} \times n = 84$

   $n = 84 \div \frac{1}{6}$

   $n = 84 \times 6$

   $n = 504$

9. $42\% = 42 \times 0.01 = 0.42$

10. $0.075 \times 72 = n$

    $5.4 = n$

11. $0.66\frac{2}{3} \times n = 105$

    $\frac{2}{3} \times n = 105$

    $n = 105 \div \frac{2}{3}$

    $n = 105 \times \frac{3}{2}$

    $n = 157.5$

12. $7.6\% = 7.6 \times 0.01 = 0.076$

13. $1.25 \times 62 = n$

    $77.5 = n$

14. $16\frac{2}{3}\% = 16\frac{2}{3} \times \frac{1}{100} = \frac{50}{3} \times \frac{1}{100} = \frac{50}{300} = \frac{1}{6}$

15. $\dfrac{n}{100} = \dfrac{40}{25}$

    $n \times 25 = 40 \times 100$

    $n \times 25 = 4000$

    $n = 4000 \div 25$

    $n = 160 = 160\%$

16. $\dfrac{20}{100} = \dfrac{15}{n}$

    $20 \times n = 100 \times 15$

    $20 \times n = 1,500$

    $n = 1,500 \div 20$

    $n = 75$

17. $0.38 \times 100\% = 38\%$

18. $0.78 \times n = 8.5$

    $n = 8.5 \div 0.78$

    $n \approx 10.89 \approx 10.9$

19. $n \times 30 = 2.2$

    $n = 2.2 \div 30$

    $n = 0.073\overline{3}$

    $n \approx 7.3\%$

20. $n \times 15 = 92$

    $n = 92 \div 15$

    $n = 6.13\overline{3}$

    $n \approx 613.3\%$

21. Strategy To find the percent of the questions answered correctly:
    • Find the number of questions answered correctly by subtracting the number missed (9) from the total number of questions (60).
    • Write and solve a proportion using n to represent the percent. The base is 60 and the amount is the number of questions answered correctly.

    Solution $60 - 9 = 51$

    $\dfrac{n}{100} = \dfrac{51}{60}$

    $n \times 60 = 51 \times 100$

    $n \times 60 = 5100$

    $n = 5100 \div 60$

    $n = 85$

    The student answered 85% of the questions correctly.

22. Strategy To find how much of the budget was spent on TV advertising, write and solve the basic percent equation using n to represent the TV advertising. The percent is 7.5% and the base is $60,000.

    Solution $7.5\% \times \$60,000 = n$

    $0.075 \times 60,000 = n$

    $4500 = n$

    The company spent $4500 on TV advertising.

23. Strategy To find the percent increase in the population of Houston:
    • Find the amount of the increase by subtracting the original population (1,600,000) from the current population (1,700,000).
    • Write and solve the basic percent equation using n to represent the unknown percent. The base is 1,600,000 and the amount is the increase in population.

    Solution
    $$\begin{array}{r} 1,700,000 \\ -\ \underline{1,600,000} \\ 100,000 \end{array}$$

    $n \times 1,600,000 = 100,000$

    $n = 100,000 \div 1,600,000$

    $n = 0.0625$

    The population of Houston increased by 6.25%.

24. Strategy   To find the total cost of the video camera:
• Find the amount of sales tax by writing and solving the basic percent equation using *n* to represent the sales tax. The percent is 6.25% and the base is $980.
• Add the sales tax to the cost of the camera ($980).

Solution
$$6.25\% \times \$980 = n \qquad \$980.00$$
$$0.0625 \times 980 = n \qquad +\ \ 61.25$$
$$61.25 = n \qquad \$1041.25$$

The total cost of the video camera is $1041.25.

25. Strategy   To find the percent of women who wore sunscreen often, write and solve the basic percent equation using *n* to represent the unknown percent. The base is 350 women and the amount is 275 women.

Solution
$$n \times 350 = 275$$
$$n = 275 \div 350$$
$$n \approx 0.7857$$

Approximately 78.6% of the women wore sunscreen often.

26. Strategy   To find the world's population in 2000, write and solve the basic percent equation using *n* to represent the population in 2000. The percent is 149% and the amount is 9,100,000,000 people.

Solution
$$149\% \times n = 9,100,000,000$$
$$1.49 \times n = 9,100,000,000$$
$$n = 9,100,000,000 \div 1.49$$
$$n \approx 6,100,000,000$$

The world's approximate population in 2000 was 6,100,000,000 people.

27. Strategy   To find the cost of the computer 4 years ago, write and solve a proportion using *n* to represent the cost 4 years ago. The percent is 60% and the amount is $1800.

Solution
$$\frac{60}{100} = \frac{1800}{n}$$
$$60 \times n = 1800 \times 100$$
$$60 \times n = 180,000$$
$$n = 180,000 \div 60$$
$$n = 3000$$

The cost of the computer 4 years ago was $3000.

28. Strategy   To find the number of points scored by the team, write and solve the basic percent equation for the number of points. The percent is 30% and the amount is 33.

Solution
$$30\% \times n = 33$$
$$0.3 \times n = 33$$
$$n = 33 \div 0.3$$
$$n = 110$$

The team scored 110 points.

CHAPTER TEST

1. $97.3\% = 97.3 \times 0.01 = 0.973$

2. $83\frac{1}{3}\% = 83\frac{1}{3} \times \frac{1}{100} = \frac{250}{3} \times \frac{1}{100} = \frac{250}{300} = \frac{5}{6}$

3. $0.3 \times 100\% = 30\%$

4. $1.63 \times 100\% = 163\%$

5. $\frac{3}{2} \times 100\% = 1.5 \times 100\% = 150\%$

6. $\frac{2}{3} \times 100\% = \frac{200}{3}\% = 66\frac{2}{3}\%$

7. $77\% \times 65 = n$
$0.77 \times 65 = n$
$50.05 = n$

8. $47.2\% \times 130 = n$
$0.472 \times 130 = n$
$61.36 = n$

9. $7\%\ \text{of}\ 120 = n$ or $76\%\ \text{of}\ 13 = n$
$\ \ 0.07 \times 120 = n \qquad\qquad 0.76 \times 13 = n$
$\qquad\quad 8.4 = n \qquad\qquad\qquad 9.88 = n$
$9.88 > 8.4$, so 76% of 13 is larger.

10. $13\%\ \text{of}\ 200 = n$ or $212\%\ \text{of}\ 12 = n$
$\ \ 0.13 \times 200 = n \qquad\qquad 2.12 \times 12 = n$
$\qquad\quad 26 = n \qquad\qquad\qquad 25.44 = n$
$25.44 < 26$, so 212% of 12 is smaller.

11. Strategy   To find the amount spent for advertising, write and solve the basic percent equation using *n* to represent the amount spent for advertising. The percent is 6% and the base is $75,000.

Solution
$$6\% \times \$75,000 = n$$
$$0.06 \times 75,000 = n$$
$$4500 = n$$

The amount spent for advertising is $4500.

12. Strategy   To find how many pounds of
vegetables were not spoiled:
• Write and solve the basic percent
equation using n to represent the
number of pounds that were spoiled.
The percent is 6.4% and the base is
1250.
• Find the number of pounds that were
not spoiled by subtracting the number
of pounds of spoiled vegetables from
the total (1250 pounds).

   Solution $6.4\% \times 1250 = n$ 1250
   $0.064 \times 1250 = n$ $-\ \ \ 80$
   $80 = n$ $\overline{1170}$

   The number of pounds of vegetables
that were not spoiled was 1170.

13. $\dfrac{440}{3000} = 0.14\overline{6} \approx 14.7\%$

14. Total number of calories $= 180 + 20 = 200$.

   The % provided $= \dfrac{200}{2200} = 9.1\%$

15. Strategy   To find what percent of the permanent
employees is hired as temporary
employees, write and solve the basic
percent using n to represent the percent
of the permanent employees. The base
is 125 and the amount is 20.

   Solution $n \times 125 = 20$
   $n = 20 \div 125$
   $n = 0.16$
   $n = 16\%$

   The percent of permanent employees
hired is 16%.

16. Strategy   To find what percent of the questions
the student answered correctly:
• Find how many questions the student
answered correctly by subtracting the
number missed (7) from the total
number of questions (80).
• Write and solve the basic percent
equation using n to represent the
percent of questions answered
correctly. The base is 80 and the
amount is the number of questions
answered correctly.

   Solution $80 - 7 = 73$
   $n \times 80 = 73$
   $n = 73 \div 80$
   $n = 0.9125$
   $n \approx 91.3\%$

   The student answered approximately
91.3% of the questions correctly.

17. $15\% \times n = 12$
$0.15 \times n = 12$
$n = 12 \div 0.15$
$n = 80$

18. $150\% \times n = 42.5$
$1.5 \times n = 42.5$
$n = 42.5 \div 1.5$
$n = 28.3$

19. Strategy   To find the number of transistors
tested, write and solve the basic
percent equation using n to represent
the number of transistors tested. The
percent is 1.2% and the amount is 384.

   Solution $1.2\% \times n = 384$
   $0.012 \times n = 384$
   $n = 384 \div 0.012$
   $n = 32,000$

20. Strategy   To find what percent the increase is of
the original price:
• Find the amount of the increase by
subtracting the original value
($95,000) from the price 5 years later
($152,000).
• Write and solve the basic percent
equation using n to represent the
percent. The base is the original price
($95,000) and the amount is the
amount of the increase.

   Solution $152,000
   $-\ \ 95,000$
   $\overline{\$57,000}$

   $n \times \$95,000 = \$57,000$
   $n = 57,000 \div 95,000$
   $n = 0.60$
   $n = 60\%$

   The increase is 60% of the original
price.

21. $\dfrac{86}{100} = \dfrac{123}{n}$
$86 \times n = 123 \times 100$
$86 \times n = 12,300$
$n = 12,300 \div 86$
$n \approx 143.02$
$n \approx 143.0$

22. $\dfrac{n}{100} = \dfrac{120}{12}$
$12 \times n = 100 \times 120$
$12 \times n = 12,000$
$n = 12,000 \div 12$
$n = 1000$
$n = 1000\%$

23. Strategy To find the dollar increase in the hourly wage:
- Write and solve a proportion to find the hourly wage last year. Let n represent last year's wage. The amount is $12.88 and the percent is 112%.
- Subtract last year's wage from this year's wage ($12.88).

Solution
$$\frac{112}{100} = \frac{12.88}{n}$$
$$112 \times n = 12.88 \times 100$$
$$112 \times n = 1288$$
$$n = 1288 \div 112$$
$$n = 11.5$$
$$\begin{array}{r} \$12.88 \\ -11.50 \\ \hline \$1.38 \end{array}$$

The dollar increase is $1.38.

24. Strategy To find what percent the population now is of the population 10 years ago, write and solve a proportion using n to represent the percent. The base is 32,500 and the amount is 71,500.

Solution
$$\frac{n}{100} = \frac{71,500}{32,500}$$
$$32,500 \times n = 71,500 \times 100$$
$$32,500 \times n = 7,150,000$$
$$n = 7,150,000 \div 32,500$$
$$n = 220$$

The population now is 220% of what is was 10 years ago.

25. Strategy To find the value of the car, write and solve a proportion using n to represent the value of the car. The percent is 1.4% and the amount is $91.

Solution
$$\frac{1.4}{100} = \frac{91}{n}$$
$$1.4 \times n = 91 \times 100$$
$$1.4 \times n = 9100$$
$$n = 9100 \div 1.4$$
$$n = 6500$$

The value of the car is $6500.

CUMULATIVE REVIEW

1. $18 \div (7-4)^2 + 2 = 18 \div (3)^2 + 2$
$$= 18 \div 9 + 2$$
$$= 2 + 2 = 4$$

2.
$$\begin{array}{c|ccc}
 & 2 & 3 & 5 \\
16 = & \boxed{2 \cdot 2 \cdot 2 \cdot 2} & & \\
24 = & 2 \cdot 2 \cdot 2 & ③ & \\
30 = & 2 & 3 & ⑤ \\
\end{array}$$
$$\text{GCF} = 2 \cdot 2 \cdot 2 \cdot 2 \cdot 3 \cdot 5 = 240$$

3.
$$\begin{array}{r}
2\frac{1}{3} = 2\frac{8}{24} \\
3\frac{1}{2} = 3\frac{12}{24} \\
+ 4\frac{5}{8} = 4\frac{15}{24} \\
\hline
9\frac{35}{24} = 10\frac{11}{24}
\end{array}$$

4.
$$\begin{array}{r}
25\frac{5}{12} = 27\frac{20}{48} = 26\frac{68}{48} \\
- 14\frac{9}{16} = 14\frac{27}{48} = 14\frac{27}{48} \\
\hline
12\frac{41}{48}
\end{array}$$

5.
$$7\frac{1}{3} \times 1\frac{5}{7} = \frac{22}{3} \times \frac{12}{7}$$
$$= \frac{22 \times 12}{3 \times 7}$$
$$= \frac{2 \cdot 11 \cdot 2 \cdot 2 \cdot \overset{1}{\cancel{3}}}{\underset{1}{\cancel{3}} \cdot 7}$$
$$= \frac{88}{7} = 12\frac{4}{7}$$

6.
$$\frac{14}{27} \div 1\frac{7}{9} = \frac{14}{27} \div \frac{16}{9}$$
$$= \frac{14}{27} \times \frac{9}{16}$$
$$= \frac{14 \times 9}{27 \times 16}$$
$$= \frac{\overset{1}{\cancel{2}} \cdot 7 \cdot \overset{1}{\cancel{3}} \cdot \overset{1}{\cancel{3}}}{3 \cdot \cancel{3} \cdot \cancel{3} \cdot 2 \cdot 2 \cdot 2 \cdot 2}$$
$$= \frac{7}{24}$$

7.
$$\left(\frac{3}{4}\right)^3 \left(\frac{8}{9}\right)^2 = \left(\frac{3}{4} \cdot \frac{3}{4} \cdot \frac{3}{4}\right)\left(\frac{8}{9} \cdot \frac{8}{9}\right)$$
$$= \frac{27}{64} \cdot \frac{64}{81}$$
$$= \frac{1}{3}$$

8.
$$\left(\frac{2}{3}\right)^2 - \left(\frac{3}{8} - \frac{1}{3}\right) \div \frac{1}{2} = \frac{4}{9} - \left(\frac{9}{24} - \frac{8}{24}\right) \div \frac{1}{2}$$
$$= \frac{4}{9} - \frac{1}{24} \div \frac{1}{2}$$
$$= \frac{4}{9} - \left(\frac{1}{24} \times \frac{2}{1}\right)$$
$$= \frac{4}{9} - \frac{1}{12}$$
$$= \frac{16}{36} - \frac{3}{36} = \frac{13}{36}$$

9.
$$\underset{\substack{\text{Given place value} \\ 9 > 5}}{3.07973}$$
$$3.08$$

10.
$$\begin{array}{r} \overset{2}{\cancel{3}}.\overset{10}{\cancel{0}}\,\overset{8}{\cancel{9}}\,\overset{9}{\overset{10}{\cancel{0}}}\,\overset{12}{\cancel{2}} \\ -\ 1.9706 \\ \hline 1.\ \ 1196 \end{array}$$

11.

$$34.28125 \approx 34.2813$$

$$\begin{array}{r} 0.032.\overline{)1.097.00000} \\ \overset{\curvearrowright}{\quad}\overset{\curvearrowright}{\quad} \\ \underline{-96} \\ 137 \\ \underline{-128} \\ 90 \\ \underline{-64} \\ 260 \\ \underline{-256} \\ 40 \\ \underline{-32} \\ 80 \\ \underline{-64} \\ 160 \\ \underline{-160} \\ 0 \end{array}$$

12. $3\dfrac{5}{8} = \dfrac{29}{8}$

$$\begin{array}{r} 3.625 \\ 8\overline{)29.000} \\ \underline{-24} \\ 50 \\ \underline{-48} \\ 20 \\ \underline{-16} \\ 40 \\ \underline{-40} \\ 0 \end{array}$$

13. $1.75 = \dfrac{175}{100} = \dfrac{7}{4} = 1\dfrac{3}{4}$

14. $\dfrac{3}{8} = 0.375$

$\dfrac{3}{8} < 0.87$

15.
$$\dfrac{3}{8} = \dfrac{20}{n}$$
$$3 \times n = 8 \times 20$$
$$3 \times n = 160$$
$$n = 160 \div 3$$
$$n \approx 53.3$$

16. $\dfrac{\$76.80}{8 \text{ hours}} = \$9.60/\text{hour}$

17. $18\dfrac{1}{3}\% = 18\dfrac{1}{3} \times \dfrac{1}{100} = \dfrac{55}{3} \times \dfrac{1}{100} = \dfrac{55}{300} = \dfrac{11}{60}$

18. $\dfrac{5}{6} \times 100\% = \dfrac{500}{6}\% = 83\dfrac{1}{3}\%$

19. $16.3\% \times 120 = n$
$0.163 \times 120 = n$
$19.56 = n$

20. $n \times 18 = 24$
$n = 24 \div 18$
$n = 1.3\overline{3}$
$n = 133\dfrac{1}{3}\%$

21. $125\% \times n = 12.4$
$1.25 \times n = 12.4$
$n = 12.4 \div 1.25$
$n = 9.92$

22. $n \times 35 = 120$
$n = 120 \div 35$
$n \approx 3.4285$
$n \approx 342.9\%$

23. Strategy To find Sergio's take-home pay:
• Find the amount deducted by multiplying the income (\$740) by $\dfrac{1}{5}$.
• Subtract the amount deducted from the income.

Solution $\dfrac{1}{5} \times \$740 = \$148$

$\$740 - \$148 = \$592$
Sergio's take-home pay is \$592.

24. Strategy To find the amount of the monthly payment:
• Find the amount that will be paid by payments by subtracting the down payment (\$1000) from the price of the car (\$8353).
• Divide the total amount remaining to be paid by the number of payments (36).

Solution
$$\begin{array}{r} \$8353 \\ -1000 \\ \hline \$7353 \end{array}$$

$$\begin{array}{r} \$204.25 \\ 36\overline{)\$7353.00} \end{array}$$

Each monthly payment is \$204.25.

25. Strategy To find the number of gallons of gasoline used during the month, divide the total paid in taxes (\$79.80) by the tax paid per gallon (\$.19).

Solution $\$79.80 \div \$.19 = 420$
The number of gallons used during the month was 420.

26. Strategy To find the real estate tax on a house valued at \$150,000, write and solve a proportion using n to represent the tax.

Solution
$$\dfrac{1440}{72,000} = \dfrac{n}{150,000}$$
$$1440 \times 150,000 = 72,000 \times n$$
$$216,000,000 = 72,000 \times n$$
$$216,000,000 \div 72,000 = n$$
$$3000 = n$$
The real estate tax is \$3000.

27. Strategy   To find what percent of the purchase price the sales tax is, write and solve the basic percent equation using n to represent the percent. The base is \$490 and the amount is \$29.40.

Solution
$$n \times \$490 = \$29.40$$
$$n = 29.40 \div 490$$
$$n = 0.06$$
$$n = 6\%$$

The sales tax is 6% of the purchase price.

28. Strategy   To find what percent of the people did not favor the candidate:
- Find the number of people who did not favor the candidate by subtracting the number of people who did favor the candidate (165) from the total surveyed (300).
- Write and solve the basic percent equation using n to represent the percent of people who did not favor the candidate. The base is 300 and the amount is the number of people who did not favor the candidate.

Solution

$$\begin{array}{r} 300 \\ -\ 165 \\ \hline 135 \end{array}$$

$$n \times 300 = 135$$
$$n = 135 \div 300$$
$$n = 0.45$$
$$n = 45\%$$

The percent of the people who did not favor the candidate was 45%.

29. Strategy   To find the average hours:
- Find the number of hours in a week by multiplying the number of hours in a day (24) by the number of days in a week (7).
- Write and solve the basic percent equation using n to represent the number of hours spent watching TV. The percent is 36.5% and the base is 168.

Solution

$$\begin{array}{r} 24 \\ \times\ 7 \\ \hline 168 \end{array}$$

$$36.5\% \times 168 = n$$
$$0.365 \times 168 = n$$
$$61.3 \approx n$$

The approximate average number of hours spent watching TV in a week is 61.3 hours.

30. Strategy   To find what percent of the children tested had levels of lead that exceeded federal standards, write and solve a proportion using n to represent the percent who had levels of lead that exceeded federal standards. The base is 5500 and the amount is 990.

Solution
$$\frac{n}{100} = \frac{990}{5500}$$
$$n \times 5500 = 990 \times 100$$
$$n \times 5500 = 99,000$$
$$n = 99,000 \div 5500$$
$$n = 18$$

The percent of the children tested who had levels of lead that exceeded federal standards was 18%.

Chapter 6: Applications for Business and Consumers

PREP TEST and GO FIGURE

1. 0.75

2. 52.05

3. 504.51

4. 9750

5. $1500 \times 0.06 \times 0.5 = 90 \times 0.5 = 45$

6. 1417.24

7.
$$\begin{array}{r} 3.33 \\ 3)\overline{10.00} \\ \underline{-9} \\ 10 \\ \underline{-9} \\ 10 \\ \underline{-9} \\ 1 \end{array}$$

8.
$$\begin{array}{r} 0.605 \\ 570)\overline{345.000} \\ \underline{-3420} \\ 300 \\ \underline{-\ 0} \\ 3000 \\ \underline{-2850} \\ 150 \end{array}$$

9. $0.379 < 0.397$

Go Figure

To find the price of the earrings including sales tax, multiply 1.04 by a number (in dollars and cents) equals a whole number. To solve for the number in dollars and cents, divide both sides by 1.04. So we want to find a whole number that when divided by 1.04 results in a terminating decimal with no more than 2 decimal places.
To anticipate a solution, the factors of 104 are 2, 2, 2, and 13. Therefore, 2, 4, or 8 divided by 104 results in a terminating decimal; 13, 26, 52, or 104 divided by 104 results in a non-terminating decimal. So the whole number cannot be 2, 4 or 8 since the result would be a non-terminating decimal.
Using trial and error,
$1 \div 1.04 = 0.961538\ldots$ (non-terminating decimal)
$2 \div 1.04 = 1.923076\ldots$ (non-terminating decimal)
$3 \div 1.04 = 2.884615\ldots$ (non-terminating decimal)
$4 \div 1.04 = 3.846153\ldots$ (non-terminating decimal)
$\vdots$
$12 \div 1.04 = 11.53846\ldots$ (non-terminating decimal)
$13 \div 1.04 = 12.5$
The earrings sold for $12.50 plus 4% tax ($.50). So a customer paid $13, including 4% tax for the pair of earrings.

SECTION 6.1

Objective A Exercises

1. **Strategy** To find the unit cost, divide the total cost ($.99) by the number of units (18).

   Solution $.99 \div 18 = 0.055$
   The unit cost is $.055 per ounce.

3. **Strategy** To find the unit cost, divide the total cost ($2.99) by the number of units (8).

   Solution $2.99 \div 8 \approx 0.3737$
   The unit cost is $.374 per ounce.

5. **Strategy** To find the unit cost, divide the total cost ($3.99) by the number of units (50).

   Solution $3.99 \div 50 = 0.0798$
   The unit cost is $.0080 per tablet.

7. **Strategy** To find the unit cost, divide the total cost ($13.95) by the number of units (2).

   Solution $13.95 \div 2 = 6.975$
   The unit price is $6.975 per clamp.

9. **Strategy** To find the unit cost, divide the total cost ($2.99) by the number of units (15).

   Solution $2.99 \div 15 \approx 0.1993$
   The unit cost is $.199 per ounce.

11. **Strategy** To find the unit cost, divide the total cost ($.95) by the number of units (8).

    Solution $0.95 \div 8 \approx 0.1187$
    The unit cost is $.119 per screw.

Objective B Application Problems

13. **Strategy** To find the more economical purchase, compare the unit costs.

    Solution Sutter Home: $3.29 \div 25.5 \approx 0.1290$
    Muir Glen: $3.79 \div 26 \approx 0.1457$
    $0.1290 < 0.1457$
    The Sutter Home pasta sauce is the more economical purchase.

15. **Strategy** To find the more economical purchase, compare the unit costs.

    Solution 20 ounces: $3.29 \div 20 = 0.1645$
    12 ounces: $1.99 \div 12 \approx 0.1658$
    $0.1645 < 0.1648$
    20 ounces is the more economical purchase.

17. Strategy   To find the more economical purchase, compare the units costs.

Solution   200 tablets: $7.39 \div 200 \approx 0.0369$
400 tablets: $12.99 \div 400 \approx 0.0324$
$0.0324 < 0.0369$
400 Golden Sun vitamin E tablets is the more economical purchase.

19. Strategy   To find the more economical purchase, compare the unit costs.

Solution   Kraft: $4.37 \div 16 \approx 0.2731$
Land to Lake: $2.29 \div 9 \approx 0.2544$
$0.2544 < 0.2731$
Land to Lake cheddar cheese is the more economical purchase.

21. Strategy   To find the more economical purchase, compare the unit price.

Solution   Maxwell House: $3.99 \div 4 = 0.9975$
Sanka: $2.39 \div 2 = 1.195$
$0.9975 < 1.195$
Maxwell House coffee is the more economical purchase.

23. Strategy   To find the more economical purchase, compare the unit price.

Solution   Purina: $4.19 \div 56 \approx 0.0748$
Friskies: $3.37 \div 50.4 \approx 0.0668$
$0.0668 < 0.0748$
Friskies Chef's Blend is the more economical purchase.

Objective C Application Problems

25. Strategy   To find the total cost, multiply the unit cost ($4.59) by the number of units (3).

Solution $4.59 \times 3 = 13.77$
The total cost is $13.77.

27. Strategy   To find the total cost, multiply the unit cost ($.23) by the number of units (8).

Solution $0.23 \times 8 = 1.84$
The total cost is $1.84.

29. Strategy   To find the total cost, multiply the unit cost ($.98) by the number of units (6.5).

Solution $0.98 \times 6.5 = 6.37$
The total cost is $6.37.

31. Strategy   To find the total cost, multiply the unit cost ($1.29) by the number of units (2.1).

Solution $1.29 \times 2.1 = 2.709$
The total cost is $2.71.

33. Strategy   To find the total cost, multiply the unit cost ($7.95) by the number of units $\left(\dfrac{3}{4}\right)$.

Solution $7.95 \times \dfrac{3}{4} = 7.95 \times 0.75 \approx 5.962$
The total cost is $5.96.

Applying the Concepts

35. Students might explain that unit pricing is used in grocery stores. The unit price of a product is the total price divided by the number of units the product contains. Consumers can use this information to determine which size of a product is the more economical purchase.

SECTION 6.2

Objective A Application Problems

1. Strategy   To find the percent increase:
• Find the amount of the increase by subtracting the enrollment for 1988 (45.4) from the projected enrollment for 2008 (54.3).
• Write and solve the basic percent equation for percent. The base is 45.4 and the amount is the amount of the increase.

Solution $54.3 - 45.4 = 8.9$
Percent $\times$ base = amount
$n \times 45.4 = 8.9$
$n = 8.9 \div 45.4$
$n \approx 0.196$
The percent increase is 19.6%.

3. Strategy   To find the percent increase:
• Subtract the number of stores before the increase (420) from the number of stores after the increase (914).
• Write and solve the basic percent equation for percent. The base is 420 and the amount is the amount of the increase.

Solution $914 - 420 = 494$
Percent $\times$ base = amount
$n \times 420 = 494$
$n = 494 \div 420$
$n \approx 0.1176$
The percent increase is 117.6%.

5. Strategy   To find the percent increase:
• Find the amount of the increase by subtracting the number of events in 1924 (14) from the number of events in 2002 (78).
• Write and solve the basic percent equation for percent. The base is 14 and the amount is the amount of the increase.

Solution $78 - 14 = 64$
Percent × base = amount
$n \times 14 = 64$
$n = 64 \div 14$
$n \approx 4.571$
The percent increase is 457.1%.

7. Strategy   To find the population 8 years later:
• Write and solve the basic percent equation for the amount of increase. The percent is 24.3% and the base is 127,000.
• Add the increase to initial population (127,000).

Solution   Percent × base = amount
$24.3\% \times 127,000 = n$
$0.243 \times 127,000 = n$
$30,861 = n$
$30,861 + 127,000 = 157,861$
The population 8 years later was 157,861 people.

9. Strategy   To find the number of hour spent online in 2000:
• Write and solve the basic percent equation for the amount of increase. The percent is 72.7% and the base is 4.4 hours.
• Add the increase to number of hours spent online in 1998 (4.4 hours).

Solution   Percent × base = amount
$72.7\% \times 4.4 = n$
$0.727 \times 4.4 = n$
$3.2 \approx n$
$3.2 + 4.4 = 7.6$
There were 7.6 hours spent online in 2000.

Objective B Application Problems

11. Strategy   To find the markup, solve the basic percent equation for amount.

Solution   Percent × base = amount
$42\% \times 85 = n$
$0.42 \times 85 = n$
$35.70 = n$
The markup is $35.70.

13. Strategy   To find the markup rate, solve the basic percent equation for percent. The base is $3250 and the amount is $975.

Solution   Percent × base = amount
$n \times 3250 = 975$
$n = 975 \div 3250$
$n = 0.30 = 30\%$
The markup rate is 30%.

15a. Strategy   To find the markup, solve the basic percent equation for amount.

Solution   Percent × base = amount
$48\% \times 162 = n$
$0.48 \times 162 = n$
$77.76 = n$
The markup is $77.76.

15b. Strategy   To find the selling price, add the markup to the cost.

Solution $77.76 + 162 = 239.76$
The selling price is $239.76.

17a. Strategy   To find the markup, solve the basic percent equation for amount.

Solution   Percent × base = amount
$55\% \times 2 = n$
$0.55 \times 2 = n$
$1.1 = n$
The markup is $1.10.

17b. Strategy   To find the selling price, add the markup to the cost.

Solution $2.00 + 1.10 = 3.10$
The selling price is $3.10.

19. Strategy   To find the selling price:
• Solve the basic percent equation for amount to find the amount of the markup.
• Add the amount of the markup to the cost ($50).

Solution   Percent × base = amount
$48\% \times 50 = n$
$0.48 \times 50 = n$
$24 = n$
$50 + 24 = 74$
The selling price is $74.

Objective C   Application Problems

21. Strategy   To find the percent decrease, solve the basic percent equation for percent. The base is $800 and the amount is $320.

Solution   Percent × base = amount
$n \times 800 = 320$
$n = 320 \div 800$
$n = 0.40 = 40\%$
The amount represents a decrease of 40%.

23. Strategy   To find the decrease in the number of employees, solve the percent equation for amount. The base is 1200 and the percent is 45%.

Solution   Percent × base = amount
$45\% \times 1200 = n$
$0.45 \times 1200 = n$
$540 = n$
The decrease in the number of employees is 540.

25a. Strategy   To find the amount of decrease, subtract the new value (39) from the original value (52).

Solution $52 - 39 = 13$
The amount of decrease is 13 minutes.

25b. Strategy   To find the percent decrease, solve the basic percent equation for percent. The base is 52 and the amount is 13.

Solution   Percent × base = amount
$n \times 52 = 13$
$n = 13 \div 52$
$n = 0.25 = 25\%$
The amount represents a decrease of 25%.

27a. Strategy   To find the amount of decrease, solve the basic percent equation for amount. The base is $1.60 and the percent is 37.5%.

Solution   Percent × base = amount
$37.5\% \times 1.60 = n$
$0.375 \times 1.60 = n$
$0.60 = n$
The amount of decrease is $.60.

27b. Strategy   To find the dividend this year, subtract the amount of the decrease ($0.60) from the dividend last year ($1.60).

Solution $1.60 - 0.60 = 1.00$
The dividend this year is $1.00.

29. Strategy   To find the percent decrease:
• Solve the basic percent equation for percent. The amount is the amount of the decrease (26) and the base is 394.

Solution   Percent × base = amount
$n \times 394 = 26$
$n = 26 \div 394$
$n \approx 0.066 = 6.6\%$
The amount represents a decrease of 6.6%.

Objective D   Application Problems

31. Strategy   To find the discount rate, solve the basic percent equation for percent. The base is $24 and the amount is $8.

Solution   Percent × base = amount
$n \times 24 = 8$
$n = 8 \div 24$
$n = 0.3\overline{3} = 33\frac{1}{3}\%$
The discount rate is $33\frac{1}{3}\%$.

33. Strategy   To find the discount, solve the basic percent equation for amount. The percent is 20% and the base is $340.

Solution   Percent × base = amount
$20\% \times 340 = n$
$0.20 \times 340 = n$
$68 = n$
The discount is $68.

35. Strategy   To find the discount rate, solve the basic percent equation for percent. The base is $140 and the amount is $42.

Solution   Percent × base = amount
$n \times 140 = 42$
$n = 42 \div 140$
$n = 0.30 = 30\%$
The discount rate is 30%.

37a. Strategy   To find the discount, solve the basic percent equation for amount. The percent is 20% and the base is $0.85.

Solution   Percent × base = amount
$20\% \times 0.85 = n$
$0.20 \times .85 = n$
$0.17 = n$
The discount is $.17 per pound.

37b. Strategy   To find the sale price, subtract the discount ($0.17) from the original price ($0.85).

Solution $0.85 - 0.17 = 0.68$
The sale price is $.68 per pound.

39a. Strategy   To find the discount, subtract the sale price ($16) from the original cost ($20).

Solution $20 - 16 = 4$
The amount of the discount is $4.

39b. Strategy   To find the discount rate, solve the basic percent equation for percent. The base is $20 and the amount is $4.

Solution   Percent $\times$ base = amount
$$n \times 20 = 4$$
$$n = 4 \div 20$$
$$n = 0.20 = 20\%$$
The discount rate is 20%.

Applying the Concepts

41. Yes.

a. $\$12 \times 0.10 + \$12 = \$1.20 + \$12 = \$13.20$

b. $\$12(1.10) = \$13.20$

43. No. Suppose the regular price is $100. Then the sale price is $75 because $100 - 0.25(100) = 100 - 25 = 75$. The promotional sale offers 25% off the sale price of $75, so the price is lowered to $56.25 because $75 - 0.25(75) = 75 - 18.75 = 56.25$. A sale that offers 50% off the regular price of $100 offers the product for $50. Because $50 < $56.25, the better price is the one that is 50% off the regular price.

SECTION 6.3

Objective A Application Problems

1a. $10,000

1b. $2050

1c. 10.25%

1d. 2 years

3. Strategy   To find the simple interest, multiply the principal by the annual interest rate by the time (in years).

Solution $8000 \times 0.09 \times 2 = 1440$
The simple interest owed is $1440.

5. Strategy   To find the simple interest, multiply the principal by the annual interest rate by the time (in years).

Solution $100,000 \times 0.09 \times \dfrac{9}{12} = 6750$
The simple interest due is $6750.

7. Strategy   To find the simple interest, multiply the principal by the annual interest rate by the time (in years).

Solution $20,000 \times 0.088 \times \dfrac{9}{12} = 1320$
The simple interest due is $1320.

9. Strategy   To find the simple interest, multiply the principal by the annual interest rate by the time (in years).

Solution $5000 \times 0.075 \times \dfrac{90}{365} \approx 92.47$
The simple interest due is $92.47.

11. Strategy   To find the simple interest, multiply the principal by the annual interest rate by the time (in years).

Solution $7500 \times 0.095 \times \dfrac{75}{365} \approx 146.40$
The simple interest due is $146.40.

13. Strategy   To find the maturity value of the loan, add the principal and the simple interest.

Solution $4800 + 320 = 5120$
The maturity value of the loan is $5120.

15. Strategy   To find the maturity value:
• Find the simple interest due by multiplying the principal by the annual interest rate by the time (in years).
• Find the maturity value by adding the principal and the simple interest.

Solution $150,000 \times 0.095 \times 1 = 14,250$
$150,000 + 14,250 = 164,250$
The maturity value is $164,250.

17. Strategy   To find the total amount due:
• Find the simple interest due by multiplying the principal by the annual interest rate by the time (in years).
• Find the total amount due by adding the principal and the simple interest.

Solution $12,500 \times 0.095 \times \dfrac{8}{12} \approx 791.67$
$12,500 + 791.67 = 13,291.67$
The total amount due on the loan is $13,291.67.

19. Strategy To find the maturity value:
• Find the simple interest due by multiplying the principal by the annual interest rate by the time (in years).
• Find the maturity value by adding the principal and the simple interest.

Solution $14,000 \times 0.1025 \times \dfrac{270}{365} \approx 1061.51$

$14,000 + 1061.51 = 15,061.51$
The maturity value is $15,061.51.

21. Strategy To find the monthly payment, divide the sum of the loan amount ($225,000) and the interest ($72,000) by the number of payments (48).

Solution $\dfrac{225,000 + 72,000}{48} = 6187.50$

The monthly payment is $6187.50.

23a. Strategy To find the simple interest charged, multiply the principal ($12,000) by the annual interest rate by the time (in years).

Solution $12,000 \times 0.045 \times 2 = 1080$
The interest charged is $1080.

23b. Strategy To find the monthly payment, divide the sum of the loan amount ($12,000) and the interest ($1080) by the number of payments (24).

Solution $\dfrac{12,000 + 1080}{24} = 545$

The monthly payment is $545.

25. Strategy To find the monthly payment:
• Find the simple interest due by multiplying the principal by the annual interest rate by the time (in years).
• Find the monthly payment by adding the interest due to the loan amount ($42,000) and dividing that sum by the number of payments (42).

Solution $42,000 \times 0.095 \times 3.5 = 13,965$
The monthly payment is
$\dfrac{42,000 + 13,965}{42} = \$1332.50$.

Objective B Application Problems

27. Strategy To find the finance charge, multiply the unpaid balance by the monthly interest rate by the number of months.

Solution $118.72 \times 0.0125 \times 1 = 1.48$
The finance charge is $1.48.

29. Strategy To find the finance charge, multiply the unpaid balance by the monthly interest rate by the number of months.

Solution $12,368.92 \times 0.015 \times 1 = 185.53$
The finance charge is $185.53.

31. Strategy To find the difference in finance charges:
• Find the difference in monthly interest rates by subtracting the smaller rate (1.15%) from the larger rate (1.85%).
• To find the difference in finance charges, multiply the unpaid balance by the difference in monthly interest rates by the number of months.

Solution $0.0185 - 0.0115 = 0.007 = 0.7\%$
$1438.20 \times 0.007 \times 1 = 10.07$
The difference in finance charges is $10.07.

Objective C Application Problems

33. Strategy To find the value of the investment in 1 year, multiply the original investment by the compound interest factor.

Solution $750 \times 1.04080 = 780.60$
The value of the investment after 1 year is $780.60.

35. Strategy To find the value of the investment after 15 years, multiply the original investment by the compound interest factor.

Solution $3000 \times 2.42726 = 7281.78$
The value of the investment after 15 years is $7281.78.

37a. Strategy To find the value of the investment in 5 years, multiply the original investment by the compound interest factor.

Solution $75,000 \times 1.48595 = 111,446.25$
The value of the investment in 5 years will be $111,446.25.

37b. Strategy To find the amount of interest that will be earned, subtract the original investment from the new value of the investment.

Solution $111,446.25 - 75,000 = 36,446.25$
The amount of interest earned will be

39. Strategy To find the amount of interest earned over a 20-year period:
• Find the value of the investment after 20 years by multiplying the original investment by the compound interest factor.
• Subtract the original value of the investment ($2500) from the new value of the investment.

Solution $2500 \times 3.31979 = 8299.48$
$8299.48 - 2500 = 5799.48$
The amount of interest earned is $5799.48.

Applying the Concepts

41. Strategy To find the simple interest owed to the credit union, multiply the principal by the monthly interest rate by the time (in months).

Solution $800 \times 0.02 \times 1 = 16$
The interest owed is $16.

b. $\$200.50\left(1+\dfrac{0.06}{12}\right)+\$100 = \$301.50$

43. You received less interest during the second month because there are fewer days in the month of September (30 days) than in the month of August (31 days). Using the simple interest formula:

$500 \times 0.05 \times \dfrac{31}{365} \approx 2.12$
$502.12 \times 0.05 \times \dfrac{30}{365} \approx 2.06$

Even though the principal is greater during the second month, the interest earned is less because there are fewer days in the month.

SECTION 6.4

Objective A Application Problems

1. Strategy To find the mortgage, subtract the down payment from the purchase price.

Solution $97,000 - 14,550 = 82,450$
The mortgage is $82,450.

3. Strategy To find the down payment, solve the basic percent equation for amount. The base is $25,000 and the percent is 30%.

Solution Percent $\times$ base = amount
$0.30 \times 25,000 = 7500$
The down payment is $7500.

5. Strategy To find the down payment, solve the basic percent equation for amount. The base is $850,000 and the percent is 25%.

Solution Percent $\times$ base = amount
$0.25 \times 850,000 = 212,500$
The down payment is $212,500.

7. Strategy To find the loan origination fee, solve the basic percent equation for amount. The base is $150,000 and the percent is $2\dfrac{1}{2}\%$.

Solution Percent $\times$ base = amount
$0.025 \times 150,000 = 3750$
The loan origination fee is $3750.

9a. Strategy To find the down payment, solve the basic percent equation for amount. The base is $150,000 and the percent is 5%.

Solution Percent $\times$ base = amount
$0.05 \times 150,000 = 7,500$
The down payment is $7,500.

9b. Strategy To find the mortgage, subtract the down payment from the purchase price.

Solution Percent $\times$ base = amount
$150,000 - 7,500 = 142,500$
The mortgage is $142,500.

11. Strategy To find the mortgage:
• Find the down payment by solving the basic percent equation for amount. The percent is 10% and the base is $210,000.
• Subtract the down payment from the purchase price.

Solution Percent $\times$ base = amount
$0.10 \times 210,000 = 21,000$
$210,000 - 21,000 = 189,000$
The mortgage is $189,000.

Objective B Application Problems

13. Strategy To find the monthly mortgage payment, multiply the mortgage by the monthly mortgage factor.

Solution $150,000 \times 0.0077182 = 1157.73$
The monthly mortgage payment is $1157.73.

15. Strategy   To determine if the couple can afford
to buy the house:
• Find the monthly mortgage payment
by multiplying the mortgage amount
by the monthly mortgage factor.
• Compare the monthly mortgage
payment with $800.

      Solution $110,000 \times 0.0073376 = 807.14$
$$\$800 < \$807.14$$
No, the couple cannot afford to buy the
house.

17. Strategy   To find the monthly property tax
payment, divide the annual property
tax by 12.

      Solution $1348.20 \div 12 = 112.35$
The monthly property tax is $112.35.

19a. Strategy   To find the monthly mortgage
payment, multiply the mortgage by the
monthly mortgage factor.

      Solution $200,000 \times 0.0083920 = 1678.40$
The monthly mortgage payment is
$1678.40.

19b. Strategy   To find how much of the monthly
mortgage payment is interest, subtract
the amount that is principal ($941.72)
from the monthly mortgage payment.

      Solution $1678.40 - 941.72 = 736.68$
The interest payment is $736.68.

21. Strategy   To find the monthly payment for
property tax and home mortgage:
• Find the monthly tax payment by
dividing the annual property tax by 12.
• Find the monthly mortgage payment
by dividing the annual mortgage
payment by 12.
• Add the monthly tax payment to the
monthly mortgage payment.

      Solution $948 \div 12 = 79$ monthly tax payment
$10,844.40 \div 12 = 903.70$
$903.70 + 79 = 982.70$
The total monthly payment for
property tax and home mortgage is
$982.70.

23. Strategy   To find the monthly mortgage
payment:
• Find the mortgage amount by
subtracting the down payment from the
purchase price.
• Multiply the mortgage amount by
the monthly mortgage factor.

      Solution $210,000 - 15,000 = 195,000$
$195,000 \times 0.0101427 = 1977.83$
The monthly mortgage payment is
$1977.83.

Applying the Concepts

25. Choice 1: 8% for 20 years
$100,000 \times 0.0083644 = 836.44$/month
Payback $= 836.44 \times 240$ months $= \$200,745.60$,
or 100,745.60 in interest
Choice 2: 8% for 30 years
$100,000 \times 0.0073376 = 733.76$
Payback $= 733.76 \times 360$ months $= \$264,153.60$
or 164,153.60 in interest
$\$164,153.60 - 100,745.60 = \$63,408$
By using the 20-year loan, the couple will save
$63,408.

SECTION 6.5

Objective A Application Problems

1. Strategy   To determine whether Amanda has
enough money for the down payment:
• Find the down payment by solving
the basic percent equation for amount.
The base is $7100 and the percent is
12%.
• Compare the required down payment
with $780.

      Solution   Percent $\times$ base $=$ amount
$0.12 \times 7100 = 852$ down payment
$$\$852 > \$780$$
No, Amanda does not have enough for
the down payment.

3. Strategy   To find how much sales tax is paid,
solve the basic percent equation for
amount. The base is $26,500 and the
percent is 4.5%.

      Solution $0.045 \times 26,500 = 1192.5$
The sales tax is $1192.50.

5. Strategy   To find the state license fee, solve the
basic percent equation for amount. The
base is $22,500 and the percent is 2%.

      Solution   Percent $\times$ base $=$ amount
$0.02 \times 22,500 = 450$
The license fee is $450.

7a. Strategy   To find the sales tax, solve the basic
percent equation for amount. The base
is $32,000 and the percent is 3.5%.

      Solution   Percent $\times$ base $=$ amount
$0.035 \times 32,000 = 1120$
The sales tax is $1120.

7b. Strategy   To find the total cost of the sales tax
and license fee, add the sales tax
($1120) and the license fee ($275).

      Solution $1120 + 275 = 1395$
The total cost of the sales tax and
license fee is $1395.

9a. Strategy   To find the down payment, solve the basic percent equation for amount. The base is $16,200 and the percent is 25%.

Solution   Percent × base = amount
$0.25 \times 16,200 = 4050$
The down payment is $4050.

9b. Strategy   To find the amount financed, subtract the down payment from the purchase price.

Solution $16,200 - 4050 = 12,150$
The amount financed is $12,150.

11. Strategy   To find the amount financed:
• Find the amount of the down payment by solving the basic percent equation for amount. The base is $35,000 and the percent is 20%.
• Subtract the down payment from the purchase price ($35,000).

Solution   Percent × base = amount
$0.20 \times 35,000 = 7000$
$35,000 - 7000 = 28,000$
The amount financed is $28,000.

Objective B Application Problems

13. Strategy   To find the monthly truck payment, multiply the amount financed by the monthly payment factor.

Solution $14,000 \times 0.0248850 = 348.39$
The monthly truck payment is $348.39.

15. Strategy   To find how much it costs to operate a car, multiply the number of miles (16,000) by the cost per mile ($0.32).

Solution $0.32 \times 16,000 = 5120$
The cost is $5120.

17. Strategy   To find the cost per mile, divide the total cost by the number of miles.

Solution $1600 \div 14,000 \approx 0.11$
The cost is about $.11 per mile.

19. Strategy   To find the amount of interest, subtract the amount of principal from the monthly car payment.

Solution $143.50 - 68.75 = 74.75$
The amount of interest is $74.75.

21a. Strategy   To find the amount financed, subtract the down payment ($5400) from the purchase price ($82,000).

Solution $82,000 - 5400 = 76,600$
The amount financed is $76,600.

21b. Strategy   To find the monthly truck payment, multiply the amount financed by the monthly payment factor.

Solution $76,600 \times 0.0207584 = 1590.09$
The monthly payment is $1590.09.

23. Strategy   To find the monthly car payment:
• Find the amount financed by subtracting the down payment from the purchase price.
• Multiply the amount financed by the monthly payment factor.

Solution $27,500 - 5500 = 22,000$
$22,000 \times 0.0322672 = 709.88$
The monthly payment is $709.88.

Applying the Concepts

25. The total loan cost for 10% loan plus application fee = $5800 × 0.0253626 × 48 + $45 − $5800 = $1305.95
The total loan cost for 11% loan with no fee = $5800 × 0.0258455 × 48 − $5800 = $1395.39
The 10% loan has the lesser loan cost.

SECTION 6.6

Objective A Application Problems

1. Strategy   To find the earnings, multiply the hourly wage by the number of hours.

Solution $9.50 \times 40 = 380$
The sales clerk earns $380.

3. Strategy   To find the commission, solve the basic percent equation for amount. The base is $131,000 and the percent is 3%.

Solution   Percent × base = amount
$0.03 \times 131,000 = 3930$
The real estate agent's commission is $3930.

5. Strategy   To find the commission, solve the basic percent equation for amount. The base is $5600 and the percent is 1.5%.

Solution   Percent × base = amount
$0.015 \times 5600 = 84$
The stockbroker's commission is $84.

7. Strategy   To find the monthly salary, divide the annual salary by 12.

Solution $38,928 \div 12 = 3244$
The teacher's monthly salary is $3244.

9. Strategy   To find the electrician's hourly wage for overtime, multiply the regular hourly wage by 2 (double time).

Solution $25.80 \times 2 = 51.60$
The overtime wage is $51.60/hour.

11. Strategy To find the commission, solve the basic percent equation for amount. The base is $450 and the percent is 25%.

Solution Percent × base = amount
$0.25 \times 450 = 112.5$
The golf pro's commission was $112.50.

13. Strategy To find the earnings, multiply the earnings per page by the number of pages.

Solution $2.75 \times 225 = 618.75$
The typist's earnings are $618.75.

15. Strategy To find the hourly wage, divide the total wage by the number of hours.

Solution $3400 \div 40 = 85$
Maxine's hourly wage is $85.

17a. Strategy To find the hourly wage on Saturday, multiply the regular hourly wage by 1.5 (time and a half).

Solution $15.90 \times 1.5 = 23.85$
Mark's hourly wage on Saturday is $23.85.

17b. Strategy To find the earnings for Saturday, multiply the hourly wage by the number of hours.

Solution $23.85 \times 8 = 190.8$
Mark earns $190.80.

19a. Strategy To find the increase in pay, solve the basic percent equation for amount. The base is $16.50 and the percent is 10%.

Solution Percent × base = amount
$0.10 \times 16.50 = 1.65$
The nurse's increase in pay is $1.65.

19b. Strategy To find the hourly wage for the night shift, add the increase in pay to the regular hourly wage.

Solution $16.50 + 1.65 = 18.15$
The nurse's hourly wage is $18.15.

21. Strategy To find the earnings for the week:
• Find the amount of sales over $1500 by subtracting $1500 from the total sales ($3000)
• Find the commission by solving the basic percent equation for amount. The base is 1500 and the percent is 15%.
• Add the commission to the weekly salary ($250).

Solution $3000 - 1500 = 1500$ sales over 1500
Percent × base = amount
$0.15 \times 1500 = 225$ commission
$250 + 225 = 475$
Nicole's earnings were $475.

Applying the Concepts

23. Let n represent the previous year's salary.
$n + 0.04n = 37,732$
$1.04n = 37,732$
$n \approx 36,281$
The industrial engineer's salary was $36,281 the previous year.

25. Liberal Arts
Last year's salary:
$n + 0.035n = 24,102$
$1.035n = 24,102$
$n \approx 23,287$
Increase in salary: $24,102 - 23,287 = 815$

Telecommunications
Last Year's salary:
$n + 0.04n = 22,447$
$1.04n = 22,447$
$n \approx 21,584$
Increase in salary: $22,447 - 21,584 = 863$

Since $815 < $863, the liberal arts major has the smaller increase in salary.

SECTION 6.7

Objective A   Application Problems

1. Strategy To find your current checking balance, add the deposit to the old balance.

Solution $342.51 + 143.81 = 486.32$
Your current checking account balance is $486.32.

3. Strategy To find the current checking account balance, subtract the amount of the check from the old balance.

Solution $2431.76 - 1209.29 = 1222.47$
The real estate firm's current checking account balance is $1222.47.

5. Strategy To find the current checking account balance, subtract the amount of each check from the original balance.

Solution
$$\begin{array}{r} 1204.63 \\ -\ 119.27 \\ \hline 1085.36 \\ -\ 260.09 \\ \hline 825.27 \end{array}$$
The nutritionist's current balance is $825.27.

7. Strategy To find the current checking account balance, add the amount of the deposit to the old balance. Then subtract the amount of each check.

Solution
3476.85
+ 1048.53
4525.38
− 848.37
3677.01
− 676.19
3000.82

The current checking account balance is $3000.82.

9. Strategy To determine if there is enough money in the account, compare $675 with the current balance after finding the current checking account balance.

Solution
404.96
+ 350.00
754.96
− 71.29
683.67

$683.67 > $675

Yes, there is enough money in the carpenter's account to purchase the refrigerator.

11. Strategy To determine if there is enough money in the account to make the two purchases, add the amounts of the two purchases and compare the total with the current checking account balance.

Solution 3500 + 2050 = 5550 total of purchases
$5550 < $5625.42
Yes, there is enough money in the account to make the two purchases.

13. Solution

Current checkbook balance:	989.86
Checks: 228	419.32
233	166.40
235	+ 288.39
	1863.97
Interest:	+ 13.22
	1877.19
Service charge:	− 0.00
	1877.19
Deposits:	− 0.00
Checkbook balance:	1877.19

Current bank balance from bank statement: $1877.19.
Checkbook balance: $1877.19.
The bank statement and checkbook balance.

15. Solution

Current checkbook balance:	1051.92
Checks: 223	414.83
224	113.37
Interest:	+ 5.15
	1585.27
Service charge:	− 0.00
	1585.27
Deposits:	− 0.00
Checkbook balance:	1585.27

Current bank balance from bank statement: $1585.27.
Checkbook balance: $1585.27.
The bank statement and checkbook balance.

Applying the Concepts

17. added

19. subtract

CHAPTER REVIEW

1. Strategy To find the unit cost, divide the total cost ($3.90) by the number of units (20).

Solution $3.90 \div 20 = 0.195$
The unit cost is $.195 per ounce or 19.5¢ per ounce.

2. Strategy To find the cost per mile:
• Find the total cost by adding the amounts spent ($1025.58, $605.82, $37.92, and $188.27).
• Divide the total cost by the number of miles (11,320).

Solution $1025.58 + 605.82 + 37.92 + 188.27 =$
1857.59
$1857.59 \div 11,320 \approx 0.164$
The cost per mile is $.164 or 16.4¢.

3. Strategy To find the percent increase:
• Find the amount of the increase by subtracting the original price ($42.375) from the increased price ($55.25).
• Solve the basic percent equation for percent. The base is $42.375 and the amount is the amount of the increase.

Solution $\$55.25 - 42.375 = 12.875$
$n \times 42.375 = 12.875$
$n = 12.875 \div 42.375 \approx 0.304 = 30.4\%$
The percent increase is 30.4%.

4. Strategy To find the markup, solve the basic percent equation for amount. The base is $180 and the percent is 40%.

Solution $0.40 \times 180 = n$
$72 = n$
The markup is $72.

5. Strategy   To find the simple interest, multiply the principal by the annual interest rate by the time (in years).

   Solution $100,000 \times 0.09 \times \dfrac{9}{12} = 6750$
   The simple interest due is $6750.

6. Strategy   To find the value of the investment in 10 years, multiply the original investment by the compound interest factor.

   Solution $25,000 \times 1.82203 = 45,550.75$
   The value of the investment after 10 years is $45,550.75.

7. Strategy   To find the percent increase:
   • Find the amount of the increase by subtracting the original amount ($4.12) from the increased amount ($4.73).
   • Solve the basic percent equation for percent. The base is $4.12 and the amount is the increased amount.

   Solution $4.73 - 4.12 = 0.61$
   $n \times 4.12 = 0.61$
   $n = 0.61 \div 4.12 \approx 0.15 = 15\%$
   The percent increase is 15%.

8. Strategy   To find the total monthly payment:
   • Find the monthly property tax by dividing the annual tax ($658.32) by 12.
   • Add the monthly property tax payment to the monthly mortgage payment ($523.67).

   Solution $658.32 \div 12 = 54.86$
   $54.86 + 523.67 = 578.53$
   The total monthly payment for the mortgage and property tax is $578.53.

9. Strategy   To find the monthly payment:
   • Find the down payment by solving the basic percent equation for amount. The percent is 8% and the base is $14,450.
   • Find the amount financed by subtracting the down payment from the purchase price ($14,450).
   • Multiply the amount financed by the monthly payment factor.

   Solution $0.08 \times 14,450 = 1156$
   $14,450 - 1156 = 13,294$
   $13,294 \times 0.0248850 = 330.82$
   The monthly payment is $330.82.

10. Strategy   To find the value of the investment in 1 year, multiply the original investment by the compound interest factor.

    Solution $50,000 \times 1.07186 = 53,593$
    The value of the investment will be $53,593.

11. Strategy   To find the down payment, solve the basic percent equation for amount. The base is $125,000 and the percent is 15%.

    Solution $0.15 \times 125,000 = 18,750$
    The down payment was $18,750.

12. Strategy   To find the total cost of the sales tax and license fee:
    • Find the sales tax by solving the basic percent equation for amount. The base is $18,500 and the percent is 6.25%.
    • Add the sales tax and the license fee ($315).

    Solution $0.0625 \times 18,500 = 1156.25$
    $1156.25 + 315 = 1471.25$
    The total cost of the sales tax and license fee is $1471.25.

13. Strategy   To find the selling price:
    • Find the markup by solving the basic percent equation for amount. The percent is 35% and the base is $1540.
    • Find the selling price by adding the markup to the cost.

    Solution $0.35 \times 1540 = 539$
    $539 + 1540 = 2079$
    The selling price is $2079.

14. Strategy   To find how much of the payment is interest, subtract the principal ($25.45) from the total payment ($122.78).

    Solution $122.78 - 25.45 = 97.33$
    The interest paid is $97.33.

15. Strategy   To find the commission, solve the basic percent equation for amount. The base is $108,000 and the percent is 3%.

    Solution $0.03 \times 108,000 = n$
    $3240 = n$
    The commission was $3240.

16. Strategy To find the sale price:
• Find the amount of the discount by solving the basic percent equation for amount. The base is $235 and the percent is 40%.
• Subtract the discount from the original price.

Solution $0.40 \times 235 = n$
$94 = n$
$235 - 94 = 141$
The discount price is $141.

17. Strategy To find the student's current checkbook balance, subtract the amount of each check and add the amount of the deposit.

Solution
```
  1568.45
- 123.76
  1444.69
- 756.45
   688.24
-  88.77
   599.47
+ 344.21
   943.68
```

The student's current checkbook balance is $943.68.

18. Strategy To find the maturity value:
• Find the simple interest due by multiplying the principal by the annual interest rate by the time (in years).
• Find the maturity value by adding the principal and the simple interest.

Solution $30,000 \times 0.08 \times \dfrac{6}{12} = 1200$
$30,000 + 1200 = 31,200$
The maturity value is $31,200.

19. Strategy To find the origination fee, solve the basic percent equation for amount. The base is $75,000 and the percent is $2\frac{1}{2}\%$.

Solution $0.025 \times 75,000 = 1875$
The origination fee is $1875.

20. Strategy To find the more economical purchase, compare the unit costs.

Solution $3.49 \div 16 \approx 0.218$
$6.99 \div 33 \approx 0.212$
The more economical purchase is 33 ounces for $6.99.

21. Strategy To find the monthly mortgage payment:
• Find the down payment by solving the basic percent equation for amount. The base is $156,000 and the percent is 10%.
• Find the amount financed by subtracting the down payment from the purchase price.
• Find the monthly mortgage payment by multiplying the amount financed by the monthly mortgage factor.

Solution $0.10 \times 156,000 = 15,600$
$156,000 - 15,600 = 140,400$
$140,400 \times 0.006653 = 934.08$
The monthly mortgage payment is $934.08.

22. Strategy To find the total income:
• Find the overtime wage by multiplying the regular wage by 1.5 (time and half).
• Find the number of overtime hours worked by subtracting the regular weekly schedule (40) from the total hours worked (48).
• Find the wages earned for overtime by multiplying the overtime wage by the number of overtime hours worked.
• Find the wages for the 40-hour week by multiplying the hourly rate ($12.60) by 40.
• Add the pay from the overtime hours to the pay from the regular week.

Solution $1.5 \times 12.60 = 18.90$
$48 - 40 = 8; 8 \times 18.90 = 151.20$
$40 \times 12.60 = 504$
$504 + 151.20 = 655.20$
The total income was $655.20.

23. Strategy To find the donut shop's current checkbook balance, subtract the amount of each check and add the amount of each deposit.

Solution
```
  9567.44
- 1023.55
  8543.89
-  345.44
  8198.45
-   23.67
  8174.78
+  555.89
  8730.67
+  135.91
  8866.58
```

The donut shop's checkbook balance is $8866.58.

24. Strategy To find the monthly payment, divide the sum of the loan amount ($55,000) and the interest ($1375) by the number of payments (4).

Solution $\dfrac{55,000+1375}{4}=14,093.75$
The monthly payment is $14,093.75.

25. Strategy To find the finance charge, multiply the unpaid balance by the monthly interest rate by the number of months.

Solution $576\times0.0125\times1=7.2$
The finance charge is $7.20.

CHAPTER TEST

1. Strategy To find the cost per foot, divide the total cost ($138.40) by the number of feet (20).

Solution $138.40\div20=6.92$
The cost per foot is $6.92.

2. Strategy To find the more economical purchase, compare the unit price of each item.

Solution $7.49\div3$ or $12.59\div5$
$7.49\div3=2.50$; $12.59\div5=2.52$
The more economical purchase is 3 pounds for $7.49.

3. Strategy To find the total cost, multiply the cost per pound ($4.15) by the number of pounds (3.5).

Solution $4.15\times3.5\approx14.53$
The total cost is $14.53.

4. Strategy To find the percent increase:
• Find the amount of the increase by subtracting the original price ($415) from the increased price ($498).
• Solve the basic percent equation for percent. The base is $415 and the amount is the amount of the increase.

Solution $498-415=83$
$n\times415=83$
$n=83\div415$
$n=0.20=20\%$
The percent increase of the cost of the exercise bicycle is 20%.

5. Strategy To find the selling price:
• Find the amount of the markup by solving the basic percent equation for amount. The percent is 40% and the base is $215.
• Add the markup to the cost ($215).

Solution $0.40\times215=86$
$215+86=301$
The selling price of a compact disc player is $301.

6. Strategy To find the percent decrease:
• Find the amount of the decrease by subtracting the decreased value ($360) from the original value ($390).
• Solve the basic percent equation for percent. The base is ($390) and the amount is the amount of the decrease.

Solution $390-360=30$
$n\times3250=975$
$n=975\div3250$
$n=0.30=30\%$
The decrease in percent is 7.7%.

7. Strategy To find the percent decrease:
• Find the amount of the decrease by subtracting the decreased value ($896) from the original value ($1120)
• Solve the basic percent equation for percent. The base is $1120 and the amount is the amount of the decrease.

Solution $1120-896=224$
$n\times1120=224$
$n=224\div1120=0.20=20\%$
The percent decrease is 20%.

8. Strategy To find the sale price:
• Find the amount of the discount by solving the basic percent equation for amount. The base is $299 and the percent is 30%.
• Subtract the amount of the discount from the regular price ($299).

Solution $0.30\times299=n$
$89.7=n$
$299-89.70=209.30$
The sale price of the corner hutch is $209.30.

9. Strategy To find the discount rate:
• Find the amount of the discount by subtracting the sale price ($2.70) from the regular price ($4.50).
• Solve the basic percent equation for percent. The base is $4.50 and the amount is the amount of the discount.

Solution $4.50-2.70=1.80$
$n\times4.50=1.80$
$n=1.80\div4.50=0.40=40\%$
The discount rate is 40%.

10. Strategy To find the simple interest due, multiply the principal by the annual interest rate by the time in years.

Solution $75,000\times0.08\times\dfrac{4}{12}=2000$

The simple interest due is $2000.

11. Strategy   To find the maturity value:
• Find the simple interest due by multiplying the principal by the annual interest rate by the time (in years).
• Find the maturity value by adding the principal and the simple interest.

Solution $25,000 \times 0.092 \times \dfrac{9}{12} = 1725$
$25,000 + 1725 = 26,725$
The maturity value is $26,725.

12. Strategy   To find the finance charge, multiply the unpaid balance by the monthly interest rate by the number of months.

Solution $374.95 \times 0.012 \times 1 = 4.50$
The finance charge is $4.50.

13. Strategy   To find the interest earned:
• Find value of the investment in 10 years, multiply the original investment by the compound interest factor.
• Find the interest earned, by subtracting the original investment from the new value of the investment.

Solution $30,000 \times 1.81402 = 54,420.6$
$54,420.60 - 30,000 = 24,420.60$
The amount of interest earned in 10 years will be $24,420.60.

14. Strategy   To find the loan origination fee, solve the basic percent equation for amount. The base is $134,000 and the percent is $2\dfrac{1}{2}\%$.

Solution $0.025 \times 134,000 = 3350$
The origination fee is $3350.

15. Strategy   To find the monthly mortgage payment, multiply the mortgage amount by the monthly mortgage factor.

Solution $222,000 \times 0.0077182 = 1713.44$
The monthly mortgage payment is $1713.44.

16. Strategy   To find the amount financed:
• Find the amount of the down payment by solving the basic percent equation for amount. The base is $23,750 and the percent is 20%.
• Subtract the down payment from the purchase price.

Solution $0.20 \times 23,750 = 4,750$
$23,750 - 4,750 = 19,000$
The amount financed is $19,000.

17. Strategy   To find the monthly car payment:
• Find the amount of the down payment by solving the basic percent equation for amount. The base is $23,714 and the percent is 15%.
• Find the amount financed by subtracting the down payment from the purchase price ($23,714).
• Multiply the amount financed by the monthly mortgage factor.

Solution $0.15 \times 23,714 = 3,557.10$
$23,714 - 3,557.10 = 20,156.90$
$20,156.90 \times 0.0239462 = 482.681$
The monthly car payment is $482.68.

18. Strategy   To find Shaney's total weekly salary:
• Find the hourly overtime wage for multiplying the hourly wage ($18.40) by 1.5 (time and a half).
• Find the salary for overtime by multiplying the number of overtime hours (15) by the hourly overtime wage.
• Find the salary for the normal hours worked by multiplying the number of hours worked (30) by the hourly rate ($13.40).
• Add the salary from the night hours to the salary from the normal hours.

Solution $18.40 \times 1.5 = 27.60$
$15 \times 27.60 = 414$
$30 \times 18.40 = 552$
$552 + 414 = 966$
Shaney earned $966.

19. Strategy   Find the current checkbook balance by subtracting the checks written and adding the deposit to the original balance.

Solution
$$\begin{array}{r} 7349.44 \\ - \underline{1349.67} \\ 5999.77 \\ - \underline{344.12} \\ 5655.65 \\ + \underline{956.60} \\ 6612.25 \end{array}$$

The current checkbook balance is $6612.25.

20. Solution

Current checkbook balance:	1106.31
Checks:	322.37
	413.45
	+ 78.20
	1920.33
Service charge:	− 0.00
	1920.33
Deposits:	− 0.00
Checkbook balance:	1920.33

Current bank balance from bank statement: $1920.33.
Checkbook balance: $1920.33
The bank statement and checkbook balance.

CUMULATIVE REVIEW

1. $12 - (10-8)^2 \div 2 + 3$

$12 - 2^2 \div 2 + 3$

$12 - 4 \div 2 + 3$

$12 - 2 + 3$

$10 + 3 = 13$

2.

$3\dfrac{1}{3} = 3\dfrac{8}{24}$

$4\dfrac{1}{8} = 4\dfrac{3}{24}$

$+ \ 1\dfrac{1}{12} = 1\dfrac{2}{24}$

$\rule{3cm}{0.4pt}$

$8\dfrac{13}{24}$

3. $12\dfrac{3}{16} = 12\dfrac{9}{48} = 11\dfrac{57}{48}$

$- \ 9\dfrac{5}{12} = 9\dfrac{20}{48} = 9\dfrac{20}{48}$

$\rule{4cm}{0.4pt}$

$2\dfrac{37}{48}$

4. $5\dfrac{5}{8} \times 1\dfrac{9}{15} = \dfrac{45}{8} \times \dfrac{24}{15}$

$= \dfrac{45 \times 24}{8 \times 15}$

$= \dfrac{\overset{1}{\cancel{5}} \cdot 3 \cdot 3 \cdot \overset{1}{\cancel{2}} \cdot \overset{1}{\cancel{2}} \cdot \overset{1}{\cancel{2}} \cdot 3}{\underset{1}{\cancel{2}} \cdot \underset{1}{\cancel{2}} \cdot \underset{1}{\cancel{2}} \cdot 3 \cdot \underset{1}{\cancel{5}}} = 9$

5. $3\dfrac{1}{2} \div 1\dfrac{3}{4} = \dfrac{7}{2} \div \dfrac{7}{4} = \dfrac{7}{2} \times \dfrac{4}{7} = \dfrac{\overset{1}{\cancel{7}} \cdot \overset{1}{\cancel{2}} \cdot 2}{\underset{1}{\cancel{2}} \cdot \underset{1}{\cancel{7}}} = 2$

6. $\left(\dfrac{3}{4}\right)^2 \div \left(\dfrac{3}{8} - \dfrac{1}{4}\right) + \dfrac{1}{2}$

$\left(\dfrac{3}{4} \cdot \dfrac{3}{4}\right) \div \left(\dfrac{3}{8} - \dfrac{2}{8}\right) + \dfrac{1}{2}$

$\dfrac{9}{16} \div \dfrac{1}{8} + \dfrac{1}{2}$

$\dfrac{9}{16} \times \dfrac{8}{1} + \dfrac{1}{2}$

$\dfrac{9}{2} + \dfrac{1}{2} = \dfrac{10}{2} = 5$

7.

$$\begin{array}{r} 52.18 \approx 52.2 \\ 0.059\overline{)3.079.20} \\ \underline{-295} \\ 129 \\ \underline{-118} \\ 112 \\ \underline{-59} \\ 530 \\ \underline{-472} \\ 58 \end{array}$$

8. $\dfrac{17}{12} = 17 \div 12$

$$\begin{array}{r} 1.4166 \approx 1.417 \\ 12\overline{)17.000} \\ \underline{-12} \\ 50 \\ \underline{-48} \\ 20 \\ \underline{-12} \\ 80 \\ \underline{-72} \\ 8 \end{array}$$

9. $\dfrac{\$410}{8 \text{ hours}} = \$51.25/\text{hour}$

10. $\dfrac{5}{n} = \dfrac{16}{35}$

$5 \times 35 = n \times 16$

$175 = n \times 16$

$175 \div 16 = n$

$10.9375 = n$

$10.94 \approx n$

11. $\dfrac{5}{8} \times 100\% = \dfrac{500}{8}\% = 62.5\%$

12. 6.5% of $420 = 0.065 \times 420 = 27.3$

13. $18.2 \times 0.01 = 0.182$

14. $n \times 20 = 8.4$

$n = 8.4 \div 20 = 0.42 = 42\%$

15. $0.12 \times n = 30$

$n = 30 \div 0.12 = 250$

16. $0.42 \times n = 65$

$n = 65 \div 0.42 \approx 154.76$

17. Strategy   To find the total rainfall for the 3 weeks, add the 3 weekly amounts $\left(3\frac{3}{4},\ 8\frac{1}{2},\ \text{and}\ 1\frac{2}{3}\ \text{inches}\right)$.

Solution
$$3\frac{3}{4}=3\frac{9}{12}$$
$$8\frac{1}{2}=8\frac{6}{12}$$
$$+\ 1\frac{2}{3}=1\frac{8}{12}$$
$$12\frac{23}{12}=13\frac{11}{12}$$

The total rainfall is $13\frac{11}{12}$ inches.

18. Strategy   Find the amount paid in taxes by multiplying the total monthly income ($4850) by the portion paid in taxes $\left(\frac{1}{5}\right)$.

Solution $4850\times\frac{1}{5}=970$

The amount paid in taxes is $970.

19. Strategy   To find the rate:
• Find the amount of the decrease by subtracting the decreased price ($30) from the original price ($75).
• Write in simplest form the ratio of the decrease to the original price.

Solution $75-30=45$
$$\frac{45}{75}=\frac{3}{5}$$

The ratio is $\frac{3}{5}$.

20. Strategy   To find the number of miles driven per gallon of gasoline, divide the number of miles driven (417.5) by the number of gallons used (12.5).

Solution $417.5\div12.5=33.4$
The number of miles driven per gallon is 33.4.

21. Strategy   To find the unit cost, divide the total cost ($12.96) by the number of pounds (14).

Solution $12.96\div14\approx0.93$
The cost per pound is $.93.

22. Strategy   To find the dividend on 200 shares, write and solve a proportion.

Solution
$$\frac{80}{112}=\frac{200}{n}$$
$$n\times80=200\times112$$
$$n\times80=22,400$$
$$n=22,400\div80$$
$$n=280$$
The dividend is $280.

23. Strategy   To find the sale price:
• Solve the basic percent equation for amount to find the amount of the discount. The base is $900 and the percent is 20%.
• Subtract the discount from the regular price.

Solution $0.20\times900=180$
$900-180=720$
The sale price is $720.

24. Strategy   To find the selling price:
• Find the amount of markup by solving the basic percent equation for amount. The base is $85 and the percent is 40%.
• Add the markup to the cost.

Solution $0.40\times85=34$
$85+34=119$
The selling price of the disc player is $119.

25. Strategy   To find the percent increase:
• Find the amount of the increase by subtracting the original value from the value after the increase.
• Solve the basic percent equation for the percent. The base is $2800 and the amount if the amount of the increase.

Solution $3024-2800=224$
$n\times2800=224$
$n=224\div2800=0.08=8\%$
The percent increase in Sook Kim's salary is 8%.

26. Strategy   To find the simple interest due, multiply the principal by the annual rate by the time (in years).

Solution $120,000\times0.10\times\frac{6}{12}=6000$
The simple interest due is $6000.

27. Strategy   To find the monthly payment:
- Find the amount financed by subtracting the down payment from the purchase price.
- Multiply the amount financed by the monthly mortgage factor.

     Solution $26,900 - 2,000 = 24,900$
$24,900 \times 0.0317997 = 791.812$
The monthly car payment is $791.81.

28. Strategy   To find the new checking account balance, add the deposit to the original balance and subtract the check amounts.

     Solution

$$\begin{array}{r} 1846.78 \\ + 568.30 \\ \hline 2415.08 \\ - 123.98 \\ \hline 2291.10 \\ - 47.33 \\ \hline 2243.77 \end{array}$$

The family's new checking account balance is $2243.77.

29. Strategy   To find the cost per mile:
- Find the total cost by adding the expenses ($840, $520, $185, and $432).
- Divide the total cost by the number of miles driven (10,000).

     Solution

$$\begin{array}{r} 840 \\ 520 \\ 185 \\ + 432 \\ \hline 1977 \end{array}$$

$1977 \div 10,000 = 0.1977$
The cost per mile is about $.20.

30. Strategy   To find the monthly mortgage payment, multiply the mortgage amount by the monthly mortgage factor.

     Solution $72,000 \times 0.0103219 \approx 743.18$
The monthly mortgage payment is $743.18.

Chapter 7: Statistics and Probability

PREP TEST and GO FIGURE

1. $\dfrac{49 \text{ billion}}{102 \text{ billion}} \approx 0.480 = 48.0\%$

2. Between 2005 and 2006
$74,418 - 70,206 = \$4212$
Between 2006 and 2007
$78,883 - 74,418 = \$4465$
Between 2007 and 2008
$83,616 - 78,883 = \$4733$
Between 2008 and 2009
$88,633 - \$83,616 = \$5017$
Between 2009 and 2010
$93,951 - 88,633 = \$5318$
The greatest cost increase is between 2009 and 2010 with an increase of $5318.

3a. $\dfrac{45 \text{ gold}}{27 \text{ silver}} = \dfrac{45}{27} = \dfrac{5}{3}$

3b. 27 silver: 27 bronze $= 27 : 27 = 1 : 1$.

4a. 3.9, 3.9, 4.2, 4.5, 5.2, 5.5, 7.1

4b. $\dfrac{3.9 + 4.5 + 4.2 + 3.9 + 5.2 + 7.1 + 5.5}{7} = 4.9$ million

5a. $90,000 \times 5\% = 90,000 \times 0.05$
$= 4500$ women

5b. $5\% = 0.05 = \dfrac{5}{100} = \dfrac{1}{20}$
OR $\dfrac{4500}{90,000} = \dfrac{1}{20}$

Go Figure

If my father's parents have 10 grandchildren, and I have 2 brothers and 1 sister, then my siblings and I account for 4 out of the 10 grandchildren. We have 6 first cousins on my father's side.
If my mother's parents have 11 grandchildren, and I have 2 brothers and 1 sister, then my siblings and I account for 4 out of the 11 grandchildren. We have 7 first cousins on my mother's side.
Adding the 6 first cousins on my father's side to the 7 first cousins on my mother's side, results in 13 first cousins.

SECTION 7.1

Objective A Exercises

1. Strategy   To find the gross revenue:
• Read the pictograph to determine the gross revenue of the four movies.
• Add the four numbers.

   Solution
   150 million
   300 million
   200 million
   +100 million
   750 million
   The gross revenue is $750 million.

3. Strategy   To find the percent, solve the basic percent equation for percent. The base is 750 million (from Exercise 1) and the amount is the revenue from the Lion King (300 million).

   Solution $n \times 750$ million $= 300$ million
   $n = 300 \div 750$
   $n = 0.40$
   The percent is 40%.

5. Strategy   To find how many more people agreed that humanity should explore planets than agreed that space exploration impacts daily life, subtract the number that agrees that space exploration impacts daily life (600) from the number that agrees that humanity should explore planets (650).

   Solution $650 - 600 = 50$
   50 more people agreed that humanity should explore space than agreed that space exploration impacts daily life.

7. Strategy   To find the number of children who said they hid vegetables under a napkin, write and solve the basic percent equation for amount. The percent is 30% and the base is 500.

   Solution $0.30 \times 500 = 150$
   150 children said they hid their vegetables under a napkin.

9. No. The sum of the percents given in the graph is only 80%, not 100%.

Objective B Exercises

11. Strategy   To find the ratio:
 • Read the circle graph to determine the units in Finance and Accounting.
 • Write in the simplest form the ratio of the numbers of units in Finance to the number of units in Accounting.

     Solution   Number of units in Finance: 15
 Number of units in Accounting: 45
 $$\frac{15}{45} = \frac{1}{3}$$
 The ratio is $\frac{1}{3}$.

13. Strategy   To find the percent, solve the basic percent equation for percent. The base is the total number of units needed (128 units) and the amount is the number of units in math (12 units).

     Solution $n \times 128 = 12$
     $n = 12 \div 128 = 0.094 = 9.4\%$
 The percent is 9.4%.

15. Strategy   To find the number of people surveyed:
 • Read the circle graph to determine the number of responses.
 • Add the five numbers.

     Solution

High ticket prices:	33
People talking:	42
Uncomfortable seats:	17
Dirty floors:	27
High food prices:	+ 31
	150

The number of people surveyed is 150 people.

17. Strategy   To find the percent, solve the basic percent equation for percent. The base is the total number of responses (150) and the amount is the number of "people talking" responses (42).

     Solution $n \times 150 = 42$
     $n = 42 \div 150 = 0.28 = 28\%$
 The percent is 28%.

19. Strategy   To find the amount of money spent:
 • Read the circle graph to find the percent of money spent on portable game machines.
 • Use the basic percent equation to find the amount.

     Solution   9% is spent on portable game machines.
 $0.09 \times 3,100,000,000 = n$
      $279,000,000 = n$
 Americans spend $279,000,000 on portable game machines.

21. Strategy   To determine if the amount spent for TV game machines is more than three times the amount spent for portable game machines:
 • Multiply by 3 the amount spent for portable game machines (Use amount from Exercise 19).
 • Compare the result with the amount spent for TV game machines. (Use amount from Exercise 18).

     Solution $3 \times 279,000,000 = 837,000,000$
 $837,000,000 < 1,085,000,000$
 Yes. The amount spent for TV game machines is more than three times the amount spent for portable game machines.

23. Strategy   To find if the number of homeless who are aged 25 to 34 is more or less than twice the number of homeless under the age of 25, read the pictograph and determine if the number of homeless aged 25 to 34 (25%) is more or less than twice the number of homeless under 25 (12%).

     Solution $2 \times 12\% = 24\%$
 $254\% > 24\%$
 The number of homeless aged 25 to 34 is more than twice the number of homeless under the age of 25.

25. Strategy   To find the amount spent for food:
 • Locate the percent of homeless over age 54.
 • Solve the basic percent equation for amount. The base is 100,000.

     Solution   Percent homeless over age 54: 8%
 Percent × base = amount
 $0.08 \times 100,000 = n$
      $8,000 = n$
 There are 8000 people over age 54 out of every 100,000 homeless people.

27. Strategy To find the difference, subtract the area of South America (6,870,000 square miles) from the area of North America (9,420,000 square miles).

Solution
$$
\begin{array}{ll}
9,420,000 & \text{North America} \\
-8,870,000 & \text{South America} \\
\hline
2,550,000 &
\end{array}
$$
North America is 2,550,000 square miles larger than South America.

29. Strategy To find the percent:
• Read the circle graph to determine the land area of Australia.
• Write and solve the basic percent equation for percent. The amount is the land area of Australia and the base is the total land area of the seven continents. (57,240,000 square miles).

Solution The area of Australia is 2,970,000 square miles.
Percent $\times$ base = amount
$n \times 57,240,000 = 2,970,000$
$n = 2,970,000 \div 57,240,000$
$n = 0.0519$
Australia is 5.2% of the total land area.

31. Strategy To find the amount spent on entertainment:
• Locate the percent of the family's income that was spent for entertainment.
• Solve the basic percent equation for amount.

Solution Spent for entertainment: 5%
Percent $\times$ base = amount
$0.05 \times 34,000 = 1,700$
The family spent $1,700 for entertainment.

33. Strategy To find out whether the amount spent for housing is more than twice the amount spent for food:
• Locate on the circle graph the percent of the income spent on food and on housing.
• Multiply the percent spent on food by two and compare with the amount spent on housing.

Solution Food: 15%
Housing: 32%
$2 \times 15\% = 30\%$
Since 32% > 30%, more than twice the amount is spent on housing than on food.

Applying the Concepts

35. Answers will vary. For example: The couple's largest single expense was rent.
Food represents approximately one-quarter of the month's expenditures.
Rent represents approximately one-third of the month's expenditures.
The expense for food is approximately the same as the expense for entertainment and transportation.
The couple spent more for transportation than for entertainment.

SECTION 7.2

Objective A Exercises

1. Strategy To find the total passenger cars produced worldwide:
• Read the bar graph to determine the number of cars produced (in millions) in each region.
• Add the five numbers.

Solution
$$
\begin{array}{ll}
15 & \text{Western Europe} \\
11 & \text{Asia} \\
8 & \text{Eastern Europe/Russia} \\
3 & \text{North America} \\
+2 & \text{Latin America} \\
\hline
39 &
\end{array}
$$
The number of passenger cars produced worldwide is 39 million.

3. Strategy To find the percent, solve the basic percent equation for percent. The base is the total number of cars produced (39 million, from Exercise 1) and the amount is the number of cars produced in Asia (11 million)

Solution $n \times 39 = 11$
$n = 11 \div 39 \approx 0.28 = 28\%$
The percent is 28%.

5. Strategy To find the difference:
• Read the double bar graph for Honda Insight to find the fuel efficiency in the city (60 MPG) and the fuel efficiency on the highway (70 MPG).
• Subtract the two numbers.

Solution
$$
\begin{array}{ll}
70 & \text{Highway} \\
-60 & \text{City} \\
\hline
10 &
\end{array}
$$
The difference is 10 miles per gallon.

7. Strategy   To estimate the difference between the maximum salaries in New York:
• Read the double-bar graph for the maximum salaries for city and suburb police officers.
• Subtract to find the difference between the two salaries.

Solution
Suburb salary: 60,000
City salary: − 44,000
                    16,000

The maximum salary of police officers in the suburbs is $16,000 higher than the maximum salary of police officers in the city.

9. Strategy   To find which city has the greatest difference between the maximum salary in the city and the suburb.
• Read the double-bar graph to find maximum salaries for the city and the suburb.
• Subtract the maximum salary in the city from the maximum salary in the suburb.

Solution
Washington D.C.
$51,000 − 41,000 = 10,000$
Detroit:
$46,000 − 38,000 = 8,000$
New York:
$60,000 − 44,000 = 16,000$
Philadelphia:
$56,000 − 38,000 = 18,000$
Los Angeles:
$52,000 − 49,000 = 3,000$
The greatest difference in salaries is in Philadelphia.

Objective B Exercises

11. Strategy   To find the amount of snowfall during January, read the broken-line graph for January.

Solution   The amount of snowfall during January was 20 inches.

13. Strategy   To find the total snowfall during March and April:
• Read the broken-line graph to find the snowfall amounts for March and April.
• Add the two amounts.

Solution
March   17
April   + 8
            25

The snowfall during March and April was 25 inches.

15. Strategy   To find the difference:
• Read the broken-line graph to find the number of Calories recommended for men and the number recommended for women 19–22 years of age.
• Subtract the number recommended for women from the number recommended for men.

Solution
For men: 2900
For women: − 2100
                 800

The difference is 800 Calories.

17. Strategy   To find the ratio:
• Read the double broken-line graph to find the number of Calories recommended for 15–18 year old women and the number recommended for 51–74 year old women.
• Write in simplest form the ratio of the number of Calories recommended for 15–18 year old women to the number recommended for 51–74 year old women.

Solution
15–18 year old women: 2,100
51–74 year old women: 1,800
$\dfrac{2,100}{1,800} = \dfrac{7}{6}$

The ratio is $\dfrac{7}{6}$.

Applying the Concepts

19. **Strategy** To create each entry of the table, read the values (in billions of dollars) from the graph of the Foreign aid and Domestic aid, and find the difference.

  Solution For 1991, the total aid:
  $2.1 − 0.8 = $1.3 billion
  For 1992, the total aid:
  $1.9 − 0.8 = $1.1 billion
  For 1993, the total aid:
  $1.7 − 0.7 = $1.0 billion
  For 1994, the total aid:
  $1.5 − 0.6 = $0.9 billion
  For 1995, the total aid:
  $1.3 − 0.5 = $0.8 billion
  For 1996, the total aid:
  $1.3 − 0.5 = $0.8 billion
  For 1997, the total aid:
  $1.8 − 0.7 = $1.1 billion
  For 1998, the total aid:
  $1.7 − 0.8 = $0.9 billion
  For 1999, the total aid:
  $2.3 − 1.3 = $1.0 billion

1991	$1.3 billion
1992	$1.1 billion
1993	$1.0 billion
1994	$0.9 billion
1995	$0.8 billion
1996	$0.8 billion
1997	$1.1 billion
1998	$0.9 billion
1999	$1.0 billion

SECTION 7.3

Objective A Exercises

1. **Strategy** Read the histogram to find the number of students that have a tuition between $3,000 and $6,000.

  Solution The number of students that have a tuition between $3,000 and $6,000 is 44.

3. **Strategy** To find the number of students that pay more than $12,000 for tuition:
  • Read the histogram to find the number of students whose tuition is between $12,000 and $15,000 and whose tuition is between $15,000 and $18,000.
  • Add the two numbers.

  Solution $12,000 − $15,000   10 students
  $15,000 − $18,000 + 8 students
  ——————————————————— 18 students

  The number of students who paid more than $12,000 tuition was 18.

5. **Strategy** To find the number of cars between 6 and 12 years old:
  • Read the histogram to find the number of cars between 6 and 9 years old and between 9 and 12 years old.
  • Add the two numbers.

  Solution 6 to 9 years: 220 cars
  9 to 12 years: + 190 cars
  ——————————————————— 410 cars

  There are 410 cars between 6 and 12 years old.

7. **Strategy** To find the number of cars more than 12 years old:
  • Read the histogram to find the number of cars 12 to 15 years old and 15 to 18 years old.
  • Add the two numbers.

  Solution 12 to 15 years: 90 cars
  15 to 18 years: + 140 cars
  ——————————————————— 230 cars

  The total number of cars more than 12 years old was 230.

9. **Strategy** To find the number of adults that spend between 1 and 2 hours at the mall, read the histogram.

  Solution The number of adults that spend between 1 and 2 hours at the mall was 54.

11. Strategy   To find the percent:
• Read the histogram to find the number of adults that spend less than 1 hour at the mall.
• Solve the basic percent equation for percent. The base is 100 and the amount is the number of adults that spend less than one hour at the mall.

Solution   Number of adults that spend less then one hour at the mall: 22
Percent × base = amount
$n \times 100 = 22$
$n = 22 \div 100$
$n = 0.22$
The percent is 22%.

Objective B Exercises

13. Strategy   To find how many teams competed:
• Read the number of entrants for each of the points on the frequency polygon.
• Add the four numbers.

Solution
  4 between 11 and 11.5 minutes
12 between 11.5 and 12 minutes
  5 between 12 and 12.5 minutes
+ 1 between 12.5 and 13 minutes
22

There were a total of 22 teams.

15. Strategy   To find the percent:
• Read the frequency polygon to find how many teams ran between 11 and 11.5 minutes and between 11.5 and 12 minutes.
• Add the two numbers.
• Solve the basic percent equation for percent. The base is 22 and the amount is the number of teams who completed the relay in less than 12 minutes.

Solution
Between 11 and 11.5 minutes   4
Between 11.5 and 12 minutes: + 12
16

Percent × base = amount
$n \times 22 = 16$
$n = 16 \div 22$
$n = 0.727$
The percent is 72.7%.

17. Strategy   To find the percent:
• Read the frequency polygon to find how many people purchased between 20 and 30 tickets.
• Solve the basic percent equation for percent. The base is 74 and the amount is the number of people who purchased between 20 and 30 tickets.

Solution   Between 20 and 30 tickets: 8
Percent × base = amount
$n \times 74 = 8$
$n = 8 \div 74$
$n \approx 0.108$
The percent is 10.8%.

19. No. A frequency polygon shows only the number of occurrences in a class. It does not show the number of occurrences for any particular value.

21. Strategy   To find the percent:
• Read the frequency polygon to find the number of students that scored between 800 and 1,000.
• Solve the basic percent equation for percent. The base is 1,080,000 and the amount is the number of students scoring between 800 and 1,000.

Solution   Number of student scoring between 800 and 1,000: 350,000
Percent × base = amount
$n \times 1,080,000 = 350,000$
$n = 350,000 \div 1,800,000$
$n = 0.3241$
The percent is 32.4%.

23. Strategy   To find the number of students:
• Read the frequency polygon to find the number of students that scored between 800 and 1,000, between 1,000 and 1,200, between 1,200 and 1,400 and between 1,400 and 1,600.
• Add the four numbers.

Solution
Between 800 and 1, 000:	350,000
Between 1, 000 and 1,200:	350,000
Between 1, 200 and 1,400:	170,000
Between 1, 400 and 1,600:	+ 30,000
	900,000

The number of students that scored over 800 was 900,000.

Applying the Concepts

25. A frequency table is another method of organizing data. In a frequency table, or a frequency distribution, data are combined into categories called classes, and the frequency, which is the number of pieces of data in each class, is shown.

SECTION 7.4

Objective A Exercises

1a. Median

1b. Mean

1c. Mode

1d. Median

1e. Mode

1f. Mean

3. Strategy To find the mean value of the number of seats occupied:
• Find the sum of the number of seats occupied.
• Divide the sum by the number of flights (16).

Solution
309
422
389
412
401
352
367
319
410
391
330
408
399
387
411
+ 398
6,105

$$16)\overline{6,105} = 381.5625$$

The mean of the number of seats filled is 381.5625.

Strategy To find the median value of the number of seats occupied, arrange the numbers in order from smallest to largest. The median is the mean of the two middle numbers.

Solution
309 ⎫
319 ⎪
330 ⎪
352 ⎬ 7 numbers
367 ⎪
387 ⎪
389 ⎭
391 ⎫
398 ⎬ Middle numbers
399 ⎫
401 ⎪
408 ⎪
410 ⎬ 7 numbers
411 ⎪
412 ⎪
422 ⎭

$$\frac{391+398}{2} = 394.5$$

The median of the number of seats filled is 394.5.

Strategy To find the mode, look at the number of seats occupied and locate the number that occurs most frequently.

Solution Since each number occurs only once, there is no mode.

5. Strategy   To find the mean cost:
- Find the sum of the costs.
- Divide the total costs by the number of purchases (8).

Solution

$45.89
$52.12
$41.43
$40.67
$48.73
$42.45
$47.81
+ $45.82
$364.92

$$\frac{45.615}{8)364.920}$$

The mean cost is $45.615.

Strategy   To find the median cost, arrange the costs in order from smallest to largest. The median is the mean of the two middle numbers.

Solution

$40.67 ⎫
$41.43 ⎬ 3 numbers
$42.45 ⎭

$45.82 ⎫
$45.89 ⎬ middle numbers

$47.81 ⎫
$48.73 ⎬ 3 numbers
$52.12 ⎭

$$\frac{\$45.82 + 45.88}{2} = \$45.855$$

The median cost is $45.855.

7. Strategy   To find the mean monthly rate:
- Find the sum of the monthly rates.
- Divide the sum by the number of plans (8).

Solution

$423
$390
$405
$396
$426
$355
$404
+ $430
$3,229

$$\frac{\$403.625}{8)\$3,229.000}$$

The mean monthly rate is $403.625.

Strategy   To find the median monthly rate, write the rates in order from smallest to largest. The median is the average of the two middle terms.

Solution

$355 ⎫
$390 ⎬ 3 numbers
$396 ⎭

$404 ⎫
$405 ⎬ middle numbers

$423 ⎫
$426 ⎬ 3 numbers
$430 ⎭

$$\frac{\$404 + \$405}{2} = \$404.50$$

The median monthly rate is $404.50.

9. Strategy   To find the mean life expectancy:
- Find the sum of the years.
- Divide the sum by the number of countries (10).

Solution

62
75
78
70
64
75
66
71
74
+ 73
708

$$\frac{70.8}{10)708.0}$$

The mean life expectancy is 70.8 years.

Strategy   To find the median life expectancy, write the years in order from lowest to highest. The median is the mean of the two middle numbers.

Solution
$$
\left.\begin{array}{l}
62 \\
64 \\
66 \\
70
\end{array}\right\} \text{4 numbers}
$$

$$
\left.\begin{array}{l}
71 \\
73
\end{array}\right\} \text{middle numbers}
$$

$$
\left.\begin{array}{l}
74 \\
75 \\
75 \\
78
\end{array}\right\} \text{4 numbers}
$$

$$\frac{71+73}{2}=72$$

The median life expectancy is 72 years.

11. No, the mean scores of the two students are not the same. The mean score of the second student is 5 points higher than the mean score of the first student.

Objective B Exercises

13a. 25%

13b. 75%

13c. 75%

13d. 25%

15. Strategy
• Read the lowest value, the highest value, the first quartile, the third quartile, the median directly from the box-and-whiskers plot.
• Find the range by subtracting the lowest from the highest.
• Interquartile range = $Q_3 - Q_1$.

Solution   Lowest is $24,880.
Highest is $47,954.
$Q_1 = \$29,778$
$Q_3 = \$36,453$
Median = $33,858
Range: $47,954 - 24,880 = \$23,074$
Interquartile range:
    $36,453 - 29,778 = \$6675$

17a. Strategy   To find the number of adults who had a cholesterol level above 217, the median, solve the basic percent equation for the amount, where the base is 80 and the percent is 50%

Solution   Percent × base = amount
$0.50 \times 80 = 40$
There were 40 adults who had cholesterol levels above 217.

17b. Strategy   To find the number of adults who had a cholesterol level below 254, the third quartile, solve the basic percent equation for the amount, where the base is 80 and the percent is 75%

Solution   Percent × base = amount
$0.75 \times 80 = 60$
There were 60 adults who had cholesterol levels below 254.

17c. Strategy   To find the number of cholesterol levels represented in each quartile, solve the basic percent equation for the amount, where the base is 80 and the percent is 25%

Solution   Percent × base = amount
$0.25 \times 80 = 20$
There are 20 cholesterol levels in each quartile.

17d. The first quartile is at 198. So 25% of the adults had cholesterol levels not more than 198

19a. Strategy
• Arrange the data from smallest to largest.
• Find the range.
• Find the median.
• Find Q_1, the median of the lower half of the data.
• Find Q_3, the median of the upper half of the data.
• Interquartile range = $Q_3 - Q_1$.

Solution

0.41	0.41	0.56	0.61	0.76
0.87	1.06	2.10	2.60	4.80

Range: $4.80 - 0.41 = 4.39$ million metric tons

median $= \dfrac{0.76 + 0.87}{2} = 0.815$ million metric tons

$Q_1 = 0.56$ million metric tons,

$Q_3 = 2.10$ million metric tons

Interquartile range $= Q_3 - Q_1$
            $= 2.10 - 0.56$
            $= 0.54$ million metric tons

19b.
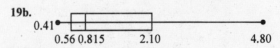

19c. 4.80

21a. Strategy To determine if the difference in means is greater than 1 inch:
• Find the sum of the rainfall in Orlando.
• Divide the sum by the number of months (12) to find the mean.
• Find the sum of the rainfall in Portland.
• Divide the sum by the number of months (12) to find the mean.
• Find the difference in the means.

Solution

Orlando		Portland	
2.1		6.2	
2.8		3.9	
3.2		3.6	
2.2		2.3	
4.0		2.1	
7.4	$\frac{4.0}{12\overline{)47.8}}$	1.5	$\frac{3.1}{12\overline{)37.5}}$
7.8		0.5	
6.3		1.1	
5.6		1.6	
2.8		3.1	
1.8		5.2	
+1.8		+6.4	
47.8		37.5	

$4.0 - 3.1 = 0.9$
No, the difference in the means is not greater than 1 inch.

21b Strategy To find the median the difference in the medians, write the rainfall in order from lowest to highest. The median is the mean of the two middle numbers. Find the difference between the Orlando median and Portland median.

Solution

Orlando		Portland	
1.8		0.5	
1.8		1.1	
2.1	5 numbers	1.5	5 numbers
2.2		1.6	
2.8		2.1	
2.8	middle	2.3	middle
3.2		3.1	
4.0		3.6	
5.6		3.9	
6.3	5 numbers	5.2	5 numbers
7.4		6.2	
7.8		6.4	

For Orlando, $\frac{2.8+3.2}{2} = 3.0$

For Portland, $\frac{2.3+3.1}{2} = 2.7$

$3.0 - 2.7 = 0.3$
The difference in medians is 0.3 inch.

21c. Strategy To draw box-and-whiskers:
• Find Q_1 and Q_3 in Orlando.
• Find Q_1 and Q_3 in Portland.

Solution For Orlando, $Q_1 = \frac{2.1+2.2}{2} = 2.15$,

$Q_3 = \frac{5.6+6.3}{2} = 5.95$

For Portland, $Q_1 = \frac{1.5+1.6}{2} = 1.55$

$Q_3 = \frac{3.9+5.2}{2} = 4.55$

21d. Answers will vary. For example, the distribution of the data is relatively similar to that in Exercise 20. However, the value of each of the 5 point on the boxplot for the Portland data is less than the corresponding value on the boxplot for the Orlando data. The average monthly rainfall in Portland is less than the average monthly rainfall in Orlando.

Applying the Concepts

23. Answers will vary. For example, 55, 55, 55, 55, 55, or 50, 55, 55, 55, 60.

25a. Q_1 is the number in a set of data that one-quarter of the data lie below.

25b. Q_3 is the number in a set of data that one-quarter of the data lie above.

25c. $\overline{x}$ is the symbol for the mean of a set of data.

27. The box does represent 50% of the data, but it provides a picture of the spread of the data. The box is not one-half the entire length of the box-and-whiskers plot because there is a greater spread of data in the interquartile range than in the first or fourth quarters of the data.

SECTION 7.5

Objective A Exercises

1. The possible outcomes of tossing a coin four times: {(HHHH), (HHHT), (HHTT), (HHTH), (HTTT), (HTHH), (HTTH), (HTHT), (TTTT), (TTTH), (TTHH), (THHH), (TTHT), (THHT), (THTT), (THTH)}.

3. The possible outcomes of tossing two tetrahedral dice: {(1, 1), (1, 2), (1, 3), (1, 4), (2, 1), (2, 2), (2, 3), (2, 4), (3, 1), (3, 2), (3, 3), (3, 4), (4, 1), (4, 2), (4, 3), (4, 4)}

5a. The sample space is {1, 2, 3, 4, 5, 6, 7, 8}.

5b. The outcomes in the event the number is less than 4 is {1, 2, 3}.

7a. Strategy   To calculate the probability:
- Count the number of possible outcomes. Refer to Table on p 317.
- Count the number of favorable outcomes.
- Use the probability formula.

Solution   There are 36 possible outcomes.
There are 4 favorable outcomes: (1, 4), (4, 1), (2, 3), (3, 2).

$$\text{Probability} = \frac{4}{36} = \frac{1}{9}$$

The probability that the sum is 5 is $\frac{1}{9}$.

7b. Strategy   To calculate the probability:
- Count the number of possible outcomes. Refer to Table on p 317.
- Count the number of favorable outcomes.
- Use the probability formula.

Solution   There are 36 possible outcomes.
There are 0 favorable outcomes.

$$\text{Probability} = \frac{0}{36} = 0$$

The probability that the sum is 15 is 0.

7c. Strategy   To calculate the probability:
- Count the number of possible outcomes. Refer to Table on p 317.
- Count the number of favorable outcomes.
- Use the probability formula.

Solution   There are 36 possible outcomes.
There are 36 favorable outcomes.

$$\text{Probability} = \frac{36}{36} = 1$$

The probability that the sum is less than 15 is 1.

7d. Strategy   To calculate the probability:
- Count the number of possible outcomes. Refer to Table on p 317.
- Count the number of favorable outcomes.
- Use the probability formula.

Solution   There are 36 possible outcomes.
There is 1 favorable outcome: (1, 1).

$$\text{Probability} = \frac{1}{36}$$

The probability that the sum is 2 is $\frac{1}{36}$.

9a. Strategy   To calculate the probability:
- Count the number of possible outcomes. Refer to Exercise 3.
- Count the number of favorable outcomes.
- Use the probability formula.

Solution   There are 16 possible outcomes.
There are 3 favorable outcomes: (1, 3), (3, 1), (2, 2).

$$\text{Probability} = \frac{3}{16}$$

The probability is $\frac{3}{16}$ that the sum of the dots on the two dice is 4.

9b. Strategy   To calculate the probability:
- Count the number of possible outcomes. Refer to Exercise 3.
- Count the number of favorable outcomes.
- Use the probability formula.

Solution   There are 16 possible outcomes.
There are 3 favorable outcomes: (2, 4), (4, 2), (3, 3).

$$\text{Probability} = \frac{3}{16}$$

The probability is $\frac{3}{16}$ that the sum of the dots on the two dice is 6.

11. Strategy   To calculate the probability:
- Count the number of possible outcomes. Refer to Table on p 317.
- Count the number of favorable outcomes.
- Use the probability formula.
- Compare the probabilities.

Solution   There are 36 possible outcomes.
There is 1 favorable outcome: (5, 5).

$$\text{Probability} = \frac{1}{36}$$

There are 4 favorable outcomes: (1, 4), (4, 1), (2, 3), (3, 2).

$$\text{Probability} = \frac{4}{36} = \frac{1}{9}$$

$$\frac{1}{9} > \frac{1}{36}$$

The probability of throwing a sum of 5 is greater.

13a. Strategy   To calculate the probability:
- Count the number of possible outcomes.
- Count the number of favorable outcomes.
- Use the probability formula.

Solution   There are 11 possible outcomes.
There are 4 favorable outcomes.

$$\text{Probability} = \frac{4}{11}$$

The probability is $\frac{4}{11}$ that the letter I is drawn.

13b. Strategy   To calculate the probability:
- Count the number of possible outcomes.
- Count the number of favorable outcomes.
- Use the probability formula.
- Compare the probabilities.

Solution   There are 11 possible outcomes.
There are 4 favorable outcomes of choosing an S.

$$\text{Probability} = \frac{4}{11}$$

There are 2 favorable outcomes of choosing a P.

$$\text{Probability} = \frac{2}{11}$$

$$\frac{4}{11} > \frac{2}{11}$$

The probability of choosing an S is greater.

15a. Strategy   To calculate the probability:
- Count the number of possible outcomes.
- Count the number of favorable outcomes.
- Use the probability formula.

Solution   There are 12 possible outcomes (3 blue + 4 green + five red).
There are 4 favorable outcomes.

$$\text{Probability} = \frac{4}{12} = \frac{1}{3}$$

The probability is $\frac{1}{3}$ that the marble chosen is green.

15b. Strategy   To calculate the probability:
- Count the number of possible outcomes.
- Count the number of favorable outcomes.
- Use the probability formula.
- Compare the probabilities.

Solution   There are 12 possible outcomes.
There are 3 favorable outcomes of choosing a blue marble.

$$\text{Probability} = \frac{3}{12} = \frac{1}{4}$$

There are 5 favorable outcomes of choosing a red marble.

$$\text{Probability} = \frac{5}{12}$$

$$\frac{5}{12} > \frac{3}{12}$$

The probability of choosing a red marble is greater.

17. Strategy   To calculate the probability:
- Count the number of possible outcomes.
- Count the number of favorable outcomes.
- Use the probability formula.

Solution   There are 47 possible outcomes (4 + 8 + 22 + 10 + 3).
There are 8 favorable outcomes.

$$\text{Probability} = \frac{8}{47}$$

The probability is $\frac{8}{47}$ that the paper has a B grade.

19. Strategy   To calculate the empirical probability, use the probability formula and divide the number of observations (587) by the total number of observations (725).

Solution $$\text{Probability} = \frac{587}{725} \approx 0.81$$

The probability is 0.81 that an employee has a group health insurance plan.

Applying the Concepts

21. No, the numbers 1 through 5 are not equally likely because the sizes of the sectors are different.

23. The mathematical probability of an event is a number from 0 to 1. Because $\frac{5}{3} = 1\frac{2}{3}, \frac{5}{3} > 1$. The probability of an even cannot be $\frac{5}{3}$ because the probability cannot be greater than 1.

CHAPTER REVIEW

1. **Strategy**   To find the amount of money:
   - Read the circle graph to determine the amounts of money spent.
   - Add the amounts.

   Solution
Defense:	$148 million
Agriculture:	$15 million
EPA:	$24 million
Commerce:	$27 million
NASA:	$31 million
Other:	+ 104 million
	$349 million

   The agencies spent $349 million in maintaining web sites.

2. **Strategy**   To find the ratio:
   - Read the circle graph to find the amount spent by the Department of Commerce and by the EPA.
   - Write in simplest form, the ratio of the amount spent by the Department of Commerce to the amount spent by the EPA.

   Solution
Commerce:	$27 million
EPA:	$24 million

   $\dfrac{\$27 \text{ million}}{\$24 \text{ million}} = \dfrac{9}{8}$

   The ratio is $\dfrac{9}{8}$.

3. **Strategy**   To find the percent, solve the basic percent equation for percent. The base is the total amount spent ($349 million) and the amount is the amount spent by NASA ($31 million)

   Solution   Percent $\times$ base = amount
   $n \times \$349 \text{ million} = \$31 \text{ million}$
   $n = \$31 \text{ million} \div \$349 \text{ million}$
   $n \approx 0.0888$
   8.9% of the total mount of money was spent by NASA.

4. **Texas**

5. **Strategy**   To find the difference in populations:
   - Read the double broken-line graph to find the Texas population and the California population in 2000.
   - Subtract the population of Texas from the population of California.

   Solution
California:	32.5
Texas:	− 20.0
	12.5

   The population of California is 12.5 million people more than the population of Texas.

6. **Strategy**   To find which 25-year period Texas had the smallest increase in population.
   - Read the double-line graph to find the population for each 25-year period.
   - Subtract the two numbers.

   Solution   1900 to 1925:
   $6 - 2.5 = 3.5$ million
   1925 to 1950:
   $8 - 6 = 2$ million
   1950 to 1975:
   $12 - 8 = 4$ million
   1975 to 2000:
   $21 - 12 = 9$ million
   The Texas population increased the least from 1925 to 1950.

7. **Strategy**   To find the number of games in which the Knicks scored fewer than 100 points:
   - Read the frequency polygon to find the number of games in which the Knicks scored between 60–70 points, 70–80 points, 80–90 points, and 90–100 points.
   - Add the four numbers.

   Solution
60 – 70 points:	1 game
70 – 80 points:	7 games
80 – 90 points:	15 games
90 – 100 points:	+ 31 games
	54

   There were 54 games in which the Knicks scored fewer than 100 points.

8. **Strategy**   To find the ratio:
   - Read the frequency polygon to find the number of games in which the Knicks scored between 90 and 100 points and between 110 and 120 points.
   - Write in simplest form the ratio of the number of games in which the Knicks scored between 90 to 100 points to the number of games in which they scored between 110 to 120 points.

   Solution
90 to 100 points:	31 games
110 to 120 points:	8 games

   $\dfrac{31 \text{ games}}{8 \text{ games}} = \dfrac{31}{8}$

   The ratio is $\dfrac{31}{8}$.

9. Strategy To find the percent:
• Read the frequency polygon to find the number of games in which the Knicks scored 110 to 120 points and 120 to 130 points.
• Add the two numbers.
• Solve the basic percent equation for percent. The base is 80 and the amount is the number of games in which more than 110 points were scored.

Solution
110–120 points	8 games
120–130 points	+ 1 game
	9 games

Percent $\times$ base = amount
$n \times 80 = 9$
$n = 9 \div 80$
$n = 0.1125$
The percent is 11.3%.

10. From the pictograph, O'Hare airport has 10 million more passengers than Los Angeles airport.

11. Strategy To find the ratio:
• Read the pictograph to find the number of passengers going through San Francisco airport and the number of passengers going through Dallas/Ft. Worth airport.
• Write in simplest form the ratio of the number of passengers going through San Francisco airport to the number of passengers going through Dallas/Ft. Worth airport.

Solution
San Francisco: 40
Dallas/Ft. Worth: 60
$\frac{40}{60} = \frac{2}{3} = 2:3$
The ratio is 2 : 3.

12. Strategy To find the difference between the total days of operation and days of full operation of the Midwest ski areas:
• Read the double-bar graph for the number of days that the resorts were open and the days of full operation.
• Subtract the two numbers.

Solution
Days open:	90
Days of full operation:	– 40
	50

The difference was 50 days.

13. Strategy To find the percent:
• Read the double-bar graph to find the number of days that the Rocky Mountain ski areas were open and the number of days of full operation.
• Solve the basic percent equation for percent. The base is the number of days open and the amount is the number of days of full operation.

Solution
Days open:	140
Days of full operation:	70

Percent $\times$ base = amount
$n \times 140 = 70$
$n = 70 \div 140$
$n = 0.5$
The percent is 50%.

14. Strategy To determine which region has the lowest number of days of full operation, read the bar graph and select the lowest graph that shows days of full operation.

Solution The Southeast had the lowest number of days of full operation; 30 days

15. Strategy To calculate the probability:
• Count the number of possible outcomes.
• Count the number of favorable outcomes.
• Use the probability formula.

Solution There are 16 possible outcomes. There are 4 favorable outcomes: THHH, HHHT, HHTH, HTHH.
Probability $= \frac{4}{16} = \frac{1}{4}$
The probability of one tail and three heads is $\frac{1}{4}$.

16. Strategy To find the number of people who slept 8 hours or more:
• Read the histogram to find the number of people who slept 8 hours, 9 hours, or more than 9 hours.
• Add the three numbers.

Solution
Slept 8 hours:	12
Slept 9 hours:	2
Slept more than 9 hours:	+ 1
	15

There were 15 people who slept more than 8 hours.

17. Strategy To find the percent:
    • Read the histogram to find the number of people who slept 7 hours.
    • Solve the basic percent equation for percent. The base is 46 and the amount is the number of people who slept 7 hours (13).

    Solution Slept 7 hours: 13
    Percent × base = amount
    $n \times 46 = 13$
    $n = 13 \div 46$
    $n \approx 0.2826$
    The percent is 28.3%.

18. Strategy To find the mean heart rates:
    • Find the sum of the heart rates.
    • Divide the sum by the number of women.

    Solution
    80
    82
    99
    91
    93
    87
    103
    94
    73
    96
    86
    80
    97
    94
    108
    81
    100
    109
    91
    84
    78
    96
    96
    + 100
    ───
    2198

    The mean heart rate is 91.6 heartbeats per minute.

    Strategy To find the median heart rate: write the heart rates in order from smallest to largest. The median is the mean of the two middle numbers.

Solution
73 ⎤
78 |
80 |
80 |
81 |
82 ⎬ 11 numbers
84 |
86 |
87 |
91 |
91 ⎦
93 ⎫ middle numbers
94 ⎭
94 ⎤
96 |
96 |
96 |
97 |
99 ⎬ 11 numbers
100 |
100 |
103 |
108 |
109 ⎦

$$\frac{93 + 94}{2} = 93.5$$

The median heart rate is 93.5 heartbeats per minute.

Strategy To find the mode, look at the heart rates and identify the number that occurs most frequently.

Solution The mode is 96 heartbeats per minute, as that is the number that occurs most frequently.

19. Strategy
• Arrange the data from smallest to largest. Then find the range.
• Find Q_1, the median of the lower half of the data.
• Find Q_3, the median of the upper half of the data.
• Interquartile range = $Q_3 - Q_1$.

Solution

73	78	80	80	81	82	84	86	87	91	91	93
94	94	96	96	96	97	99	10	100	10	10	109

Range = 109 − 73 = 36 heartbeats per minute

$Q_1 = \dfrac{82 + 84}{2} = 83$

$Q_3 = \dfrac{97 + 99}{2} = 98$

$Q_3 - Q_1 = 98 - 83 = 15$

Interquartile range = 15 heartbeats per minute

CHAPTER TEST

1. Strategy
To find the number of students that spent between $15 and $25 each week:
• Read the frequency polygon to find the number of students that spent between $15 and $20 and between $20 and $25.
• Add the two numbers

Solution
Number between $15–$20:   12
Number between $20–$25:   +7
                              19

Nineteen students spent between $15 and $25 each week.

2. Strategy
To find the ratio:
• Read the frequency polygon to find the number of students who spend between $10 and $15 and the number who spend between $15 and $20.
• Write in simplest form the ratio of the number of students who spend between $10 and $15 to the number of students who spend between $15 and $20.

Solution
Between $10 and $15: 8 students
Between $15 and $20: 12 students
$\dfrac{8 \text{ students}}{12 \text{ students}} = \dfrac{2}{3}$
The ratio is $\dfrac{2}{3}$.

3. Strategy
To find the percent:
• Read the frequency polygon to find the number of students who spend between $0 to $5, between $5 and $10, and between $10 and $15 each week.
• Add the three numbers.
• Solve the basic percent equation for percent. The base is 40 and the amount is the number of students who spend less than $15 per week.

Solution
Between $0 and $5:   4 students
Between $5 and $10:   6 students
Between $10 and $15: + 8 students
                           18 students

Percent × base = amount
$n \times 40 = 18$
$n = 18 \div 40$
$n = 0.45$
The percent is 45%.

4. Strategy
To find the number of people surveyed:
• Read the pictograph to determine the number of people for each letter grade.
• Add the four numbers.

Solution
Number of A grades: 21
Number of B grades: 10
Number of C grades: 4
Number of D grades: + 1
                           36

There were 36 people that were surveyed for the Gallup poll.

5. Strategy
To find the ratio:
• Read the pictograph to find the number of people who gave their marriage a B grade and the number who gave their marriage a C grade.
• Write in simplest form the ratio of the number of people who gave their marriage a B grade to the number of people who gave their marriage a C grade.

Solution
Number of B grades: 10 people
Number of C grades: 4 people
$\dfrac{10 \text{ people}}{4 \text{ people}} = \dfrac{5}{2}$
The ratio is $\dfrac{5}{2}$.

6. Strategy   To find the percent:
• Read the pictograph to find the number of people who gave their marriage an A grade.
• Solve the basic percent equation for percent. The base is 36 (from Exercise 4) and the amount is the number of people who gave their marriage an A grade.

Solution   Number of A grades: 21 people
Percent × base = amount
$n \times 36 = 21$
$n = 21 \div 36$
$n \approx 0.5833$
The percent is 58.3%.

7. Strategy   Read the bar graph to find the two consecutive years that the number of fatalities were the same.

Solution   During 1995 and 1996, the number of fatalities was the same.

8. Strategy   To find the total fatalities on amusement rides during 1991 to 1999:
• Read the bar graph to determine the number of fatalities for each year.
• Add the nine numbers.

Solution

3	1991
2	1992
4	1993
2	1994
3	1995
3	1996
4	1997
5	1998
+6	1999
32	

There were 32 fatal accidents from 1991 to 1999.

9. Strategy   To find how many more fatalities in 1995 to 1998 than 1991 to 1994:
• Add the number of fatalities for 1995 to 1998.
• Add the number of fatalities for 1991 to 1994.
• Subtract the two numbers.

Solution

3	1995	3	1991
3	1996	2	1992
4	1997	4	1993
+5	1998	+2	1994
15		11	

$15 - 11 = 4$
There were 4 more fatalities from 1995 to 1998.

10. Strategy   To find how many more R-rated films than PG:
• Read the circle graph to find the number of films rated R and PG.
• Subtract the two numbers.

Solution   R: 427
PG: − 72
     355

There were 355 more films rated R.

11. Strategy   To find how many times more PG-13 films were released than NC-17:
• Read the circle graph to find the number of films rated PG-13 and NC-17.
• Divide the two numbers.

Solution $7\overline{)112}$ with quotient 16

There were 16 times more films rated PG-13.

12. Strategy   To find the percent of films rated G:
• Read the circle graph to find the number of films for each rating.
• Add the five numbers.
• Write and solve the basic percent equation for the percent. The base is the total number of films and the amount is the number of G rated films (37).

Solution

G:	37
PG:	72
PG-13:	112
R:	427
NC-17	+ 7
	655

Percent × base = amount
$n \times 655 = 37$
$n = 37 \div 655$
$n \approx 0.056$
The percent of films rated G was 5.6%.

13. Strategy   To find the number of states with per capita income between $16,000 and $22,000.
• Read the histogram to find the number of states with per capita income between $16,000 and $19,000 and between $19,000 and $22,000.
• Add the two numbers.

Solution $16,000 to $19,000:   15 states
$19,000 to $22,000: + 19 states
                  34 states

There are 34 states that have a per capita income between $16,000 and $22,000.

14. Strategy To find the percent of the states with a per capita income between $19,000 and $22,000, solve the basic percent equation for percent. The base is 50 and the amount is the number of states with a per capita income between $19,000 and $22,000 (19).

Solution Percent × base = amount
$$n \times 50 = 19$$
$$n = 19 \div 50$$
$$n = 0.38$$
The percent is 38%.

15. Strategy To find the percent:
• read the histogram to find the number of states that have a per capita income between $25,000 and $28,000 and between $28,000 and $31,000.
• Add the two numbers.
• Solve the basic percent equation for percent. The base is 50 and the amount is the number of states with a per capita income above $25,000.

Solution $25,000 – $28,000: 5 states
$28,000 – $31,000: + 2 states
                    7 states

Percent × base = amount
$$n \times 50 = 7$$
$$n = 7 \div 50$$
$$n = 0.14$$
The percent is 14%.

16. Strategy To calculate the probability:
• Count the number of possible outcomes.
• Count the number of favorable outcomes.
• Use the probability formula.

Solution There are 50 possible outcomes. There are 15 favorable outcomes.
$$\text{Probability} = \frac{15}{50} = \frac{3}{10}$$
The probability is $\frac{3}{10}$ that the ball chosen is red.

17. Strategy To find which decade had the smallest increase in enrollment.
• Read the line graph to find the enrollment for each decade.
• Subtract the two numbers.

Solution 1960 to 1970:
$8 – 4 = 4$ million
1970 to 1980:
$12 – 8 = 4$ million
1980 to 1990:
$14 – 12 = 2$ million
1990 to 2000:
$15 – 14 = 1$ million
The student enrollment increased the least during the 1990's.

18. Strategy To approximate the increase in enrollment:
• Read the enrollment for 1960 and 2000.
• Subtract the two numbers.

Solution 2000:   15 million
1960: − 4 million
            11 million

The increase in enrollment was 11 million students.

19. Strategy To find the mean lifetime of the batteries:
• Find the sum of the times.
• Divide the sum by the number of batteries tested (20).

Solution
2.9
2.4
3.1
2.5
2.6
2.0
3.0
2.3
2.4
2.7
2.0
2.4
2.6
2.7   2.53
2.1   20$\overline{)50.60}$
2.9
2.8
2.4
2.0
+ 2.8
50.6

The mean time is 2.53 days.

Strategy   To find the median lifetime of the batteries, write the times in order from lowest to highest. The median is the mean of the two middle numbers.

Solution

$$
\begin{array}{l}
2.0 \\
2.0 \\
2.0 \\
2.1 \\
2.3 \\
2.4 \\
2.4 \\
2.4 \\
2.4
\end{array} \Big\} \text{9 numbers}
$$

$$
\begin{array}{l}
2.5 \\
2.6
\end{array} \Big\} \text{middle numbers}
$$

$$
\begin{array}{l}
2.6 \\
2.7 \\
2.7 \\
2.8 \\
2.8 \\
2.9 \\
2.9 \\
3.0 \\
3.1
\end{array} \Big\} \text{9 numbers}
$$

$$\frac{2.5+2.6}{2}=2.55$$

The median time is 2.55 days.

20. Strategy   • Arrange the data from smallest to largest. The median is 2.55 (Exercise 19).
• Find Q_1, the median of the lower half of the data.
• Find Q_3, the median of the upper half of the data.
• Draw the box-and-whiskers plot.

Solution

2.0	2.0	2.	2.1	2.	2.	2.4	2.4	2.4	2.5
2.6	2.6	2.	2.7	2.	2.	2.9	2.9	3.0	3.1

$$Q_1=\frac{2.3+2.4}{2}=2.35$$

$$Q_3=\frac{2.8+2.8}{2}=2.8$$

2.0   2.35 2.55   2.8   3.1

CUMULATIVE REVIEW

1. $2^2 \cdot 3^3 \cdot 5 = (2\cdot 2)\cdot (3\cdot 3\cdot 3)\cdot (5)$
$= 4\cdot 27\cdot 5 = 540$

2. $3^2 \cdot(5-2)\div 3+5$
$9\cdot(3)\div 3+5$
$27\div 3+5$
$9+5$
14

3.
$$
\begin{array}{r}
24 = \\
40 =
\end{array}
\begin{array}{|c|c|c|}
\hline
2\cdot 2\cdot 2 & 3 & \\
\hline
2\cdot 2\cdot 2 & & 5 \\
\hline
\end{array}
$$
$\text{GCF}=2\cdot 2\cdot 2\cdot 3\cdot 5=120$

4. $\dfrac{60}{144}=\dfrac{2\cdot 2\cdot 3\cdot 5}{2\cdot 2\cdot 2\cdot 2\cdot 3\cdot 3}=\dfrac{5}{12}$

5.
$$4\frac{1}{2}=4\frac{60}{120}$$
$$2\frac{3}{8}=2\frac{45}{120}$$
$$+\;5\frac{3}{15}=5\frac{24}{120}$$
$$\rule{3cm}{0.4pt}$$
$$11\frac{129}{120}=12\frac{9}{120}=12\frac{3}{40}$$

6.
$$12\frac{5}{8}=12\frac{15}{24}=11\frac{39}{24}$$
$$-\;7\frac{11}{12}=7\frac{22}{24}=7\frac{22}{24}$$
$$\rule{3cm}{0.4pt}$$
$$4\frac{17}{24}$$

7. $\dfrac{5}{8}\times 3\frac{1}{5}=\dfrac{5}{8}\times\dfrac{16}{5}$
$=\dfrac{5\cdot 16}{8\cdot 5}$
$=\dfrac{5\cdot 2\cdot 2\cdot 2\cdot 2}{2\cdot 2\cdot 2\cdot 5}=2$

8. $3\frac{1}{5}\div 4\frac{1}{4}=\dfrac{16}{5}\div\dfrac{17}{4}=\dfrac{16}{5}\times\dfrac{4}{17}=\dfrac{16\cdot 4}{5\cdot 17}=\dfrac{64}{85}$

9. $\dfrac{5}{8}\div\left(\dfrac{3}{4}-\dfrac{2}{3}\right)+\dfrac{3}{4}=\dfrac{5}{8}\div\left(\dfrac{9}{12}-\dfrac{8}{12}\right)+\dfrac{3}{4}$
$=\dfrac{5}{8}\div\dfrac{1}{12}+\dfrac{3}{4}$
$=\dfrac{5}{8}\times\dfrac{12}{1}+\dfrac{3}{4}$
$=\dfrac{5\cdot 2\cdot 2\cdot 3}{2\cdot 2\cdot 2}+\dfrac{3}{4}$
$=\dfrac{15}{2}+\dfrac{3}{4}=\dfrac{30}{4}+\dfrac{3}{4}=\dfrac{33}{4}=8\frac{1}{4}$

10. 209.305

11.
$$
\begin{array}{r}
4.092 \\
\times\; 0.69 \\
\hline
36828 \\
24552 \\
\hline
2.82348
\end{array}
$$

12. $16\frac{2}{3} = \frac{50}{3}$

$$\begin{array}{r} 16.66\overline{6} \approx 16.67 \\ 3\overline{)50.000} \\ \underline{-3} \\ 20 \\ \underline{-18} \\ 20 \\ \underline{-18} \\ 20 \\ \underline{-18} \\ 20 \\ \underline{-18} \\ 20 \end{array}$$

13. $\dfrac{330 \text{ miles}}{12.5 \text{ gal}} = 26.4 \text{ mpg}$

14. $\dfrac{n}{5} = \dfrac{16}{25}$

$n \times 25 = 5 \times 16$

$n \times 25 = 80$

$n = 80 \div 25 = 3.2$

15. $\dfrac{4}{5} \times 100\% = 80\%$

16. $10\% \times n = 8$

$0.10 \times n = 8$

$n = 8 \div 0.10 = 80$

17. $38\% \times 43 = n$

$0.38 \times 43 = n$

$16.34 = n$

18. $n \times 75 = 30$

$n = 30 \div 75 = 0.40 = 40\%$

19. **Strategy**   To find the income for the week:
• Find the commission earned on sales by solving the basic percent equation for amount. The base is $27,500 and the percent is 2%.
• Find the total income by adding the base salary ($100) to the commission.

    Solution $2\% \times 27,500 = n$

$0.02 \times 27,500 = n$

$550 = n$

$100 + 550 = 650$

The salesperson's income for the week was $650.

20. **Strategy**   To find the cost, write and solve a proportion.

    Solution

$$\frac{4.15}{1000} = \frac{n}{50,000}$$

$4.15 \times 50,000 = n \times 1000$

$207,500 = n \times 1000$

$207,500 \div 1000 = n$

$207.50 = n$

The cost is $207.50.

21. **Strategy**   To find the interest due, multiply the principal by the annual interest rate and the time (in years).

    Solution $125,000 \times 0.11 \times \dfrac{6}{12} = 6875$

The interest due is $6875.

22. **Strategy**   To find the markup rate of the compact disc player:
• Find the markup amount by subtracting the cost ($180) from the selling price ($279).
• Solve the basic percent equation for percent. The base is $180 and the amount is the amount of the markup.

    Solution $279 - 180 = 99$

$n \times 180 = 99$

$n = 99 \div 180 = 0.55 = 55\%$

The markup rate is 55%.

23. **Strategy**   To find how much is budgeted for food:
• Read the circle graph to find what percent of the budget is spent on food.
• Solve the basic percent equation for amount. The base is $3000 and the rate is the percent of the budget that is spent on food.

    Solution   Amount spent on food: 19%

$n \times \text{base} = \text{amount}$

$19\% \times 3000 = \text{amount}$

$0.19 \times 3000 = 570$

The amount budgeted for food is $570.

24. **Strategy**   To find the difference:
• Read the double-broken-line graph to find the number of problems student 1 answered correctly on test 1 and the number of problems student 2 answered correctly on test 1.
• Subtract the student 1 total from the student 2 total to find the difference.

    Solution

$$\begin{array}{ll} \text{student 2:} & 27 \text{ answered correctly} \\ \text{student 1:} & \underline{-15 \text{ answered correctly}} \\ & 12 \text{ answered correctly} \end{array}$$

The difference in the number of problems answered correctly is 12.

25. Strategy   To find the mean high temperature:
                • Find the sum of the high temperatures.
                • Divide the sum of the high
                temperatures by the number of
                temperatures (7).

     Solution

$$
\begin{array}{r}
56° \\
72° \\
80° \\
75° \\
68° \\
62° \\
+\,74° \\
\hline
487° \text{ sum of high temperatures}
\end{array}
$$

$$
\begin{array}{r}
69.57 \\
7)\overline{487.00}
\end{array}
$$

                The mean high temperatures is 69.6°.

26. Strategy   To calculate the probability:
                • Count the number of possible
                outcomes.
                • Count the number of favorable
                outcomes.
                • Use the probability formula.

     Solution   There are 36 possible outcomes.
                There are 5 favorable outcomes: (2, 6),
                (6, 2), (3, 5), (5, 3), (4, 4).

$$
\text{Probability} = \frac{5}{36}
$$

                The probability is $\dfrac{5}{36}$ that the sum of
                the dots on the two dice is 8.

Chapter 8: U.S. Customary Units of Measurement

PREP TEST and GO FIGURE

1. 702

2. 58

3. 4

4. $\dfrac{5}{3} \times 6 = \dfrac{5 \cdot 2 \cdot \cancel{6}^{\,1}}{\cancel{3}_{\,1}} = 10$

5. $400 \times \dfrac{1}{8} \times \dfrac{1}{2} = \dfrac{\cancel{2} \cdot \cancel{2} \cdot \cancel{2} \cdot \cancel{2} \cdot 5 \cdot 5}{\cancel{2} \cdot \cancel{2} \cdot \cancel{2} \cdot \cancel{2}} = 25$

6. $5\dfrac{3}{4} \times 8 = \dfrac{23}{4} \times 8 = \dfrac{23 \cdot \cancel{2} \cdot \cancel{2} \cdot 2}{\cancel{2} \cdot \cancel{2}} = 46$

7. $3\overline{)714}$ quotient 238

8. $12\overline{)18.0}$ quotient 1.5

Go Figure

To go from the bank to the bookstore takes 10 minutes. She goes a distance of
$\dfrac{3}{4} - \dfrac{1}{3} = \dfrac{9}{12} - \dfrac{4}{12} = \dfrac{5}{12}$.

We want to find how long it takes her to reach the halfway mark on her way to work. The distance between the bank and the halfway point is
$\dfrac{1}{2} - \dfrac{1}{3} = \dfrac{3}{6} - \dfrac{2}{6} = \dfrac{1}{6} = \dfrac{2}{12}$.

We can say that $\dfrac{1}{12} = 1$ unit. This is similar to

the inches and foot relationship. Using a proportion: it takes 10 minutes to go 5 units, how many minutes does it take to go 2 units?

$\dfrac{10 \text{ min}}{5 \text{ units}} = \dfrac{n \text{ min}}{2 \text{ units}}$
$10 \times 2 = 5 \times n$
$20 = 5 \times n$
$20 \div 5 = n$
$4 = n$

It takes Mandy 4 minutes to go 2 units. It takes 4 minutes to go from the bank to the halfway point.
7:52AM + 4 minutes = 7:56AM
So she reaches the halfway point at 7:56AM.

SECTION 8.1

Objective A Exercises

1. $6 \text{ ft} = 6 \cancel{\text{ ft}} \times \dfrac{12 \text{ in.}}{1 \cancel{\text{ ft}}} = 72 \text{ in.}$

3. $30 \text{ in.} = 30 \cancel{\text{ in.}} \times \dfrac{1 \text{ ft}}{12 \cancel{\text{ in.}}} = 2\dfrac{1}{2} \text{ ft}$

5. $13 \text{ yd} = 13 \cancel{\text{ yd}} \times \dfrac{3 \text{ ft}}{1 \cancel{\text{ yd}}} = 39 \text{ ft}$

7. $16 \text{ ft} = 16 \cancel{\text{ ft}} \times \dfrac{1 \text{ yd}}{3 \cancel{\text{ ft}}} = 5\dfrac{1}{3} \text{ yd}$

9. $2\dfrac{1}{3} \text{ yd} = 2\dfrac{1}{3} \cancel{\text{ yd}} \times \dfrac{36 \text{ in.}}{1 \cancel{\text{ yd}}} = 84 \text{ in.}$

11. $120 \text{ in.} = 120 \cancel{\text{ in.}} \times \dfrac{1 \text{ yd}}{36 \cancel{\text{ in.}}} = 3\dfrac{1}{3} \text{ yd}$

13. $2 \text{ mi} = 2 \cancel{\text{ mi}} \times \dfrac{5280 \text{ ft}}{1 \cancel{\text{ mi}}} = 10{,}560 \text{ ft}$

15. $7\dfrac{1}{2} \text{ in.} = 7\dfrac{1}{2} \cancel{\text{ in.}} \times \dfrac{1 \text{ ft}}{12 \cancel{\text{ in.}}} = \dfrac{5}{8} \text{ ft}$

Objective B Exercises

17.
```
        1 mi 1120 ft
5280 ) 6400
      −5280
       1120
```
6400 ft = 1 mi 1120 ft

19.
```
  6 ft  7 in.
+ 3 ft  4 in.
  9 ft 11 in.
```

21.
```
   4 ft 15 in.
   5 ft  3 in.
 − 2 ft  6 in.
   2 ft  9 in.
```

23.
```
      2 ft  5 in.
  ×          6
  12 ft 30 in. = 14 ft 6 in.
```

25.
```
       2 ft 8 in.
  2 ) 5 ft 4 in.
     − 4 ft
       1 ft = 12 in.
             16 in.
           − 16 in.
                 0
```

27. $4\dfrac{2}{3} \text{ ft} + 6\dfrac{1}{2} \text{ ft} = 4\dfrac{4}{6} \text{ ft} + 6\dfrac{3}{6} \text{ ft}$
$= 10\dfrac{7}{6} \text{ ft}$
$= 11\dfrac{1}{6} \text{ ft}$

29.
```
   1 mi 4200 ft
 + 2 mi 3600 ft
   3 mi 7800 ft = 4 mi 2520 ft
```

Objective C   Application Problems

31. Strategy   To find how many 4-in. tiles can be placed along the counter:
• Convert the length (4 ft 8 in.) to inches.
• Divide the total length by the length of one tile (4 in.).

Solution $4 \text{ ft} = 4 \text{ ft} \times \dfrac{12 \text{ in.}}{1 \text{ ft}} = 48 \text{ in.}$

4 ft 8 in. = 48 in. + 8 in. = 56 in.

$56 \div 4 = 14$

The number of tiles is 14.

33. Strategy   To find the missing dimension:
• Find the total of the two given dimensions by adding the 2 lengths $\left(1\dfrac{1}{3}\text{ ft and }1\dfrac{1}{3}\text{ ft}\right)$.
• Subtract the total of the two given dimensions from the entire length $\left(4\dfrac{1}{2}\text{ ft}\right)$.

Solution $1\dfrac{1}{3}\text{ ft} + 1\dfrac{1}{3}\text{ ft} = 2\dfrac{2}{3}\text{ ft}$

$$4\dfrac{1}{2}\text{ ft} = 4\dfrac{3}{6} = 3\dfrac{9}{6}$$
$$-2\dfrac{2}{3}\text{ ft} = 2\dfrac{4}{6} = 2\dfrac{4}{6}$$
$$\overline{\qquad\qquad\qquad 1\dfrac{5}{6}}$$

The missing dimension is $1\dfrac{5}{6}$ ft.

35. Strategy   To find the length of material needed, add the diameters of the two holes (3 in. each) and the lengths of the three spaces left in between.

Solution
$$3 \text{ in.}$$
$$3 \text{ in.}$$
$$\dfrac{1}{2} \text{ in.}$$
$$\dfrac{1}{2} \text{ in.}$$
$$+\ \dfrac{1}{2} \text{ in.}$$
$$\overline{\qquad\qquad}$$
$$6\dfrac{3}{2} \text{ in.} = 7\dfrac{1}{2} \text{ in.}$$

The length of material needed is $7\dfrac{1}{2}$ in.

37. Strategy   To find the length of each piece, divide the total length $\left(6\dfrac{2}{3}\text{ ft}\right)$ by the number of equal pieces (4).

Solution $6\dfrac{2}{3}\text{ ft} \div 4 = \dfrac{20}{3} \div 4$

$$= \dfrac{20}{3} \times \dfrac{1}{4} = \dfrac{2 \cdot 2 \cdot 5}{3 \cdot 2 \cdot 2}$$
$$= \dfrac{5}{3}\text{ ft} = 1\dfrac{2}{3}\text{ ft}$$

The length of each piece is $1\dfrac{2}{3}$ ft.

39. Strategy   To find the length of framing needed, add the lengths of the four sides of the frame (1 ft 9 in., 1 ft 6 in., 1 ft 9 in., 1 ft 6 in.).

Solution
$$1 \text{ ft } 9 \text{ in.}$$
$$1 \text{ ft } 6 \text{ in.}$$
$$1 \text{ ft } 9 \text{ in.}$$
$$+\ 1 \text{ ft } 6 \text{ in.}$$
$$\overline{4 \text{ ft } 30 \text{ in.} = 6 \text{ ft } 6 \text{ in.}}$$

The length of framing needed is 6 ft 6 in.

41. Strategy   To find the total length of the wall in feet:
• Multiply the length of each brick (9 in.) by the number of bricks (45) to find the total length in inches.
• Convert the total length in inches to feet.

Solution
$$\begin{array}{r} 9 \text{ in.} \\ \times\ 45 \\ \hline 405 \text{ in.} \end{array}$$
$405 \text{ in.} \times \dfrac{1 \text{ ft}}{12 \text{ in.}} = 33\dfrac{3}{4}\text{ ft}$

The total length of the wall is $33\dfrac{3}{4}$ ft.

Applying the Concepts

43. $\dfrac{19 \text{ in.}}{1} \times \dfrac{1 \text{ ft}}{12 \text{ in.}} \times \dfrac{1 \text{ mi}}{5280 \text{ ft}} \times 2{,}000{,}000{,}000$
$\approx 599{,}747 \text{ mi}$

Since 599,747 > 25,000 the line would reach around the Earth at the equator.

SECTION 8.2

Objective A Exercises

1. $64 \text{ oz} = 64 \text{ oz} \times \dfrac{1 \text{ lb}}{16 \text{ oz}} = 4 \text{ lb}$

3. $8 \text{ lb} = 8 \text{ lb} \times \dfrac{16 \text{ oz}}{1 \text{ lb}} = 128 \text{ oz}$

5. $3200 \text{ lb} = 3200 \text{ lb} \times \dfrac{1 \text{ ton}}{2000 \text{ lb}} = 1\dfrac{3}{5} \text{ tons}$

7. $6 \text{ tons} = 6 \text{ tons} \times \dfrac{2000 \text{ lb}}{1 \text{ ton}} = 12{,}000 \text{ lb}$

9. $66 \text{ oz} = 66 \text{ oz} \times \dfrac{1 \text{ lb}}{16 \text{ oz}} = 4\dfrac{1}{8} \text{ lb}$

11. $1\dfrac{1}{2} \text{ lb} = 1\dfrac{1}{2} \text{ lb} \times \dfrac{16 \text{ oz}}{1 \text{ lb}} = 24 \text{ oz}$

13. $1\dfrac{3}{10} \text{ tons} = 1\dfrac{3}{10} \text{ tons} \times \dfrac{2000 \text{ lb}}{1 \text{ ton}} = 2600 \text{ lb}$

15. $500 \text{ lb} = 500 \text{ lb} \times \dfrac{1 \text{ ton}}{2000 \text{ lb}} = \dfrac{1}{4} \text{ ton}$

17. $180 \text{ oz} = 180 \text{ oz} \times \dfrac{1 \text{ lb}}{16 \text{ oz}} = 11\dfrac{1}{4} \text{ lb}$

Objective B Exercises

19.
```
       4 tons 1000 lb
2000 )9000
      −8000
       1000
```
9000 lb = 4 tons 1000 lb

21.
```
     2 lb 8 oz
16 )40
   −32
     8
```
40 oz = 2 lb 8 oz

23.
```
   1 ton    800 lb
 + 3 tons 1600 lb
   4 tons 2400 lb = 5 tons 400 lb
```

25.
```
    2    2500
    3 tons 500 lb
 −1 ton    800 lb
   1 ton 1700 lb
```

27. $5\dfrac{1}{2} \text{ lb} \times 6 = \dfrac{11}{2} \text{ lb} \times 6$

$= \dfrac{66}{2} \text{ lb}$

$= 33 \text{ lb}$

29. $4\dfrac{2}{3} \text{ lb} \times 3 = \dfrac{14}{3} \text{ lb} \times 3$

$= \dfrac{42}{3} \text{ lb}$

$= 14 \text{ lb}$

31.
```
    6 1/2 oz
 + 2 1/2 oz
   8 2/2 oz = 9 oz
```

33.
```
      1 lb  7 oz
 4 )5 lb 12 oz
   −4 lb
    1 lb = 16 oz
         28 oz
        −28 oz
          0
```

Objective C Application Problems

35. **Strategy** To find the weight of the load, multiply the number of bricks (800) by the weight of one brick $\left(2\dfrac{1}{2} \text{ lb}\right)$.

Solution $800 \times 2\dfrac{1}{2} \text{ lb} = 800 \times \dfrac{5}{2} \text{ lb}$

$= \dfrac{4000}{2} \text{ lb}$

$= 2000 \text{ lb}$

The load of bricks weighs 2000 lb.

37. **Strategy** To find the weight of the package in pounds:
   • Multiply the number of tiles (144) by the weight of one tile (7 oz).
   • Convert the number of ounces to pounds.

Solution
```
      144
   ×   7 oz
     1008 oz
```
$1008 \text{ oz} \times \dfrac{1 \text{ lb}}{16 \text{ oz}} = 63 \text{ lb}$

The package of tiles weighs 63 lb.

39. **Strategy** To find the weight of the case in pounds:
   • Find the weight in ounces by multiplying the weight of each can (6 oz) by the number of cans (24).
   • Convert the weight in ounces to pounds.

Solution
```
       6 oz
   ×   24
     144 oz
```
$144 \text{ oz} \times \dfrac{1 \text{ lb}}{16 \text{ oz}} = 9 \text{ lb}$

The weight of the case of soft drinks is 9 lb.

41. Strategy To find how much shampoo is in each container, divide the total weight of the shampoo (5 lb 4 oz) by the number of containers (4).

Solution

$$
\begin{array}{r}
1\text{ lb }\ \ 5\text{ oz} \\
4\overline{)5\text{ lb }\ 4\text{ oz}} \\
\underline{-4\text{ lb}} \\
1\text{ lb}=\underline{16\text{ oz}} \\
20\text{ oz} \\
\underline{-20\text{ oz}} \\
0
\end{array}
$$

Each container holds 1 lb 5 oz of shampoo.

43. Strategy To find the cost of the ham roast:
• Convert 5 lb 10 oz to pounds.
• Multiply the number of pounds by the price per pound ($3.20).

Solution $5\text{ lb }10\text{ oz}=5\dfrac{5}{8}\text{ lb}=5.625\text{ lb}$

$$
\begin{array}{r}
\$3.20 \\
\times\ \ 5.625 \\
\hline
\$18.00000
\end{array}
$$

The ham roast costs $18.

45. Strategy To find the cost of mailing the manuscript:
• Convert 2 lb 3 oz to ounces.
• Multiply the number of ounces by the postage rate per ounce ($.25)

Solution $2\text{ lb }3\text{ oz}=35\text{ oz}$

$$
\begin{array}{r}
\$0.25 \\
\times\ \ \ 35 \\
\hline
\$8.75
\end{array}
$$

The cost of mailing the manuscript is $8.75.

Applying the Concepts

47. Answers will vary.

SECTION 8.3

Objective A Exercises

1. $60\text{ fl oz}=60\ \cancel{\text{fl oz}}\times\dfrac{1\text{ c}}{8\ \cancel{\text{fl oz}}}=7\dfrac{1}{2}\text{ c}$

3. $3\text{ c}=3\ \cancel{\text{c}}\times\dfrac{8\text{ fl oz}}{1\ \cancel{\text{c}}}=24\text{ fl oz}$

5. $8\text{ c}=8\ \cancel{\text{c}}\times\dfrac{1\text{ pt}}{2\ \cancel{\text{c}}}=4\text{ pt}$

7. $3\dfrac{1}{2}\text{ pt}=3\dfrac{1}{2}\ \cancel{\text{pt}}\times\dfrac{2\text{ c}}{1\ \cancel{\text{pt}}}=7\text{ c}$

9. $22\text{ qt}=22\ \cancel{\text{qt}}\times\dfrac{1\text{ gal}}{4\ \cancel{\text{qt}}}=5\dfrac{1}{2}\text{ gal}$

11. $2\dfrac{1}{4}\text{ gal}=2\dfrac{1}{4}\ \cancel{\text{gal}}\times\dfrac{4\text{ qt}}{1\ \cancel{\text{gal}}}=9\text{ qt}$

13. $7\dfrac{1}{2}\text{ pt}=7\dfrac{1}{2}\ \cancel{\text{pt}}\times\dfrac{1\text{ qt}}{2\ \cancel{\text{pt}}}=3\dfrac{3}{4}\text{ qt}$

15. $20\text{ fl oz}=20\ \cancel{\text{fl oz}}\times\dfrac{1\ \cancel{\text{c}}}{8\ \cancel{\text{fl oz}}}\times\dfrac{1\text{ pt}}{2\ \cancel{\text{c}}}=\dfrac{5}{4}\text{ pt}=1\dfrac{1}{4}\text{ pt}$

17. $17\text{ c}=17\ \cancel{\text{c}}\times\dfrac{1\ \cancel{\text{pt}}}{2\ \cancel{\text{c}}}\times\dfrac{1\text{ qt}}{2\ \cancel{\text{pt}}}=4\dfrac{1}{4}\text{ qt}$

Objective B Exercises

19.
$$
\begin{array}{r}
3\text{ gal }2\text{ qt} \\
4\overline{)14}\ \ \ \ \ \ \ \\
\underline{-12}\ \ \ \ \ \ \ \\
2\ \ \ \ \ \ \
\end{array}
$$
$14\text{ qt}=3\text{ gal }2\text{ qt}$

21.
$$
\begin{array}{r}
2\text{ qt }1\text{ pt} \\
2\overline{)5}\ \ \ \ \ \ \ \\
\underline{-4}\ \ \ \ \ \ \ \\
1\ \ \ \ \ \ \
\end{array}
$$
$5\text{ pt}=2\text{ qt }1\text{ pt}$

23.
$$
\begin{array}{r}
4\text{ qt }1\text{ pt} \\
+\ 2\text{ qt }1\text{ pt} \\
\hline
6\text{ qt }2\text{ pt}=7\text{ qt}
\end{array}
$$

25.
$$
\begin{array}{r}
2\text{ c }11\text{ fl oz} \\
\cancel{3\text{ c}\ \ \ 3\text{ fl oz}} \\
-\ 2\text{ c }\ \ 5\text{ fl oz} \\
\hline
6\text{ fl oz}
\end{array}
$$

27. $3\dfrac{1}{2}\text{ pt}\times5=\dfrac{7}{2}\text{ pt}\times5=\dfrac{35}{2}\text{ pt}=17\dfrac{1}{2}\text{ pt}$

29. $3\dfrac{1}{2}\text{ gal}\div4=3\dfrac{1}{2}\text{ gal}\times\dfrac{1}{4}$

$\qquad\qquad\ =\dfrac{7}{2}\text{ gal}\times\dfrac{1}{4}$

$\qquad\qquad\ =\dfrac{7}{8}\text{ gal}$

31.
$$
\begin{array}{r}
3\text{ gal }3\text{ qt} \\
+\ 1\text{ gal }2\text{ qt} \\
\hline
4\text{ gal }5\text{ qt}=5\text{ gal }1\text{ qt}
\end{array}
$$

33.
$$
\begin{array}{r}
2\text{ gal }4\text{ qt} \\
\cancel{3\text{ gal}}\ \ \ \ \ \ \ \\
-\ 1\text{ gal }2\text{ qt} \\
\hline
1\text{ gal }2\text{ qt}
\end{array}
$$

35. $4\dfrac{1}{2}\text{ gal}-1\dfrac{3}{4}\text{ gal}=4\dfrac{2}{4}\text{ gal}-1\dfrac{3}{4}\text{ gal}$

$\qquad\qquad\qquad\qquad\ =3\dfrac{6}{4}\text{ gal}-1\dfrac{3}{4}\text{ gal}$

$\qquad\qquad\qquad\qquad\ =2\dfrac{3}{4}\text{ gal}$

Objective C Application Problems

37. Strategy To find how many gallons of coffee should be prepared:
• Find how many cups of coffee should be prepared by multiplying the number of adults attending (60) by the number of cups each adult will drink (2).
• Convert the number of cups to gallons.

Solution $2 \text{ c} \times 60 = 120 \text{ c}$

$$120 \, \text{c} \times \frac{1 \, \text{pt}}{2 \, \text{c}} \times \frac{1 \, \text{qt}}{2 \, \text{pt}} \times \frac{1 \, \text{gal}}{4 \, \text{qt}} = \frac{120 \, \text{gal}}{16}$$
$$= 7\frac{1}{2} \, \text{gal}$$

The number of gallons of coffee that should be prepared is $7\frac{1}{2}$.

39. Strategy To find the number of quarts of final solution:
• Find the total number of ounces in the solution by adding the number of ounces in the three components (72, 16, and 48 oz).
• Convert the number of ounces to quarts.

Solution
$$\begin{array}{r} 72 \text{ oz} \\ 16 \text{ oz} \\ + 48 \text{ oz} \\ \hline 136 \text{ oz} \end{array}$$

$$136 \, \text{oz} \times \frac{1 \, \text{c}}{8 \, \text{oz}} \times \frac{1 \, \text{pt}}{2 \, \text{c}} \times \frac{1 \, \text{qt}}{2 \, \text{pt}} = \frac{136}{32} \, \text{qt}$$
$$= 4\frac{1}{4} \, \text{qt}$$

The number of quarts of final solution is $4\frac{1}{4}$.

41. Strategy To find the number of gallons of oil the farmer used:
• Multiply 5 qt by 7 to find the number of quarts used.
• Convert the number of quarts to gallons.

Solution
$$\begin{array}{r} 5 \text{ qt} \\ \times \quad 7 \\ \hline 35 \text{ qt of oil} \end{array}$$

$$35 \text{ qt} = 35 \, \text{qt} \times \frac{1 \, \text{gal}}{4 \, \text{qt}} = 8\frac{3}{4} \, \text{gal}$$

The farmer used $8\frac{3}{4}$ gal of oil.

43. Strategy To find the more economical purchase:
• Convert 1 qt to ounces.
• Compare the price per ounce of each brand of orange juice.

Solution $1 \text{ qt} = 1 \, \text{qt} \times \dfrac{2 \, \text{pt}}{1 \, \text{qt}} \times \dfrac{2 \, \text{c}}{1 \, \text{pt}} \times \dfrac{8 \, \text{oz}}{1 \, \text{c}} = 32 \text{ oz}$

First brand: $32 \overline{)\$1.5900}$ $\$0.0497$

Second brand: $24 \overline{)\$1.250}$ $\$0.052$

$\$0.0497 < \$0.052$
The more economical purchase is 1 qt for $1.59 (the first brand).

45. Strategy To find the profit:
• Convert 5 qt to fluid ounces.
• Divide the number of fluid ounces by 8 to find the number of bottles.
• Multiply the number of bottles by $4.25 to find the total income.
• Subtract the original cost ($41.50) from the total income to find the profit.

Solution $5 \text{ qt} = 5 \, \text{qt} \times \dfrac{2 \, \text{pt}}{1 \, \text{qt}} \times \dfrac{2 \, \text{c}}{1 \, \text{pt}} \times \dfrac{8 \, \text{oz}}{1 \, \text{c}}$
$= 160 \text{ fl oz}$

$8 \overline{)160 \text{ fl oz}}$ 20 number of bottles

$$\begin{array}{r} \$4.25 \\ \times \quad 20 \\ \hline \$85.00 \text{ total income} \end{array}$$

$$\begin{array}{r} \$85.00 \\ - \quad 41.50 \\ \hline \$43.50 \end{array}$$

The profit made is $43.50.

Applying the Concepts

47. **Grain:** A unit of weight in the U.S. Customary System; an avoirdupois unit equal to 0.002286 ounce or 0.036 dram.
Dram: A unit of weight in the U.S. Customary System; an avoirdupois unit equal to 0.0625 ounce or 27.344 grains
Furlong: A unit of measuring distance, equal to 0.125 mile, or 220 yards.
Rod: A linear measure equal to 5.5 yards, to 16.5 feet, and to 5.03 meters.

SECTION 8.4

Objective A Exercises

1. $25 \text{ Btu} = 25 \, \text{Btu} \times \dfrac{778 \, \text{ft} \cdot \text{lb}}{1 \, \text{Btu}}$
$= 19{,}450 \text{ ft} \cdot \text{lb}$

3. $25{,}000 \text{ Btu} = 25{,}000 \, \text{Btu} \times \dfrac{778 \, \text{ft} \cdot \text{lb}}{1 \, \text{Btu}}$
$= 19{,}450{,}000 \text{ ft} \cdot \text{lb}$

5. Energy $= 150 \text{ lb} \times 10 \text{ ft}$
$= 1500 \text{ ft} \cdot \text{lb}$

7. Energy $= 3300 \text{ lb} \times 9 \text{ ft}$
$= 29,700 \text{ ft} \cdot \text{lb}$

9. 3 tons $= 6000 \text{ lb}$
Energy $= 6000 \text{ lb} \times 5 \text{ ft}$
$= 30,000 \text{ ft} \cdot \text{lb}$

11. $850 \times 3 \text{ lb} = 2550 \text{ lb}$
Energy $= 2550 \text{ lb} \times 10 \text{ ft} = 25,500 \text{ ft} \cdot \text{lb}$

13. $45,000 \text{ Btu} = 45,000 \cancel{\text{Btu}} \times \dfrac{778 \text{ ft} \cdot \text{lb}}{1 \cancel{\text{Btu}}}$
$= 35,010,000 \text{ ft} \cdot \text{lb}$

15. $12,000 \text{ Btu} = 12,000 \cancel{\text{Btu}} \times \dfrac{778 \text{ ft} \cdot \text{lb}}{1 \cancel{\text{Btu}}}$
$= 9,336,000 \text{ ft} \cdot \text{lb}$

Objective B Exercises

17. $\dfrac{1100}{550} = 2 \text{ hp}$

19. $\dfrac{4400}{550} = 8 \text{ hp}$

21. $5 \times 550 \dfrac{\text{ft} \cdot \text{lb}}{\text{s}} = 2750 \dfrac{\text{ft} \cdot \text{lb}}{\text{s}}$

23. $7 \times 550 \dfrac{\text{ft} \cdot \text{lb}}{\text{s}} = 3850 \dfrac{\text{ft} \cdot \text{lb}}{\text{s}}$

25. Power $= \dfrac{125 \text{ lb} \times 12 \text{ ft}}{3 \text{ s}}$
$= 500 \dfrac{\text{ft} \cdot \text{lb}}{\text{s}}$

27. Power $= \dfrac{3000 \text{ lb} \times 40 \text{ ft}}{25 \text{ s}} = 4800 \dfrac{\text{ft} \cdot \text{lb}}{\text{s}}$

29. Power $= \dfrac{180 \text{ lb} \times 40 \text{ ft}}{5 \text{ s}} = 1440 \dfrac{\text{ft} \cdot \text{lb}}{\text{s}}$

31. $\dfrac{4950}{550} = 9 \text{ hp}$

33. $\dfrac{6600}{550} = 12 \text{ hp}$

CHAPTER REVIEW

1. $4 \text{ ft} = 4 \cancel{\text{ft}} \times \dfrac{12 \text{ in.}}{1 \cancel{\text{ft}}} = 48 \text{ in.}$

2.
$$
\begin{array}{r}
2 \text{ ft } 6 \text{ in.} \\
3 \overline{)7 \text{ ft} \quad\; 6 \text{ in.}} \\
\underline{-6 \text{ ft}} \\
1 \text{ ft} = \underline{12 \text{ in.}} \\
18 \text{ in.} \\
\underline{-18 \text{ in.}} \\
0
\end{array}
$$

3. Energy $= 200 \text{ lb} \times 8 \text{ ft} = 1600 \text{ ft} \cdot \text{lb}$

4. $2\frac{1}{2} \text{ pt} = 2\frac{1}{2} \cancel{\text{pt}} \times \dfrac{2 \cancel{\text{c}}}{1 \cancel{\text{pt}}} \times \dfrac{8 \text{ fl oz}}{1 \cancel{\text{c}}}$
$= 2\frac{1}{2} \times 16 \text{ fl oz}$
$= \dfrac{5}{2} \times 16 \text{ fl oz}$
$= 40 \text{ fl oz}$

5. $14 \text{ ft} = 14 \cancel{\text{ft}} \times \dfrac{1 \text{ yd}}{3 \cancel{\text{ft}}}$
$= \dfrac{14}{3} \text{ yd}$
$= 4\frac{2}{3} \text{ yd}$

6. $2400 \text{ lb} = 2400 \cancel{\text{lb}} \times \dfrac{1 \text{ ton}}{2000 \cancel{\text{lb}}}$
$= \dfrac{2400}{2000} \text{ tons} = 1\frac{1}{5} \text{ tons}$

7.
$$
\begin{array}{r}
2 \text{ lb } \quad 7 \text{ oz} \\
3 \overline{)7 \text{ lb} \quad\; 5 \text{ oz}} \\
\underline{-6 \text{ lb}} \\
1 \text{ lb} = \underline{16 \text{ oz}} \\
21 \text{ oz} \\
\underline{-21 \text{ oz}} \\
0
\end{array}
$$

8. $3\frac{3}{8} \text{ lb} = 3\frac{3}{8} \cancel{\text{lb}} \times \dfrac{16 \text{ oz}}{1 \cancel{\text{lb}}}$
$= 3\frac{3}{8} \times 16 \text{ oz}$
$= \dfrac{27}{8} \times 16 \text{ oz} = 54 \text{ oz}$

9.
$$
\begin{array}{r}
3 \text{ ft } \quad 9 \text{ in.} \\
+ 5 \text{ ft } \quad 6 \text{ in.} \\
\hline
8 \text{ ft } 15 \text{ in.} = 9 \text{ ft } 3 \text{ in.}
\end{array}
$$

10.
$$
\begin{array}{r}
\overset{2}{3} \text{ tons } \overset{2500}{500} \text{ lb} \\
\underline{-1 \text{ ton } 1500 \text{ lb}} \\
1 \text{ ton } 1000 \text{ lb}
\end{array}
$$

11.
$$
\begin{array}{r}
5 \text{ lb } 11 \text{ oz} \\
+ 3 \text{ lb } \;\, 8 \text{ oz} \\
\hline
8 \text{ lb } 19 \text{ oz} = 9 \text{ lb } 3 \text{ oz}
\end{array}
$$

12.
$$
\begin{array}{r}
\overset{4}{5} \text{ yd } \overset{4}{1} \text{ ft} \\
\underline{-3 \text{ yd } 2 \text{ ft}} \\
1 \text{ yd } 2 \text{ ft}
\end{array}
$$

13. $12 \cancel{\text{c}} \times \dfrac{1 \cancel{\text{pt}}}{2 \cancel{\text{c}}} \times \dfrac{1 \text{ qt}}{2 \cancel{\text{pt}}} = \dfrac{12}{4} \text{ qt} = 3 \text{ qt}$

14.
$$
\begin{array}{r}
2 \text{ ft } 8 \text{ in.} \\
\times \qquad 5 \\
\hline
10 \text{ ft } 40 \text{ in.} = 13 \text{ ft } 4 \text{ in.}
\end{array}
$$

15. $2.5 \text{ hp} \times 550 \dfrac{\text{ft} \cdot \text{lb}}{\text{s}} = 1375 \dfrac{\text{ft} \cdot \text{lb}}{\text{s}}$

16.
$$
\begin{array}{r}
5 \text{ lb } 8 \text{ oz} \\
\times \qquad 8 \\
\hline
40 \text{ lb } 64 \text{ oz} = 44 \text{ lb}
\end{array}
$$

17. $50 \text{ Btu} = 50 \cancel{\text{Btu}} \times \dfrac{778 \text{ ft} \cdot \text{lb}}{1 \cancel{\text{Btu}}} = 38,900 \text{ ft} \cdot \text{lb}$

18. $\dfrac{3850}{550} = 7$ hp

19. Strategy   To find the length of the remaining piece of board, subtract the length of the piece cut (6 ft 11 in.) from the total length (10 ft 5 in.).

   Solution
$$
\begin{array}{r}
\overset{9}{\cancel{10}} \text{ ft } \overset{17}{\cancel{5}} \text{ in.}\\
- \text{ 6 ft 11 in.}\\
\hline
3 \text{ ft } 6 \text{ in.}
\end{array}
$$
The length of the remaining piece is 3 ft 6 in.

20. Strategy   To find the cost of mailing the book:
• Find the weight of the book in ounces.
• Multiply the weight of the book in ounces by the price per ounce for postage ($0.18).

   Solution   2 lb 3 oz = 35 oz
$$
\begin{array}{r}
\$0.18\\
\times \quad 35\\
\hline
\$6.30
\end{array}
$$
The cost of mailing the book is $6.30.

21. Strategy   To find the number of quarts in a case:
• Find the number of ounces in a case by multiplying the number of ounces in a can (18 fl oz) by the number of cans in a case (24).
• Convert the number of ounces to quarts.

   Solution
$$
\begin{array}{r}
18 \text{ fl oz}\\
\times \quad 24\\
\hline
432 \text{ fl oz}
\end{array}
$$
$$
432 \text{ fl oz} \times \frac{1 \text{ c}}{8 \text{ fl oz}} \times \frac{1 \text{ pt}}{2 \text{ c}} \times \frac{1 \text{ qt}}{2 \text{ pt}}
$$
$$
= \frac{432}{32} \text{ qt} = 13\frac{1}{2} \text{ qt}
$$
the number of quarts in a case is $13\frac{1}{2}$.

22. Strategy   To find how many gallons of milk were sold:
• Find the number of cups sold by multiplying the number of cartons (256) by the number of cups per carton (1).
• Convert the number of cups to gallons.

   Solution   256 cartons × 1 c = 256 c
$$
256 \text{ c} = 256 \text{ c} \times \frac{1 \text{ pt}}{2 \text{ c}} \times \frac{1 \text{ qt}}{2 \text{ pt}} \times \frac{1 \text{ gal}}{4 \text{ qt}}
$$
$$
= \frac{256}{16} \text{ gal} = 16 \text{ gal}
$$
The number of gallons of milk sold that day was 16.

23. $35,000$ Btu $= 35,000 \text{ Btu} \times \dfrac{778 \text{ ft} \cdot \text{lb}}{1 \text{ Btu}}$
$= 27,230,000 \text{ ft} \cdot \text{lb}$

24. Power $= \dfrac{800 \text{ lb} \times 15 \text{ ft}}{25 \text{ s}} = 480 \dfrac{\text{ft} \cdot \text{lb}}{\text{s}}$

CHAPTER TEST

1. $2\dfrac{1}{2}$ ft $= 2\dfrac{1}{2}$ ft $\times \dfrac{12 \text{ in.}}{1 \text{ ft}} = 2\dfrac{1}{2} \times 12$ in.
$= \dfrac{5}{2} \times 12$ in.
$= 30$ in.

2.
$$
\begin{array}{r}
\overset{3}{\cancel{4}} \text{ ft } \overset{14}{\cancel{2}} \text{ in.}\\
- 1 \text{ ft } 9 \text{ in.}\\
\hline
2 \text{ ft } 5 \text{ in.}
\end{array}
$$

3. Strategy   To find the length of each equal piece, divide the total length $\left(6\dfrac{2}{3} \text{ ft}\right)$ by the number of pieces (5).

   Solution $6\dfrac{2}{3}$ ft $\div 5 = \dfrac{20}{3}$ ft $\div 5$
$$
= \dfrac{20}{3} \text{ ft} \times \dfrac{1}{5}
$$
$$
= \dfrac{4}{3} \text{ ft} = 1\dfrac{1}{3} \text{ ft}
$$

4. Strategy   To find the length of the wall in feet:
• Find the length of the wall in inches by multiplying the length of one brick (8 in.) by the number of bricks (72).
• Convert the length in inches to feet.

   Solution
$$
\begin{array}{r}
8 \text{ in.}\\
\times \quad 72\\
\hline
576 \text{ in.}
\end{array}
$$
576 in. $= 576$ in. $\times \dfrac{1 \text{ ft}}{12 \text{ in.}}$
$= 48$ ft
The wall is 48 ft long.

5. $2\dfrac{7}{8}$ lb $= 2\dfrac{7}{8}$ lb $\times \dfrac{16 \text{ oz}}{1 \text{ lb}}$
$= 2\dfrac{7}{8} \times 16 \text{ oz} = \dfrac{23}{8} \times 16 \text{ oz}$
$= 46$ oz

6.
$$
\begin{array}{r}
2 \text{ lb } 8 \text{ oz}\\
16 \overline{)40}\\
\underline{-32}\\
8
\end{array}
$$
40 oz = 2 lb 8 oz

7.
$$
\begin{array}{r}
9 \text{ lb } 6 \text{ oz}\\
+ 7 \text{ lb } 11 \text{ oz}\\
\hline
16 \text{ lb } 17 \text{ oz} = 17 \text{ lb } 1 \text{ oz}
\end{array}
$$

8.
$$
\begin{array}{r}
1 \text{ lb } \quad 11 \text{ oz}\\
4 \overline{)6 \text{ lb } 12 \text{ oz}}\\
\underline{-4 \text{ lb}}\\
2 \text{ lb} = \underline{32 \text{ oz}}\\
44 \text{ oz}\\
\underline{-44 \text{ oz}}\\
0
\end{array}
$$

9. Strategy To find the total weight of the workbooks in pounds:
• Find the total weight of the workbooks in ounces by multiplying the number of workbooks (1000) by the weight per workbook (12 oz).
• Convert the weight in ounces to pounds.

Solution $1000 \times 12 \text{ oz} = 12{,}000 \text{ oz}$

$12{,}000 \text{ oz} = 12{,}000 \text{ oz} \times \dfrac{1 \text{ lb}}{16 \text{ oz}} = 750 \text{ lb}$

The total weight of the workbooks is 750 lb.

10. Strategy To find the amount received for recycling the cans:
• Find the weight in ounces of the cans by solving a proportion.
• Convert the weight in ounces to pounds.
• Multiply the weight in pounds by the price paid per pound.

Solution $\dfrac{4 \text{ cans}}{3 \text{ oz}} = \dfrac{800 \text{ cans}}{n}$

$\quad 4 \times n = 3 \times 800$
$\quad 4 \times n = 2400$
$\qquad n = 2400 \div 4 = 600$

The cans weigh 600 oz.

$600 \text{ oz} = 600 \text{ oz} \times \dfrac{1 \text{ lb}}{16 \text{ oz}} = 37.5 \text{ lb}$

$37.5 \text{ lb} \times \$0.75 = \$28.13$

The amount the class received for recycling was \$28.13.

11. $13 \text{ qt} = 13 \text{ qt} \times \dfrac{1 \text{ gal}}{4 \text{ qt}} = \dfrac{13}{4} \text{ gal} = 3\dfrac{1}{4} \text{ gal}$

12. $3\dfrac{1}{2} \text{ gal} = 3\dfrac{1}{2} \text{ gal} \times \dfrac{4 \text{ qt}}{1 \text{ gal}} \times \dfrac{2 \text{ pt}}{1 \text{ qt}}$

$\qquad = 3\dfrac{1}{2} \times 8 \text{ pt} = 28 \text{ pt}$

13. $1\dfrac{3}{4} \text{ gal} \times 7 = \dfrac{7}{4} \text{ gal} \times 7 = \dfrac{49}{4} \text{ gal}$

$\qquad = 12\dfrac{1}{4} \text{ gal}$

14. $\begin{aligned} 5 \text{ gal } 2 \text{ qt} \\ + 2 \text{ gal } 3 \text{ qt} \\ \hline 7 \text{ gal } 5 \text{ qt} = 8 \text{ gal } 1 \text{ qt} \end{aligned}$

15. Strategy To find the number of cups of grapefruit juice in a case:
• Find the number of ounces of juice in a case by multiplying the number of cans in a case (24) by the number of ounces in a can (20).
• Convert the number of ounces to cups.

Solution $24 \times 20 \text{ oz} = 480 \text{ oz}$

$480 \text{ oz} = 480 \text{ oz} \times \dfrac{1 \text{ c}}{8 \text{ oz}} = 60 \text{ c}$

The number of cups in a case is 60.

16. Strategy To find the profit:
• Convert 40 gal to quarts.
• To find the total income for the sale of the oil, multiply the number of quarts by the sale price per quart (\$2.15).
• To find the profit, subtract the price the mechanic pays for the oil (\$200) from the total income.

Solution $40 \text{ gal} = 40 \text{ gal} \times \dfrac{4 \text{ qt}}{1 \text{ gal}} = 160 \text{ qt}$

$160 \text{ qt} \times \$2.15 = \$344 \text{ total income}$
$\$344 - \$200 = \$144$
Nick's profit is \$144.

17. Energy $= 250 \text{ lb} \times 15 \text{ ft} = 3750 \text{ ft} \cdot \text{lb}$

18. $40{,}000 \text{ Btu} = 40{,}000 \text{ Btu} \times \dfrac{778 \text{ ft} \cdot \text{lb}}{1 \text{ Btu}}$
$\qquad = 31{,}120{,}000 \text{ ft} \cdot \text{lb}$

19. Power $= \dfrac{200 \text{ lb} \times 20 \text{ ft}}{25 \text{ s}} = 160 \dfrac{\text{ft} \cdot \text{lb}}{\text{s}}$

20. $\dfrac{2200}{550} = 4 \text{ hp}$

CUMULATIVE REVIEW

1.

	2	3	5
9 =		3·3	
12 =	2·2	3	
15 =		3	5

LCM $= 2 \cdot 2 \cdot 3 \cdot 3 \cdot 5 = 180$

2. $\dfrac{43}{8} = 8\overline{)43} \quad 5\dfrac{3}{8}$
$\qquad \dfrac{-40}{3}$

3. $5\dfrac{7}{8} = 5\dfrac{21}{24}$
$-2\dfrac{7}{12} = 2\dfrac{14}{24}$
$\qquad\qquad 3\dfrac{7}{24}$

4. $5\dfrac{1}{3} \div 2\dfrac{2}{3} = \dfrac{16}{3} \div \dfrac{8}{3} = \dfrac{16}{3} \times \dfrac{3}{8} = 2$

5. $\dfrac{5}{8} \div \left(\dfrac{3}{8} - \dfrac{1}{4}\right) - \dfrac{5}{8}$

$\dfrac{5}{8} \div \left(\dfrac{3}{8} - \dfrac{2}{8}\right) - \dfrac{10}{16}$

$\dfrac{5}{8} \div \dfrac{1}{8} - \dfrac{10}{16}$

$\dfrac{5}{8} \times \dfrac{8}{1} - \dfrac{10}{16}$

$5 - \dfrac{10}{16}$

$4\dfrac{16}{16} - \dfrac{10}{16} = 4\dfrac{6}{16} = 4\dfrac{3}{8}$

6. $\quad$ Given place value
2.0972
$\qquad$ 7 > 5
2.10

7. $\quad$ 0.0792
$\times \quad$ 0.49
$\quad$ 7128
$\quad$ 3168
0.038808

8. $\dfrac{n}{12} = \dfrac{44}{60}$

$n \times 60 = 12 \times 44$

$n \times 60 = 528$

$n = 528 \div 60 = 8.8$

9. $2\dfrac{1}{2}\% \times 50 = n$

$0.025 \times 50 = n$

$1.25 = n$

10. $42\% \times n = 18$

$0.42 \times n = 18$

$n = 18 \div 0.42$

$n = 42.86$

11. $29.88 \div 7.2 \text{ lb} = 4.15/\text{lb}$

12. $3\dfrac{2}{5}$ in. $= 3\dfrac{6}{15}$ in.

$+ 5\dfrac{1}{3}$ in. $= 5\dfrac{5}{15}$ in.

$\qquad\qquad 8\dfrac{11}{15}$ in.

13. $\quad$ 1 lb 8 oz
16)24 $\qquad$ 24 oz = 1 lb 8 oz
$\underline{-16}$
$\quad$ 8

14. $\quad$ 3 lb 8 oz
$\times \qquad 9$
27 lb 72 oz = 31 lb 8 oz

15. $4\dfrac{1}{3}$ qt $= 4\dfrac{2}{6}$ qt $= 3\dfrac{8}{6}$ qt

$-1\dfrac{5}{6}$ qt $= 1\dfrac{5}{6}$ qt $= 1\dfrac{5}{6}$ qt

$\qquad\qquad 2\dfrac{3}{6}$ qt $= 2\dfrac{1}{2}$ qt

16. $\quad$ 3 $\quad$ 22
$\quad$ 4 lb 6 oz
$\underline{- 2 \text{ lb } 10 \text{ oz}}$
$\quad$ 1 lb 12 oz

17. Strategy $\quad$ To find the dividend, solve a proportion.

Solution $\quad \dfrac{\$56}{40 \text{ shares}} = \dfrac{n}{200 \text{ shares}}$

$56 \times 200 = n \times 40$

$11,200 = n \times 40$

$11,200 \div 40 = n$

$280 = n$

The dividend would be $280.

18. Strategy $\quad$ To find Anna's checking balance, subtract the amounts of the checks and add the deposit.

Solution $\quad$ 578.56
$\underline{- 216.98}$
$\quad$ 361.58
$\underline{- 34.12}$
$\quad$ 327.46
$\underline{+ 315.33}$
$\quad$ 642.79

Anna's balance is $642.79.

19. Strategy $\quad$ To find the executive's total monthly income:
• Find the amount of sales over $25,000 by subtracting $25,000 from the total sales ($140,000).
• Find the amount of the commission by solving the basic percent equation for amount. The base is the amount of sales over $25,000 and the percent is 2%.
• Add the amount of commission to the salary ($800).

Solution $\quad$ $140,000 - \$25,000 = \$115,000$
Percent $\times$ base = amount
$2\% \times 115,000 = n$
$0.02 \times 115,000 = n$
$2300 = n$
$\$2300 + \$800 = \$3100$
The executive's monthly income is $3100.

20. Strategy   To find the amount of carrots that could be sold:
• Find the amount of spoiled carrots by solving the basic percent equation for amount. The percent is 3% and the base is 2500.
• Subtract the amount of spoiled carrots from the total amount of the shipment (2500 lb).

Solution   Percent × base = amount
$3\% \times 2500 = n$
$0.03 \times 2500 = n$
$75 = n$
2500 lb – 75 lb = 2425 lb
The amount of carrots that could be sold is 2425 lb.

21. Strategy   To find the percent:
• Find the total number of students who took the final exam by reading the histogram and adding the frequencies.
• Find the number of students who received a score between 80% and 90% by reading the histogram.
• Solve the basic percent equation for percent. The base is the total number of students who took the exam and the amount is the number of students with scores between 80% and 90%.

Solution

Score: 40–50	2 students
50–60:	1 student
60–70:	5 students
70–80:	7 students
80–90:	4 students
90–100:	3 students
Total number of students:	22

Percent × base = amount
$n \times 22 = 4$
$n = 4 \div 22 = 0.181 \approx 18\%$
The percent is 18%.

22. Strategy   To find the selling price:
• Find the amount of the markup by solving the basic percent equation for amount. The base is $220 and the percent is 40%.
• Add the markup to the cost ($220).

Solution   Percent × base = amount
$40\% \times 220 = n$
$0.40 \times 220 = n$
$88 = n$
$220 + $88 = $308
The selling price of a compact disc player is $308.

23. Strategy   To find the interest paid, multiply the principal ($200,000) by the annual interest rate (11%) by the time (8 months) in years.

Solution $200,000 \times 0.11 \times \dfrac{8}{12} = 14,666.67$
The interest paid on the loan is $14,666.67.

24. Strategy   To find how much each student received:
• Convert 1 lb 3 oz to ounces.
• Find the total value of the gold by multiplying the number of ounces by the price per ounce ($400).
• Divide the total value by the number of students (6).

Solution   1 lb 3 oz = 19 oz
$19 \times \$400 = \$7600 =$ total value
$\$7600 \div 6 \approx \$1267$
Each student received $1267.

25. Strategy   To find the cost of mailing the books:
• Find the total weight of the books by adding the 4 weights (1 lb 3 oz, 13 oz, 1 lb 8 oz, and 1 lb).
• Convert the total weight to ounces.
• Find the cost by multiplying the total number of ounces by the price per ounce ($0.28).

Solution

```
  1 lb  3 oz
       13 oz
  1 lb  8 oz
+ 1 lb
  3 lb 24 oz = 72 oz
```

```
  $0.28
×    72
 $20.16
```

The cost of mailing the books is $20.16.

26. Strategy   To find the better buy:
• Find the unit price for each brand.
• Compare unit prices.

Solution $.79 for 8 oz $2.98 for 36 oz
$\dfrac{0.79}{8} = 0.09875$ $\dfrac{2.98}{36} \approx 0.08278$
$0.08278 < $0.09875
The better buy is $2.98 for 36 oz.

27. Strategy   To calculate the probability:
   - Count the number of possible outcomes.
   - Count the number of favorable outcomes.
   - Use the probability formula.

   Solution   There are 36 possible outcomes. There are 4 favorable outcomes: (3, 6), (4, 5), (5, 4), (6, 3).

   $$\text{Probability} = \frac{4}{36} = \frac{1}{9}$$

   The probability is $\frac{1}{9}$ that the sum of the dots on the two dice is 9.

28. $\text{Energy} = 400 \text{ lb} \times 8 \text{ ft} = 3200 \text{ ft} \cdot \text{lb}$

29. $\text{Power} = \dfrac{600 \text{ lb} \times 8 \text{ ft}}{12 \text{ s}} = 400 \dfrac{\text{ft} \cdot \text{lb}}{\text{s}}$

Chapter 9: The Metric System of Measurement

PREP TEST and GO FIGURE

1. 37,320

2. 659,000

3. 0.04107

4. $28,496 \div 10^3 = 28,496 \div 1000 = 28.496$

5. 5.125

6. 5.96

7. 0.13

8. $35 \times \dfrac{1.61}{1} = 35 \times 1.61 = 56.35$

9. $1.67 \times \dfrac{1}{3.34} = 1.67 \div 3.34$

$$3.34\overline{)1.67.0}^{\;0.5}$$

10. $4\dfrac{1}{2} \times 150 = \dfrac{9}{2} \times 150$

$$= \frac{3 \cdot 3 \cdot \cancel{2} \cdot 3 \cdot 5 \cdot 5}{\cancel{2}} = 675$$

Go Figure

Adding an even amount (6) of odd numbers results in an even number. Using that rule eliminates 15, 29, and 31 as possible solutions. Also, 4 cannot be a solution since the lowest possible score is 6. The highest possible score is 56, so 58 can also be eliminated as a solution. A possible score is 28, $28 = 9 + 7 + 5 + 5 + 3$.

SECTION 9.1

Objective A Exercises

1. 42 cm = 420 mm

3. 81 mm = 8.1 cm

5. 6804 m = 6.804 km

7. 2.109 km = 2109 m

9. 432 cm = 4.32 m

11. 0.88 m = 88 cm

13. 7038 m = 7.038 km

15. 3.5 km = 3500 m

17. 260 cm = 2.60 m

19. 1.685 m = 168.5 cm

21. 14.8 cm = 148 mm

23. 62 m 7 cm = 62 m + 0.07 m = 62.07 m

25. 31 cm 9 mm = 31 cm + 0.9 cm = 31.9 cm

27. 8 km 75 m = 8 km + 0.075 km = 8.075 km

Objective B Application Problems

29. **Strategy** To find the missing dimension:
• Find the sum of the given dimensions (4 cm and 15 cm 6 mm).
• Subtract the sum of the given dimensions from the entire length (27 cm 4 mm).

Solution 15 cm 6 mm = 15.6 cm

$$\begin{array}{r} 4 \text{ cm} \\ + \ 15.6 \text{ cm} \\ \hline 19.6 \text{ cm} \end{array}$$

27 cm 4 mm = 27.4 cm

$$\begin{array}{r} 27.4 \text{ cm} \\ - 19.6 \text{ cm} \\ \hline 7.8 \text{ cm} \end{array}$$

The missing dimension is 7.8 cm.

31. **Strategy** To find the distance between the rivets, divide the total length of the plate (3 m 40 cm) by the number of spaces between the rivets (19).

Solution 3 m 40 cm = 340 cm

$$19\overline{)340}^{\;17.89 \approx 17.9}$$

The distance between the rivets is 17.9 cm.

33. **Strategy** To find how much fencing is left on the roll:
• Convert the length and width of the dog run to meters.
• Add the four lengths of fencing to make the dog run.
• Subtract the total fencing used from the length of the full roll (50 m).

Solution

340 cm	3.40 m
1380 cm	13.80 m
340 cm	3.40 m
1380 cm	+ 13.80 m
	34.40 m

$50 - 34.40 = 15.6$

The amount of fencing left on the roll is 15.6 m.

35. Strategy   To find the time for light to travel to the Earth from the sun:
• Convert the distance light travels in one second (300,000,000 m) to kilometers.
• Divide the distance from the sun to the earth (150,000,000 km) by the distance light travels in one second.

Solution   300,000,000 m = 300,000 km
150,000,000 km ÷ 300,000 km/s
    = 500 s
It takes 500 s for light to travel from the sun to the Earth.

37. Strategy   To find the distance that light travels in one day:
• Find the number of seconds in one day.
• Multiply the distance that light travels in one second (300,000 km) by the number of seconds in one day.

Solution   1 day
$= 1 \text{ day} \times \dfrac{24 \text{ h}}{1 \text{ day}} \times \dfrac{60 \text{ min}}{1 \text{ h}} \times \dfrac{60 \text{ s}}{1 \text{ min}}$
= 86,400 s (in one day).
300,000 × 86,400
    = 25,920,000,000 km
Light travels 25,920,000,000 km in one day.

Applying the Concepts

39. A very brief history of the metric system appears on page 363. You might decide whether you want students to write about its early development at the end of the 18th century or about its more recent history. For example, in 1975, Congress passed the Metric Conversion Act, which encourages voluntary use of the metric system in the United States.

SECTION 9.2

Objective A

1. 420 g = 0.420 kg

3. 127 mg = 0.127 g

5. 4.2 kg = 4200 g

7. 0.45 g = 450 mg

9. 1856 g = 1.856 kg

11. 4057 mg = 4.057 g

13. 1.37 kg = 1370 g

15. 0.0456 g = 45.6 mg

17. 18,000 g = 18.000 kg

19. 3 kg 922 g = 3 kg + 0.922 kg = 3.922 kg

21. 7 g 891 mg = 7 g + 0.891 g = 7.891 g

23. 4 kg 63 g = 4 kg + 0.063 kg = 4.063 kg

Objective B   Application Problems

25. Strategy   To find the number of grams in one serving of Quaker Oats:
• Convert the amount of Quaker Oats (1.19 kg) to grams.
• Divide the amount of Quaker Oats by the number of servings (30).

Solution   1.19 kg = 1190 g
$\begin{array}{r} 39.67 \\ 30\overline{)1190.0} \end{array}$
There are 40 g in 1 serving.

27a. Strategy   To find the number of grams of cholesterol in one dozen eggs:
• Convert the amount of cholesterol that one egg contains (274 mg) to grams.
• Multiply the number of grams of cholesterol in one egg by the number of eggs (12).

Solution   274 mg = 0.274 g
12 × 0.274 = 3.288 g
There are 3.288 g of cholesterol in 12 eggs.

27b. Strategy   To find the number of grams of cholesterol in 4 glasses of milk:
• Convert the amount of cholesterol in one glass of milk (33 mg) to grams.
• Multiply the number of grams in one glass of milk by the number of glasses of milk (4).

Solution   33 mg = 0.033 g
4 × 0.033 = 0.132
There are 0.132 g of cholesterol in 4 glasses of milk.

29a. Strategy   To find the weight of the package in kilograms:
• Convert the weight of one serving (31 g) to kilograms.
• Multiply the weight of one serving by the number of servings (6).

Solution   31 g = 0.031 kg
0.031 × 6 = 0.186
There are 0.186 kg of mix in the package.

29b. Strategy To find the number of grams of sodium contained in two servings:
• Convert the weight of sodium in one servings (210 mg) to grams.
• Multiply the weight of sodium in one servings by the number of servings (2).

Solution 210 mg = 0.210 g
$0.210 \times 2 = 0.420$
There are 0.42 g of sodium in two servings.

31. Strategy To find the amount of grass seed:
• Convert 80 g to kilograms.
• Write and solve a proportion.

Solution 80 g = 0.08 kg
$$\frac{0.08 \text{ kg}}{100 \text{ m}^2} = \frac{n}{2000 \text{ m}^2}$$
$0.08 \times 2000 = 100 \times n$
$160 = 100 \times n$
$160 \div 100 = n$
$1.6 = n$
The amount of seed needed is 1.6 kg.

33. Strategy To find the profit:
• Convert the weight of a 10-kg container to grams.
• Find the number of bags of nuts in a 10-kg container by dividing the total weight in grams by the weight of one bag (200 g).
• Find the cost of the bags by multiplying the number of bags by $.04.
• Add the cost of the bags to the cost of a 10-kg container ($75) to find the total cost.
• Multiply the number of bags by $2.89 to find the total revenue.
• Subtract the total cost from the revenue to find the profit.

Solution 10 kg = 10,000 g
10,000 g ÷ 200 g = 50 (bags of nuts)
$50 \times \$0.04 = \$2$ (cost of the bags)
$\$75 + \$2 = \$77$ (total cost)
$50 \times \$2.89 = \$144.50$ (total revenue)
$\$144.50 - \$77.00 = \$67.50$ (profit)
The profit from repackaging the nuts is $67.50.

35. Strategy To find the percent of corn:
• Add the amount of exports of wheat (37,141 million kg), rice (2,680 million kg), and corn (40,365 million kg).
• Solve the basic percent equation for percent. The base is the total amount of exports and the amount is the amount of corn (40,365 million kg).

Solution
   37,141 million kg
    2,680 million kg
 + 40,365 million kg
   80,186 million kg
Percent × base = amount
$n \times 80,186 = 40,365$
$n = 40,365 \div 80,186$
$n \approx 0.5034$
Corn was 50.3% of the total exports.

Applying the Concepts

37. Students might list familiarity among the advantages of the U.S. Customary System and difficulty in converting units among the disadvantages. They might list ease of conversion among the advantages of the metric system, as well as the fact that international trade is based on the metric system.
A disadvantage for Americans is that they are unfamiliar with metric units. Another disadvantage is related to American industry: If forced to change to the metric system, companies would face the difficulty and expense of altering the present dimensions of machinery, tools, and products..

SECTION 9.3

Objective A Exercises

1. 4200 ml = 4.200 L

3. 3.42 L = 3420 ml

5. 423 ml = 423 cm^3

7. 642 cm^3 = 642 ml

9. 42 cm^3 = 42 ml = 0.042 L

11. 0.435 L = 435 ml = 435 cm^3

13. 4.62 kl = 4620 L

15. 1423 L = 1.423 kl

17. 1.267 L = 1267 cm^3

19. 3 L 42 ml = 3 L + 0.042 L = 3.042 L

21. 3 kl 4 L = 3 kl + 0.004 kl = 3.004 kl

23. 8 L 200 ml = 8 L + 0.200 L = 8.200 L

Objective B Application Problems

25a. Strategy To determine if the amount of oxygen in 50 L of air is more or less than 25 L, note the percent of air that is oxygen.

Solution Because oxygen makes up only 21% of air, which is much less than $\frac{1}{2}$, there could not be 25 L of oxygen in 50 L of air.

25b. Strategy To find the amount of oxygen, solve the basic percent equation for the amount. The percent is 21% and the base is 50 L.

Solution Percent × base = amount
$21\% \times 50 = n$
$0.21 \times 50 = n$
$10.5 = n$
There are 10.5 L of oxygen in 50 L of air.

27. Strategy To find the amount of chlorine used in one month:
• Convert 800 ml to liters.
• Multiply the amount of chlorine used in a day by the number of days in a month (30).

Solution $800 \text{ ml} = 0.8 \text{ L}$
$0.8 \times 30 = 24 \text{ L}$
24 L of chlorine were used in one month.

29. Strategy To find how many patients can be immunized:
• Convert 3 cm^3 to liters
• Divide the total number of liters of flu vaccine (12) by the number of liters of vaccine each person receives.

Solution $3 \text{ cm}^3 = 3 \text{ ml} = 0.003 \text{ L}$
$12 \div 0.003 = 4000$
4000 patients can be immunized.

31. Strategy To determine the better buy:
• Find the unit cost (cost per liter) of the 12 one-liter bottles by dividing the cost ($19.80) by the amount of apple juice (12 L).
• Find the unit cost (cost per liter) of the 24 cans by converting the amount to liters and then dividing $14.50 by the amount of juice.

Solution The case of 12 one-liter bottles:
$19.80 \div 12 = 1.65$
The unit cost is $1.65 per liter.
The case of 24 cans:
$24 \times 340 \text{ ml} = 8160 \text{ ml} = 8.16 \text{ L}$
$14.50 \div 8.16 \approx 1.78$
The unit cost is $1.78 per liter.
Since $1.65 < $1.78, the 12-one liter bottles are the better buy.

33. Strategy To find the profit:
• Convert 85 kl to liters.
• Find the income by multiplying the price per liter ($.379) by the number of liters sold.
• Subtract the cost ($23,750) from the total income.

Solution $85 \text{ kl} = 85,000 \text{ L}$

$$\begin{array}{rr} 85,000 \text{ L} & \$32,215 \\ \times\ \$.379 & -23,750 \\ \hline \text{Income: } \$32,215 & \$8,465 \end{array}$$

The profit on the gasoline is $8465.

Applying the Concepts

35. $3 \text{ L} - 280 \text{ ml} = 3 \text{ L} - 0.280 \text{ L} = 2.72 \text{ L}$
$2.72 \text{ L} = 2720 \text{ ml}$
$2.72 \text{ L} = 2 \text{ L } 720 \text{ml}$

SECTION 9.4

Objective A Application Problems

1. Strategy To find the number of Calories that can be omitted from your diet, multiply the number of Calories omitted each day (110) by the number of days (30).

Solution $110 \times 30 = 3300$
3300 Calories can be omitted from your diet.

3a. Strategy • From the nutrition label find the number of Calories per serving.
• Multiply the number of Calories per serving by $1\frac{1}{2}$.

Solution There are 60 Calories per serving.
$60 \times 1\frac{1}{2} = 90$

There are 90 Calories in $1\frac{1}{2}$ servings.

3b. Strategy
- From the nutrition label find the serving size and the number of Calories from fat.
- Determine how many servings are in 6 slices of bread.
- Multiply the number of fat Calories in a serving by the number of servings.

Solution   2 slices of bread is one serving.
10 fat Calories are in one serving.
$6 \div 2 = 3$ (number of servings)
$10 \times 3 = 30$
There are 30 fat Calories in 6 slices of bread.

5. Strategy   To find how many Calories a 135-lb person would need to maintain body weight, multiply the body weight (135 lb) by the number of Calories per pound needed (15).

Solution $135 \times 15 = 2025$
The number of Calories needed would be 2025.

7. Strategy   To find how many Calories you burn up playing tennis:
- Convert 45 min to hours.
- Find how many hours of tennis are played by multiplying the number of days (30) by the time per day in hours.
- Multiply the number of hours played by the Calories burned per hour (450).

Solution $45 \text{ min} = 45 \text{ min} \times \dfrac{1 \text{ h}}{60 \text{ min}} = \dfrac{3}{4} \text{ h}$

$30 \times \dfrac{3}{4} \text{ h} = \dfrac{90}{4} \text{ h} = 22.5 \text{ h}$

$450 \text{ cal} \times 22.5 = 10{,}125 \text{ cal}$
The number of Calories you burn up is 10,125.

9. Strategy   To find how many hours you would have to hike:
- Add to find the total number of Calories consumed ($375 + 150 + 280$).
- Divide the sum by the number of Calories used in 1 h (315).

Solution
375
150
$+\ 280$
805 number of Calories consumed

$805 \div 315 \approx 2.6$

You would have to hike for 2.6 h.

11. Strategy   To find the energy used, multiply the number of watts (500) by the number of hours $\left(2\dfrac{1}{2}\right)$.

Solution $500 \times 2\dfrac{1}{2} = 1250$ watt-hours

13. Strategy   To find the number of kilowatt-hours used:
- Find the number of Watt-hours used in standby mode.
- Find the number of Watt-hours used when in operation
- Add the two numbers.
- Convert Watt-hours to kilowatt hours.

Solution
$39 \times\ 9 =\ \ \ 351$
$6 \times 36 = \underline{+\ 216}$
567 Wh

$567 \text{ W} = 0.567 \text{ kWh}$
The fax machine used 0.567 kWh.

15. Strategy   To find the cost:
- Convert 2200 W to kilowatts.
- Multiply the kilowatts by the number of hours (8) to find the number of kilowatt-hours.
- Multiply the number of kilowatt-hours by the price per kilowatt-hour (9¢).

Solution $2200 \text{ W} = 2.2 \text{ kW}$
$2.2 \text{ kW} \times 8 \text{ h} = 17.6 \text{ kWh}$
$17.6 \text{ kWh} \times \$0.09 = \$1.584 \approx \$1.58$
The cost of running an air conditioner is \$1.58.

17a. Strategy   To determine if the light output of the Energy Saver Bulb is more or less than half the energy of the Long Life Soft White bulb, compare the light output of each bulb.

Solution   Energy Saver - 400 lumens
Soft White - 835 lumens
400 lumens is less than half the output of the Soft White bulb.

17b. Strategy   To find the cost for each bulb:
- Find the number of watt-hours by multiplying the number of watts by the number of hours.
- Convert watt-hours to kilowatt-hours.
- Multiply the number of kilowatt-hours by the cost per kilowatt hour.
- Find the difference in cost.

Solution   Sylvania Long Life Bulb:
$60 \times 150 = 9000$ Wh
9000 Wh $= 9$ kWh
$9 \times \$.108 = \$.972$
Energy Saver Soft White Bulb:
$34 \times 150 = 5100$ Wh
5100 Wh $= 5.1$ kWh
$5.1 \times \$.108 = \$.5508$
$\$.972 - \$.5508 = \$.4212$
The energy saver bulb costs \$.42 less to operate.

19. Strategy To find the cost:
• Multiply to find the total number of hours the welder is used.
• Multiply the number of hours by the number of kilowatts used each hour.
• Multiply the number of kilowatt-hours by the cost per kilowatt-hour.

Solution $30 \times 6 = 180$ h
$180 \times 6.5 = 1170$ kWh
$1170 \times 0.094 = \$109.98$
The cost of using the welder is $109.98.

Applying the Concepts

21.

Age	Weight	Calories per Day to Maintain Weight	Calories per Day to lose 1 lb/Week
men 11–14	99	2500	2000
15–18	145	3000	2500
19–24	160	2900	2400
25–50	174	2900	2400
51+	170	2300	1800
women 11–14	101	2200	1700
15–18	120	2200	1700
19–24	128	2200	1700
25–50	138	2200	1700
51+	143	1900	1400

SECTION 9.5

Objective A Exercises

1. $100 \text{ yd} = 100 \text{ yd} \times \dfrac{0.9174 \text{ m}}{1 \text{ yd}} = 91.74 \text{ m}$

3. 5 ft 8 in. = 60 in. + 8 in. = 68 in.
$68 \text{ in.} \times \dfrac{2.54 \text{ cm}}{1 \text{ in.}} \approx 172.72 \text{ cm}$
$172.72 \text{ cm} = 1.7272 \text{ m} \approx 1.73 \text{ m}$

5. $15 \text{ lb} = 15 \text{ lb} \times \dfrac{1 \text{ kg}}{2.2 \text{ lb}} \approx 6.82 \text{ kg}$

7. $1 \text{ c} = 1 \text{ c} \times \dfrac{1 \text{ pt}}{2 \text{ c}} \times \dfrac{1 \text{ qt}}{2 \text{ pt}} \times \dfrac{1 \text{ L}}{1.06 \text{ qt}}$
$\approx 0.23585 \text{ L} \approx 235.85 \text{ ml}$

9. $65\dfrac{\text{mi}}{\text{h}} = 65\dfrac{\text{mi}}{\text{h}} \times \dfrac{1.61 \text{ km}}{1 \text{ mi}} = 104.65\dfrac{\text{km}}{\text{h}}$

11. $\dfrac{\$3.49}{\text{lb}} \approx \dfrac{\$3.49}{\text{lb}} \times \dfrac{2.2 \text{ lb}}{1 \text{ kg}} \approx \$7.68/\text{kg}$

13. $\dfrac{\$1.47}{\text{gal}} = \dfrac{\$1.47}{\text{gal}} \times \dfrac{1 \text{ gal}}{4 \text{ qt}} \times \dfrac{1.06 \text{ qt}}{1 \text{ L}} \approx \$.39/\text{L}$

15. $24,887 \text{ mi} = 24,887 \text{ mi} \times \dfrac{1.61 \text{ km}}{1 \text{ mi}}$
$= 40,068.07 \text{ km}$

Objective B Exercises

17. $100 \text{ m} = 100 \text{ m} \times \dfrac{3.28 \text{ ft}}{1 \text{ m}} = 328 \text{ ft}$

19. $6 \text{ L} = 6 \text{ L} \times \dfrac{1.06 \text{ qt}}{1 \text{ L}} \times \dfrac{1 \text{ gal}}{4 \text{ qt}} = 1.58 \text{ gal}$

21. $1500 \text{ m} = 1500 \text{ m} \times \dfrac{3.28 \text{ ft}}{1 \text{ m}} = 4920 \text{ ft}$

23. $24 \text{ L} = 24 \text{ L} \times \dfrac{1 \text{ gal}}{3.79 \text{ L}} \approx 6.33 \text{ gal}$

25. $\dfrac{80 \text{ km}}{\text{h}} \approx \dfrac{80 \text{ km}}{\text{h}} \times \dfrac{1 \text{ mi}}{1.61 \text{ km}} = 49.69 \text{ mi/hr}$

27. $\dfrac{\$.385}{\text{L}} \approx \dfrac{\$.385}{\text{L}} \times \dfrac{3.79 \text{ L}}{\text{gal}} = \$1.46/\text{gal}$

29. $2.1 \text{ kg} \approx 2.1 \text{ kg} \times \dfrac{2.2 \text{ lb}}{\text{kg}} = 4.62 \text{ lb}$

31. Strategy To find the number of pounds lost:
• Multiply to find the number of hours spent hiking.
• Multiply the number of hours spent hiking by the number of extra Calories used in hiking to find the total number of extra Calories used.
• Multiply the number of extra Calories consumed each day by the number of days.
• Subtract to find the difference between the number of Calories used in hiking and the number of extra Calories consumed.
• Divide the difference by 3500.

Solution $5 \times 5 = 25$ h
$25 \times 320 = 8000$ Calories
$5 \times 900 = 4500$ Calories
$8000 - 4500 = 3500$
$\dfrac{3500}{3500} = 1$ lb
Gary will lose 1 lb.

33. $\dfrac{300,000 \text{ km}}{\text{s}} \approx \dfrac{300,000 \text{ km}}{\text{s}} \times \dfrac{1 \text{ mi}}{1.61 \text{ km}}$
$\approx 186,335.4 \text{ mi}/\text{s}$

Applying the Concepts

35. a. False.
 b. False.
 c. True.
 d. True.
 e. False.

CHAPTER REVIEW

1. 1.25 km = 1250 m

2. 0.450 g = 450 mg

3. 0.0056 L = 5.6 ml

4. $1000 \text{ m} \approx 1000 \text{ m} \times \dfrac{1.09 \text{ yd}}{1 \text{ m}} = 1090 \text{ yd}$

5. 79 mm = 7.9 cm

6. 5 m 34 cm = 5 m + 0.34 m = 5.34 m

7. 990 g = 0.990 kg

8. 2550 ml = 2.550 L

9. 4870 m = 4.870 km

10. 0.37 cm = 3.7 mm

11. 6 g 829 mg = 6 g + 0.829 g = 6.829 g

12. $1.2 \text{ L} = 1200 \text{ cm}^3$

13. 4.050 kg = 4050 g

14. 8.7 m = 870 cm

15. $192 \text{ ml} = 192 \text{ cm}^3$

16. 356 mg = 0.356 g

17. 372 cm = 3.72 m

18. 8.3 kl = 8300 L

19. 2 L 89 ml = 2 L + 0.089 L = 2.089 L

20. $5410 \text{ cm}^3 = 5.410 \text{ L}$

21. 3792 L = 3.792 kl

22. $468 \text{ cm}^3 = 468 \text{ ml}$

23. Strategy   To find the amount of the wire left on the roll:
    • Convert the lengths of the three pieces cut from the roll to meters.
    • Add the three numbers.
    • Subtract the sum from the length of the original roll (50 m).

    Solution
    240 cm =   2.40 m
    560 cm =   5.60 m
    480 cm = + 4.80 m
    —————————————————
                12.80 m

        50.0 m
    –   12.8 m
    —————————
        37.2 m

    There are 37.2 m of wire left on the roll.

24. Strategy   To find the total cost:
    • Convert the weights of the packages to kilograms.
    • Add the weights.
    • Multiply the total weight by the cost per kilogram ($5.59).

    Solution
    790 g =   0.790 kg
    830 g =   0.830 kg
    655 g = + 0.655 kg
    —————————————————
              2.275 kg

    2.275 × $5.59 = $12.71725
    The total cost of the chicken is $12.72.

25. $\dfrac{\$3.40}{\text{lb}} \approx \dfrac{\$3.40}{\text{lb}} \times \dfrac{2.2 \text{ lb}}{1 \text{ kg}} = \$7.48/\text{kg}$

26. Strategy   To find how many liters of coffee should be prepared:
    • Convert 400 ml to liters.
    • Multiply the number of guests expected to attend (125) by the number of liters per guest.

    Solution   400 ml = 0.4 L
    0.4 L × 125 = 50 L
    The amount of coffee that should be prepared is 50 L.

27. Strategy   To find the number of Calories that can be eliminated, multiply the number of Calories in one egg (90) by the number of days it is eliminated (30).

    Solution   90 cal × 30 = 2700 Cal
    The number of Calories that can be eliminated is 2700.

28. Strategy   To find the cost of running the TV set:
    • Find the number of hours the TV is used each month by multiplying the number of hours per day (5) by the number of days (30).
    • Find the number of watt-hours by multiplying the number of watts per hour (240) by the total number of hours.
    • Convert watt-hours to kilowatt-hours.
    • Multiply the number of kilowatt-hours by the cost per kilowatt-hour (9.5¢).

    Solution   5 h × 30 = 150 h
    150 h × 240 W = 36,000 Wh
    36,000 Wh = 36 kWh
    36 kWh × ($.095) = $3.42
    The cost of running the TV set is $3.42.

29. $1.90 \text{ kg} = 1.90 \text{ kg} \times \dfrac{2.2 \text{ lb}}{1 \text{ kg}} = 4.18 \text{ lb}$

30. Strategy To find how many hours of cycling are necessary to lose 1 lb, divide 1 lb (3500 cal) by the number of Calories cycling burns per hour (400).

Solution $\dfrac{3500}{400} = 8.75$

The number of hours of cycling needed is 8.75.

31. Strategy To find the profit:
• Convert the amount of soap purchased (6 L) to milliliters:
• Divide the volume of one plastic container (150 ml) into the amount of soap purchased to determine the number of containers of soap for sale.
• Multiply the number of containers by the cost per container ($.26) to find the cost of the containers.
• Multiply the number of liters of soap (6) by the cost per liter ($11.40) to find the cost of the soap.
• Add the cost of the soap and the cost of the containers to find the total cost.
• Multiply the number of containers by $3.29 to find the total revenue.
• Subtract the total cost from the total revenue to find the profit.

Solution 6 L = 6000 ml (Amount of soap)
6000 ÷ 150 = 40 (Number of containers)
40 × $.26 = $10.40 (Cost of containers)
6 × $11.40 = $68.40 (Cost of soap)
$10.40 + $68.40 = $78.80 (Total cost)
40 × $3.29 = $131.60 (Revenue)
$131.60 − $78.80 = $52.80
The profit was $52.80.

32. Strategy To find the number of kilowatt-hours of energy used:
• Multiply 80 W times 2 h times 7 days to find the number of watt-hours used.
• Convert the watt-hours to kilowatt-hours.

Solution 80 × 2 × 7 = 1120 Wh
1120 Wh = 1.120 kWh
The color TV used 1.120 kWh of electricity.

33. Strategy To find the amount of fertilizer:
• Multiply the number of trees (500) by the amount of fertilizer per tree (250 g).
• Convert the grams to kilograms.

Solution 500 × 250 = 125,000 g
125,000 g = 125 kg
The amount of fertilizer used was 125 kg.

CHAPTER TEST

1. 2.96 km = 2960 m

2. 0.378 g = 378 mg

3. 0.046 L = 46 ml

4. 919 cm^3 = 919 ml

5. 42.6 mm = 4.26 cm

6. 7 m 96 cm = 7 m + 0.96 m = 7.96 m

7. 847 g = 0.847 kg

8. 3920 ml = 3.920 L

9. 5885 m = 5.885 km

10. 1.5 cm = 15 mm

11. 3 g 89 mg = 3 g + 0.089 g = 3.089 g

12. 1.6 L = 1600 cm^3

13. 3.29 kg = 3290 g

14. 4.2 m = 420 cm

15. 96 ml = 96 cm^3

16. 1375 mg = 1.375 g

17. 402 cm = 4.02 m

18. 8.92 kl = 8920 L

19. 5 km + 38 m = 5000 m + 38 m = 5038 m

20. 6020 L = 6.020 kl

21. Strategy To find the length of the rafters:
• Multiply the number of rafters (30) by the length of each rafter (380 cm).
• Convert the length in centimeters to meters.

Solution 30 × 380 = 11,400 cm
11,400 cm = 114 m
The length of the rafters is 114 m.

22. Strategy To find the weight of the box of tiles, multiply the weight of one tile (250 g) by the number of tiles in the box (144).

Solution
$$\begin{array}{r} 250 \text{ g} \\ \times\ 144 \\ \hline 36{,}000 \text{ g} = 36 \text{ kg} \end{array}$$
The weight of the box is 36 kg.

23. Strategy To find how many liters of vaccine are needed:
• Multiply the number of people (2600) by the amount of vaccine per flu shot (2 cm^3).
• Convert the total amount of vaccine to liters.

Solution
$$\begin{array}{r} 2600 \\ \times\ \ 2 \text{ cm}^3 \\ \hline 5200 \text{ cm}^3 = 5.2 \text{ L} \end{array}$$
The amount of vaccine needed is 5.2 L.

24. $35 \text{ mi/h} \approx \dfrac{35 \text{ mi}}{\text{h}} \times \dfrac{1.61 \text{ km}}{1 \text{ mi}} \approx 56.4 \text{ km/h}$

25. Strategy To find the distance between the rivets:
• Convert the length of the plate (4.20 m) to centimeters.
• Divide the length of the plate by the number of spaces (24).

Solution
4.20 m = 420 cm
420 ÷ 24 = 17.5 cm
The distance between the rivets is 17.5 cm.

26. Strategy To find how much it costs to fertilize the orchard:
• Find out how much fertilizer is needed by multiplying the number of trees in the orchard (1200) by the amount of fertilizer for each tree (200 g).
• Convert the total amount of fertilizer to kilograms.
• Multiply the number of kilograms of fertilizer by the cost per kilogram ($2.75).

Solution
1200 × 200 = 240,000 g
240,000 g = 240 kg
240 × $2.75 = $660
The cost to fertilize the trees is $660.

27. Strategy To find the cost of the electricity:
• Determine the amount of electricity used by multiplying 1600 W times the hours used per day (4) times the number of days (30).
• Convert the watt-hours to kilowatt hours.
• Multiply the kilowatt-hours by the cost per kilowatt hour ($.085).

Solution
1600 × 4 × 30 = 192,000 Wh
192,000 Wh = 192 kWh
192 × $.085 = $16.32
The total cost is $16.32.

28. Strategy To find how much acid should be ordered:
• Find the amount of acid needed by multiplying the number of classes (3) times the number of students in each class (40) times the amount of acid needed by each student (90).
• Convert the amount to liters.

Solution
3 × 40 × 90 = 10,800 ml
10.8 L
The assistant should order 11 L of the acid.

29. $137 \text{ m} \approx 137 \ \cancel{\text{m}} \times \dfrac{3.28 \text{ ft}}{1 \ \cancel{\text{m}}} = 449.36 \text{ ft}$

30. $110 \text{ in.} \approx 110 \ \cancel{\text{in.}} \times \dfrac{2.54 \text{ cm}}{1 \ \cancel{\text{in.}}} = 279.4 \text{ cm}$

CUMULATIVE REVIEW

1.
$$\begin{aligned} 12 - 8 \div (6-4)^2 \cdot 3 &= 12 - 8 \div 2^2 \cdot 3 \\ &= 12 - 8 \div 4 \cdot 3 \\ &= 12 - 2 \cdot 3 \\ &= 12 - 6 \\ &= 6 \end{aligned}$$

2.
$$\begin{array}{r} 5\dfrac{3}{4} = 5\dfrac{27}{36} \\ 1\dfrac{5}{6} = 1\dfrac{30}{36} \\ +\ 4\dfrac{7}{9} = 4\dfrac{28}{36} \\ \hline 10\dfrac{85}{36} = 12\dfrac{13}{36} \end{array}$$

3.
$$\begin{array}{r} 4\dfrac{2}{9} = 4\dfrac{8}{36} = 3\dfrac{44}{36} \\ -\ 3\dfrac{5}{12} = 3\dfrac{15}{36} = 3\dfrac{15}{36} \\ \hline \dfrac{29}{36} \end{array}$$

4. $5\dfrac{3}{8} \div 1\dfrac{3}{4} = \dfrac{43}{8} \div \dfrac{7}{4}$

$= \dfrac{43}{8} \times \dfrac{4}{7}$

$= \dfrac{43 \cdot \overset{1}{2} \cdot \overset{1}{2}}{\underset{1}{2} \cdot \underset{1}{2} \cdot 2 \cdot 7} = \dfrac{43}{14} = 3\dfrac{1}{14}$

5. $\left(\dfrac{2}{3}\right)^4 \cdot \left(\dfrac{9}{4}\right)^2 = \left(\dfrac{2}{3} \cdot \dfrac{2}{3} \cdot \dfrac{2}{3} \cdot \dfrac{2}{3}\right)\left(\dfrac{9}{4} \cdot \dfrac{9}{4}\right) = \dfrac{16}{81} \cdot \dfrac{81}{16} = 1$

6. $\begin{array}{r} 12.0072 \\ -\ 9.937 \\ \hline 2.0702 \end{array}$

7. $\dfrac{5}{8} = \dfrac{n}{50}$

$5 \times 50 = 8 \times n$

$250 = 8 \times n$

$250 \div 8 = n$

$n = 31.3$

8. $1\dfrac{3}{4} = \dfrac{7}{4} \times 100\% = \dfrac{700}{4}\% = 175\%$

9. $4.2\% \times n = 6.09$

$0.042 \times n = 6.09$

$n = 6.09 \div 0.042 = 145$

10. $18\ \cancel{pt} \times \dfrac{1\ \cancel{qt}}{2\ \cancel{pt}} \times \dfrac{1\ \text{gal}}{4\ \cancel{qt}} = \dfrac{18}{8}\ \text{gal} = 2.25\ \text{gal}$

11. $875\ \text{cm} = 8.75\ \text{m}$

12. $3420\ \text{m} = 3.420\ \text{km}$

13. $5.05\ \text{kg} = 5050\ \text{g}$

14. $3\ \text{g}\ 672\ \text{mg} = 3\ \text{g} + 0.672\ \text{g} = 3.672\ \text{g}$

15. $6\ \text{L} = 6000\ \text{ml}$

16. $2.4\ \text{kl} = 2400\ \text{L}$

17. Strategy   To find how much money is left after the rent is paid:
• Find the amount that is paid in rent by multiplying $\dfrac{1}{4}$ by the total monthly income ($3244).
• Subtract the amount paid in rent from the total monthly income.

Solution $\dfrac{1}{4} \times \$3244 = \dfrac{\$3244}{4} = \$811$

$\begin{array}{r} \$3244 \\ -\ 811 \\ \hline \$2433 \end{array}$

$2433 is left after the rent is paid.

18. Strategy   To find the amount of income tax paid:
• Find the amount of income tax paid on the profit by multiplying 0.08 by the profit ($82,340).
• Add $620 to the amount of income tax paid on the profit.

Solution $0.08 \times \$82{,}340 = \$6587.20$

$\begin{array}{r} \$6587.20 \\ +\ 620.00 \\ \hline \$7207.20 \end{array}$

The business paid $7207.20 in income tax.

19. Strategy   To find the property tax, solve a proportion.

Solution $\dfrac{\$900}{\$45{,}000} = \dfrac{n}{\$75{,}000}$

$900 \times 75{,}000 = 45{,}000 \times n$

$67{,}500{,}000 = 45{,}000 \times n$

$67{,}500{,}000 \div 45{,}000 = n$

$1500 = n$

The property tax is $1500.

20. Strategy   To find the rebate, solve the basic percent equation for amount. The base is $23,500 and the rate is 12%.

Solution   Percent × base = amount

$12\% \times 23{,}500 = n$

$0.12 \times 23{,}500 = n = 2820$

The car buyer will receive a rebate of $2820.

21. Strategy   To find the percent, solve the basic percent equation for percent. The base is $8200 and the amount is $533.

Solution   Percent × base = amount

$n \times 8200 = 533$

$n = 533 \div 8200$

$n = 0.065 = 6.5\%$

The percent is 6.5%.

22. Strategy   To find your average grade, find the sum of the grades and divide the sum by the number of grades (5).

Solution
$\begin{array}{r} 78 \\ 92 \\ 45 \\ 80 \\ +\ 85 \\ \hline 380 \end{array}$ sum of grades

$5\overline{)380}\ \ \ 76$

Your average grade is 76.

23. Strategy   To find what the salary will be next
year:
• Find the amount of the increase by
solving the basic percent equation for
amount. The base is $22,500 and the
percent is 12%.

Solution   Percent × base = amount
$12\% \times 22,500 = n$ $22,500
$0.12 \times 22,500 = n$ + 2,700
$2700 = n$ $25,200
Karla's salary next year will be
$25,200.

24. Strategy   To find the discount rate:
• Find the amount of the discount by
subtracting the sale price ($62.40)
from the original price ($80).
• Solve the basic percent equation for
percent. The base is the original price
($80) and the amount is the amount of
the discount.

Solution $80.00
− 62.40
$17.60
Percent × base = amount
$n \times 80 = 17.60$
$n = 17.60 \div 80$
$n = 0.22 = 22\%$
The discount rate is 22%.

25. Strategy   To find the length of the wall:
• Convert 9 in. to feet.
• Multiply the length of one block (9
in.) by the number of blocks (48).

Solution $9 \text{ in.} = 9 \text{ in.} \times \dfrac{1 \text{ ft}}{12 \text{ in.}} = \dfrac{9}{12} \text{ ft} = 0.75 \text{ ft}$

48
× 0.75 ft
36 ft
The length of the wall is 36 ft.

26. Strategy   To find the number of quarts:
• Find the total amount of juice by
multiplying the amount in a jar (24 oz)
by the number of jars in a case (16).
• Convert the ounce amount of juice to
quarts.

Solution   24 oz
× 16
384 oz

$384 \text{ oz} \times \dfrac{1 \text{ c}}{8 \text{ oz}} \times \dfrac{1 \text{ pt}}{2 \text{ c}} \times \dfrac{1 \text{ qt}}{2 \text{ pt}}$

$= \dfrac{384}{32} \text{ qt} = 12 \text{ qt}$

There are 12 qt of apple juice in the
case.

27. Strategy   To find the profit:
• Convert the amount of oil to quarts.
• Find the cost by multiplying the
number of gallons (40) by the cost per
gallon ($4.88).
• Find the revenue by multiplying the
number of quarts by the selling price
per quart ($1.99).
• Subtract the cost from the revenue.

Solution   40 gal = 160 quarts
$40 \times \$4.88 = \$195.20$ (cost)
$160 \times \$1.99 = \$318.40$ (Revenue)
$\$318.40 - \$195.20 = \$123.20$
The profit was $123.20.

28. Strategy   To find the amount of chlorine used:
• Convert the amount of chlorine used
to liters.
• Multiply the amount used each day
by the number of days (20).

Solution   1200 ml = 1.2 L
$1.2 \text{ L} \times 20 = 24 \text{ L}$
24 L of chlorine was used.

29. Strategy   To find how much it costs to operate
the hairdryer:
• Find how many hours the hair dryer
is used by multiplying the amount used
each day $\left(\dfrac{1}{2} \text{ h} \right)$ by the number of
days (30).
• Find the watt-hours by multiplying
the number of watts (1200) by the
number of hours.
• Convert watt-hours to kilowatt-
hours.
• Multiply the number of kilowatt-
hours by the cost per kilowatt-hour
(10.5¢).

Solution $30 \times \dfrac{1}{2} \text{ h} = 15 \text{ h}$

$1200 \text{ W} \times 15 \text{ h} = 18,000 \text{ Wh}$
$18,000 \text{ Wh} = 18 \text{ kWh}$
$18 \text{ kWh} \times \$0.105 = \$1.89$
The total cost of operating the hair
dryer is $1.89.

30. $\dfrac{60 \text{ mi}}{1 \text{ h}} = \dfrac{1.61 \text{ km}}{1 \text{ mi}} = 96.6 \text{ km/h}$

Chapter 10: Rational Numbers

PREP TEST and GO FIGURE

1. $54 > 45$

2. 4

3. $15,847$

4. 3779

5. $26,432$

6. $\dfrac{144}{24} = \dfrac{2 \cdot \overset{1}{\cancel{2}} \cdot \overset{1}{\cancel{2}} \cdot \overset{1}{\cancel{2}} \cdot 3 \cdot \overset{1}{\cancel{3}}}{\underset{1}{\cancel{2}} \cdot \underset{1}{\cancel{2}} \cdot \underset{1}{\cancel{2}} \cdot \underset{1}{\cancel{3}}} = 6$

7. $\dfrac{2}{3} + \dfrac{3}{5} = \dfrac{10}{15} + \dfrac{9}{15}$
   $= \dfrac{19}{15} = 1\dfrac{4}{15}$

8. $\dfrac{3}{4} - \dfrac{5}{16} = \dfrac{12}{16} - \dfrac{5}{16}$
   $= \dfrac{7}{16}$

9. 11.058

10. 3.781

11. $\dfrac{3}{4} \times \dfrac{8}{15} = \dfrac{\overset{1}{\cancel{3}} \cdot \overset{1}{\cancel{2}} \cdot \overset{1}{\cancel{2}} \cdot 2}{\underset{1}{\cancel{2}} \cdot \underset{1}{\cancel{2}} \cdot \underset{1}{\cancel{3}} \cdot 5} = \dfrac{2}{5}$

12. $\dfrac{5}{12} \div \dfrac{3}{4} = \dfrac{5}{12} \cdot \dfrac{4}{3}$
    $= \dfrac{5 \cdot \overset{1}{\cancel{2}} \cdot \overset{1}{\cancel{2}}}{\underset{1}{\cancel{2}} \cdot \underset{1}{\cancel{2}} \cdot 3 \cdot 3} = \dfrac{5}{9}$

13. 9.4

14. $2.4\,\overline{)0.9{,}6}\;\;^{0.4}$

15. $(8-6)^2 + 12 \div 4 \cdot 3^2 = 2^2 + 12 \div 4 \cdot 9$
    $= 4 + 3 \cdot 9$
    $= 4 + 27$
    $= 31$

Go Figure

If it takes one loaf 30 minutes to fill the oven, doubling in volume each minute, then at 29 minutes, the oven would be half filled. At 28 minutes, the oven would be one quarter filled with one loaf. So for the oven to be half filled with two loaves of bread, would be at the 28-minute mark.

SECTION 10.1

Objective A Exercises

1. -120 ft

3. $+2$ dollars

5.

7.

9. $-2 > -5$

11. $-16 < 1$

13. $3 > -7$

15. $-11 < -8$

17. $35 > 28$

19. $-42 < 27$

21. $21 > -34$

23. $-27 > -39$

25. $-87 < 63$

27. $86 > -79$

29. $-62 > -84$

31. $-131 < 101$

Objective B Exercises

33. -4

35. 2

37. -22

39. 31

41. -70

43. $|2| = 2$

45. $|-6| = 6$

47. $|8| = 8$

49. $|-9| = 9$

51. $-|-1| = -1$

53. $-|0| = 0$

55. $|19| = 19$

57. $|-22| = 22$

59. $-|20| = -20$

61. $-|-18| = -18$

63. $|-23| = 23$

65. $-|27| = -27$

67. $|25| = 25$

69. $|-74| = 74$

71. $-|88| = -88$

Applying the Concepts

73. New York

75. a. −6 is halfway between −7 and −5.

   b. −8 is halfway between −10 and −6.

   c. −9 and −6 are both one-third of the way between −12 and −3.

SECTION 10.2

Objective A Exercises

1. −14, −364

3. $3 + (-5) = -2$

5. $8 + 12 = 20$

7. $-3 + (-8) = -11$

9. $-4 + (-5) = -9$

11. $6 + (-9) = -3$

13. $-6 + 7 = 1$

15. $2 + (-3) + (-4) = -1 + (-4) = -5$

17. $-3 + (-12) + (-15) = -15 + (-15) = -30$

19. $-17 + (-3) + 29 = -20 + 29 = 9$

21. $-3 + (-8) + 12 = -11 + 12 = 1$

23. $13 + (-22) + 4 + (-5) = -9 + 4 + (-5)$
$= -5 + (-5) = -10$

25. $-22 + 10 + 2 + (-18) = -12 + 2 + (-18)$
$= -10 + (-18) = -28$

27. $-16 + (-17) + (-18) + 10 = -33 + (-18) + 10$
$= -51 + 10 = -41$

29. $-126 + (-247) + (-358) + 339$
$= -373 + (-358) + 339$
$= -731 + 339$
$= -392$

31. $-12 + (-8) = -20$

33. $-7 + (-16) = -23$

35. $-4 + 2 = -2$

37. $-2 + 8 + (-12) = 6 + (-12) = -6$

39. $2 + (-3) + 8 + (-13) = -1 + 8 + (-13)$
$= 7 + (-13) = -6$

Objective B Exercises

41. Negative six minus positive four

43. Positive six minus negative four

45. $9 + 5$

47. $1 + (-8)$

49. $16 - 8 = 16 + (-8) = 8$

51. $7 - 14 = 7 + (-14) = -7$

53. $-7 - 2 = -7 + (-2) = -9$

55. $7 - (-29) = 7 + 29 = 36$

57. $-6 - (-3) = -6 + 3 = -3$

59. $6 - (-12) = 6 + 12 = 18$

61. $-4 - 3 - 2 = -4 + (-3) + (-2)$
$= -7 + (-2) = -9$

63. $12 - (-7) - 8 = 12 + 7 + (-8)$
$= 19 + (-8) = 11$

65. $4 - 12 - (-8) = 4 + (-12) + 8$
$= -8 + 8 = 0$

67. $-6 - (-8) - (-9) = -6 + 8 + 9$
$= 2 + 9 = 11$

69. $-30 - (-65) - 29 - 4 = -30 + 65 + (-29) + (-4)$
$= 35 + (-29) + (-4)$
$= 6 + (-4) = 2$

71. $-16 - 47 - 63 - 12 = -16 + (-47) + (-63) + (-12)$
$= -63 + (-63) + (-12)$
$= -126 + (-12)$
$= -138$

73. $47 - (-67) - 13 - 15 = 47 + 67 + (-13) + (-15)$
$= 114 + (-13) + (-15)$
$= 101 + (-15) = 86$

75. $167 - 432 - (-287) - 359$
$= 167 + (-432) + 287 + (-359)$
$= -265 + 287 + (-359)$
$= 22 + (-359)$
$= -337$

77. $-4 - (-8) = -4 + 8 = 4$

79. $-8 - 4 = -8 + (-4) = -12$

81. $-4 - 8 = -4 + (-8) = -12$

83. $1 - (-2) = 1 + 2 = 3$

Objective C Exercises

85. Strategy   To find the temperature, add the increase (7) to the previous temperature (−8).

   Solution $-8 + 7 = -1$
   The temperature is −1°C.

87. **Strategy** To find Nick's score, subtract 26 points from his original score (11).

Solution $11 - 26 = 11 + (-26) = -15$
Nick's score was -15 points after his opponent shot the moon.

89. **Strategy** To find the price of Byplex stock add the change in price for each day of the week.

Solution $-2 + (-3) + (-1) + (-2) + (-1)$
$= -5 + (-1) + (-2) + (-1)$
$= -6 + (-2) + (-1)$
$= (-8) + (-1)$
$= -9$
The change in the price of the stock is -9 dollars.

91. **Strategy** To find the difference in temperature, subtract the temperature in Earth's stratosphere (-70) from the temperature of Earth's surface (45).

Solution $45 - (-70) = 45 + 70 = 115$
The difference is $115°$F.

93. **Strategy** To find the difference in elevation, subtract the elevation of Valdes Peninsula (-131) from the elevation of Mt. Aconcagua (22,834).

Solution $22,834 - (-131) = 22,834 + 131$
$= 22,965$
The difference in elevation is 22,965 ft.

95. **Strategy** To find the difference between the highest and lowest temperature in Africa, subtract the lowest temperature (-24) from the highest temperature (58).

Solution $58 - (-24) = 58 + 24 = 82$
The difference is $82°$C.

97. **Strategy** To find the difference, subtract the lowest temperature in Asia (-68) from the lowest temperature in Europe (-55).

Solution $-55 - (-68) = -55 + 68 = 13$
The difference in temperature is $13°$C.

Applying the Concepts

99.

-3	2	1
4	0	-4
-1	-2	3

101. Students should include in their description the fact that "minus" refers to the operation of subtraction, whereas "negative" refers to the sign of a number.

SECTION 10.3

Objective A Exercises

1. Subtraction

3. Multiplication

5. $14 \times 3 = 42$

7. $-4 \cdot 6 = -24$

9. $-2 \cdot (-3) = 6$

11. $(9)(2) = 18$

13. $5(-4) = -20$

15. $-8(2) = -16$

17. $(-5)(-5) = 25$

19. $(-7)(0) = 0$

21. $-24 \times 3 = -72$

23. $6(-17) = -102$

25. $-4(-35) = 140$

27. $-6 \cdot (38) = -228$

29. $8(-40) = -320$

31. $-4(39) = -156$

33. $5 \times 7 \times (-2) = 35 \times (-2) = -70$

35. $(-9)(-9)(2) = 81(2) = 162$

37. $-5(8)(-3) = -40(-3) = 120$

39. $-1(4)(-9) = -4(-9) = 36$

41. $4(-4) \cdot 6(-2) = -16 \cdot 6(-2) = -96(-2) = 192$

43. $-9(4) \cdot 3(1) = -36 \cdot 3(1) = -108(1) = -108$

45. $(-6) \cdot 7 \cdot (-10)(-5) = -42 \cdot (-10)(-5)$
$= 420(-5)$
$= -2100$

47. $-5(-4) = 20$

49. $-8(6) = -48$

51. $-4(7)(-5) = 140$

Objective B Exercises

53. $3(-12) = -36$

55. $-5(11) = -55$

57. $12 \div (-6) = -2$

59. $(-72) \div (-9) = 8$

61. $0 \div (-6) = 0$

63. $45 \div (-5) = -9$

65. $-36 \div 4 = -9$

67. $-81 \div (-9) = 9$

69. $72 \div (-3) = -24$

71. $(-60) \div 5 = -12$

73. $-93 \div (-3) = 31$

75. $(-85) \div (-5) = 17$

77. $120 \div 8 = 15$

79. $78 \div (-6) = -13$

81. $-72 \div 4 = -18$

83. $-114 \div (-6) = 19$

85. $-104 \div (-8) = 13$

87. $57 \div (-3) = -19$

89. $-136 \div (-8) = 17$

91. $-130 \div (-5) = 26$

93. $(-92) \div (-4) = 23$

95. $-150 \div (-6) = 25$

97. $204 \div (-6) = -34$

99. $-132 \div (-12) = 11$

101. $-182 \div 14 = -13$

103. $143 \div 11 = 13$

105. $-180 \div (-15) = 12$

107. $154 \div (-11) = -14$

109. $\dfrac{182}{-13} = -14$

111. $\dfrac{144}{-24} = -6$

113. $\dfrac{-88}{22} = -4$

Objective C Application Problems

115. Strategy   To find the average daily high temperature:
- Add the seven temperature readings.
- Divide by 7.

Solution $-6 + (-11) + 1 + 5 + (-3) + (-9) + (-5)$
$= -17 + 1 + 5 + (-3) + (-9) + (-5)$
$= -16 + 5 + (-3) + (-9) + (-5)$
$= -11 + (-3) + (-9) + (-5)$
$= -14 + (-9) + (-5)$
$= -23 + (-5)$
$= -28$

$-28 \div 7 = -4$
The average high temperature was $-4°$.

117. Strategy   To find the average score, divide the combined scores (-20) by the number of golfers (10).

Solution $-20 \div 10 = -2$
The average score was -2.

119. Strategy   To find the wind chill factor multiply the wind chill factor at 25°F with a 35 mph wind (-12) by 5.

Solution $-12 \times 5 = -60$
The wind chill factor is $-60°$F.

Applying the Concepts

121. a. The largest possible product is that of (-5) and (-5). $(-5)(-5) = 25$
All other combinations, (-1) and (-9), (-2) and (-8), (-3) and (-7), and (-4) and (-6) have a product less than 25.

     b. The smallest possible sum is -17.
$(-1) + (-16) = -17$ (the sum)
$-1(-16) = 16$ (the product)
Other combinations are (-2) and (-8), and (-4) and (-4), which have a product of 16 and a sum greater than -17.

123. a. True

     b. True

SECTION 10.4

Objective A Exercises

1. $\dfrac{5}{8} - \dfrac{5}{6} = \dfrac{15}{24} - \dfrac{20}{24}$
$= \dfrac{15}{24} + \dfrac{(-20)}{24}$
$= \dfrac{15 + (-20)}{24}$
$= -\dfrac{5}{24}$

3. $-\dfrac{5}{12}-\dfrac{3}{8}=\dfrac{-10}{24}-\dfrac{9}{24}$

$\quad=\dfrac{-10}{24}+\dfrac{(-9)}{24}$

$\quad=\dfrac{-10+(-9)}{24}=\dfrac{-19}{24}=-\dfrac{19}{24}$

5. $-\dfrac{6}{13}+\dfrac{17}{26}=\dfrac{-12}{26}+\dfrac{17}{26}$

$\quad=\dfrac{-12+17}{26}=\dfrac{5}{26}$

7. $-\dfrac{5}{8}-\left(-\dfrac{11}{12}\right)=\dfrac{-15}{24}-\left(\dfrac{-22}{24}\right)$

$\quad=\dfrac{-15}{24}+\dfrac{22}{24}=\dfrac{-15+22}{24}=\dfrac{7}{24}$

9. $\dfrac{5}{12}-\dfrac{11}{15}=\dfrac{25}{60}-\dfrac{44}{60}$

$\quad=\dfrac{25}{60}+\dfrac{(-44)}{60}=\dfrac{25+(-44)}{60}=\dfrac{-19}{60}=-\dfrac{19}{60}$

11. $-\dfrac{3}{4}-\dfrac{5}{8}=\dfrac{-6}{8}-\dfrac{5}{8}$

$\quad=\dfrac{-6}{8}+\dfrac{(-5)}{8}$

$\quad=\dfrac{-6+(-5)}{8}=\dfrac{-11}{8}=-1\dfrac{3}{8}$

13. $-\dfrac{5}{2}-\left(-\dfrac{13}{4}\right)=\dfrac{-10}{4}-\left(\dfrac{-13}{4}\right)$

$\quad=\dfrac{-10}{4}+\dfrac{13}{4}=\dfrac{-10+13}{4}=\dfrac{3}{4}$

15. $-\dfrac{3}{8}-\dfrac{5}{12}-\dfrac{3}{16}=\dfrac{-18}{48}-\dfrac{20}{48}-\dfrac{9}{48}$

$\quad=\dfrac{-18}{48}+\dfrac{(-20)}{48}+\dfrac{(-9)}{48}$

$\quad=\dfrac{-18+(-20)+(-9)}{48}=\dfrac{-47}{48}=-\dfrac{47}{48}$

17. $\dfrac{1}{2}-\dfrac{3}{8}-\left(-\dfrac{1}{4}\right)=\dfrac{4}{8}-\dfrac{3}{8}-\left(\dfrac{-2}{8}\right)$

$\quad=\dfrac{4}{8}+\dfrac{(-3)}{8}+\dfrac{2}{8}$

$\quad=\dfrac{4+(-3)+2}{8}=\dfrac{3}{8}$

19. $\dfrac{1}{3}-\dfrac{1}{4}-\dfrac{1}{5}=\dfrac{20}{60}-\dfrac{15}{60}-\dfrac{12}{60}$

$\quad=\dfrac{20}{60}+\dfrac{(-15)}{60}+\dfrac{(-12)}{60}$

$\quad=\dfrac{20+(-15)+(-12)}{60}=\dfrac{-7}{60}=-\dfrac{7}{60}$

21. $\dfrac{1}{2}+\left(-\dfrac{3}{8}\right)+\dfrac{5}{12}=\dfrac{12}{24}+\dfrac{(-9)}{24}+\dfrac{10}{24}$

$\quad=\dfrac{12+(-9)+10}{24}=\dfrac{13}{24}$

23. $3.4+(-6.8)=-3.4$

25. $-8.32+(-0.57)=-8.89$

27. $-4.8+(-3.2)=-8.0$

29. $-4.6+3.92=-0.68$

31. $-45.71+(-135.8)=-181.51$

33. $4.2+(-6.8)+5.3=-2.6+5.3=2.7$

35. $-4.5+3.2+(-19.4)=-1.3+(-19.4)=-20.7$

37. $-18.39+4.9-23.7=-18.39+4.9+(-23.7)$

$\qquad=-13.49+(-23.7)=-37.19$

39. $-3.09-4.6-27.3=-3.09+(-4.6)+(-27.3)$

$\qquad=-7.69+(-27.3)=-34.99$

41. $-4.02+6.809-(-3.57)-(-0.419)$

$=-4.02+6.809+3.57+0.419$

$=2.789+3.57+0.419$

$=6.359+0.419=6.778$

43. $0.27+(-3.5)-(-0.27)+(-5.44)$

$=0.27+(-3.5)+0.27+(-5.44)$

$=-3.23+0.27+(-5.44)$

$=-2.96+(-5.44)$

$=-8.4$

Objective B Exercises

45. $-\dfrac{2}{9}\times\left(-\dfrac{3}{14}\right)=\dfrac{2\cdot3}{9\cdot14}=\dfrac{6}{126}=\dfrac{2}{42}=\dfrac{1}{21}$

47. $\left(-\dfrac{3}{4}\right)\left(-\dfrac{8}{27}\right)=\dfrac{3\cdot8}{4\cdot27}=\dfrac{24}{108}=\dfrac{2}{9}$

49. $\dfrac{5}{12}\times\left(-\dfrac{8}{15}\right)=-\left(\dfrac{5\cdot8}{12\cdot15}\right)=-\dfrac{40}{180}=-\dfrac{2}{9}$

51. $\left(\dfrac{3}{8}\right)\left(-\dfrac{15}{41}\right)=-\left(\dfrac{3\cdot15}{8\cdot41}\right)=-\dfrac{45}{328}$

53. $\left(-\dfrac{5}{7}\right)\left(-\dfrac{14}{15}\right)=\dfrac{5\cdot14}{7\cdot15}=\dfrac{70}{105}=\dfrac{2}{3}$

55. $\left(\dfrac{1}{2}\right)\left(-\dfrac{3}{4}\right)\left(-\dfrac{5}{8}\right)=\dfrac{1\cdot3\cdot5}{2\cdot4\cdot8}=\dfrac{15}{64}$

57. $-\dfrac{3}{8}\div\dfrac{7}{8}=-\dfrac{3}{8}\times\dfrac{8}{7}=-\dfrac{3}{7}$

59. $\dfrac{5}{6}\div\left(-\dfrac{3}{4}\right)=\dfrac{5}{6}\times\left(-\dfrac{4}{3}\right)$

$\quad=-\left(\dfrac{5\cdot4}{6\cdot3}\right)=-\dfrac{20}{18}=-\dfrac{10}{9}=-1\dfrac{1}{9}$

61. $-\dfrac{5}{16}\div\left(-\dfrac{3}{8}\right)=-\dfrac{5}{16}\times-\dfrac{8}{3}=\dfrac{5\cdot8}{16\cdot3}=\dfrac{40}{48}=\dfrac{5}{6}$

63. $-\dfrac{8}{19}\div\dfrac{7}{38}=-\dfrac{8}{19}\times\dfrac{38}{7}$

$\quad=-\left(\dfrac{8\cdot38}{19\cdot7}\right)=-\dfrac{304}{133}=-\dfrac{16}{7}=-2\dfrac{2}{7}$

65. $-6 \div \dfrac{4}{9} = -\dfrac{6}{1} \times \dfrac{9}{4}$

$= -\left(\dfrac{6 \cdot 9}{1 \cdot 4}\right) = -\dfrac{54}{4} = -\dfrac{27}{2} = -13\dfrac{1}{2}$

67. $-8.9 \times (-3.5) = 8.9 \times 3.5 = 31.15$

$$
\begin{array}{r}
8.9 \\
\times\ 3.5 \\
\hline
445 \\
267 \\
\hline
31.15
\end{array}
$$

69. $-14.3 \times 7.9 = -(14.3 \times 7.9) = -112.97$

$$
\begin{array}{r}
14.3 \\
\times\ 7.9 \\
\hline
1287 \\
1001 \\
\hline
112.97
\end{array}
$$

71. $(-1.21)(-0.03) = (1.21)(0.03) = 0.0363$

$$
\begin{array}{r}
1.21 \\
\times\ 0.03 \\
\hline
0.0363
\end{array}
$$

73. $-77.6 \div (-0.8) = 77.6 \div 0.8 = 97$

$$
\begin{array}{r}
97. \\
0.8\overline{)77.6} \\
\underline{72} \\
56 \\
\underline{-56} \\
0
\end{array}
$$

75. $(-7.04) \div (-3.2) = 7.04 \div 3.2 = 2.2$

$$
\begin{array}{r}
2.2 \\
3.2\overline{)7.04} \\
\underline{-64} \\
64 \\
\underline{-64} \\
0
\end{array}
$$

77. $-3.312 \div (0.8) = -(3.312 \div 0.8) = -4.14$

$$
\begin{array}{r}
4.14 \\
0.8\overline{)3.312} \\
\underline{-3\,2} \\
11 \\
\underline{-8} \\
32 \\
\underline{-32} \\
0
\end{array}
$$

79. $26.22 \div (-6.9) = -(26.22 \div 6.9) = -3.8$

$$
\begin{array}{r}
3.8 \\
6.9\overline{)26.22} \\
\underline{-20\,7} \\
552 \\
\underline{-552} \\
0
\end{array}
$$

81. $21.792 \div (-0.96) = -(21.792 \div 0.96) = -22.70$

$$
\begin{array}{r}
22.70 \\
0.96\overline{)21.7920} \\
\underline{-19\,2} \\
2\,59 \\
\underline{-192} \\
672 \\
\underline{-672} \\
0 \\
\underline{-0} \\
0
\end{array}
$$

83. $-3.171 \div (-45.3) = 3.171 \div 45.3 = 0.07$

$$
\begin{array}{r}
.07 \\
45.3\overline{)3.171} \\
\underline{-3\,171} \\
0
\end{array}
$$

85. $(-13.97) \div (-25.4) = 13.97 \div 25.4 = 0.55$

$$
\begin{array}{r}
0.55 \\
25.4\overline{)13.970} \\
\underline{-1270} \\
1270 \\
\underline{-1270} \\
0
\end{array}
$$

Objective C Application Problems

87. Strategy To find the amount the temperature fell from 9:00 A.M. subtract the temperature at 9:27 A.M. from the temperature at 9:00 A.M.

Solution $12.22 - (-20) = 32.22$
The temperature fell 32.22°C in 27 minutes.

89. Strategy To find the difference, subtract the melting point (–218.4) from the boiling point of oxygen (–182.962).

Solution $-182.962 - (-218.4)$
$= -182.962 + 218.4$
$= 35.438$
The difference between the boiling point and the melting point of oxygen is 35.438°C.

91a. Strategy To find the closing price on the previous day, subtract the change in price $\left(-\frac{1}{2}\right)$ from the closing price on October 6 $\left(24\frac{13}{16}\right)$.

Solution

$$24\frac{13}{16} = 24\frac{13}{16} = 24\frac{13}{16}$$
$$-\left(-\frac{1}{2}\right) = +\frac{8}{16} = \quad +\frac{8}{16}$$
$$\overline{\qquad\qquad\qquad 24\frac{21}{16} = 25\frac{5}{16}}$$

The closing price on October 5 was $\$25\frac{5}{16}$.

91b. Strategy To find the closing price on the previous day, subtract the change in price $\left(+\frac{5}{8}\right)$ from the closing price on October 6 $\left(27\frac{1}{4}\right)$.

Solution

$$27\frac{1}{4} = 27\frac{2}{8} = 26\frac{10}{8}$$
$$-\frac{5}{8} = \quad -\frac{5}{8} = \quad -\frac{5}{8}$$
$$\overline{\qquad\qquad\qquad\qquad 26\frac{5}{8}}$$

The closing price on October 5 was $\$26\frac{5}{8}$.

Applying the Concepts

93. $-\frac{17}{24}$ is one example

95. Given any two different rational numbers, it is always possible to find a rational number between them. One method is to add the two numbers and divide by 2.
Another method is to add the numerators and add the denominators. For example, given the fractions $\frac{2}{5}$ and $\frac{3}{4}$, $\frac{2+3}{5+4} = \frac{5}{9}$ and $\frac{2}{5} < \frac{5}{9} < \frac{3}{4}$.

SECTION 10.5

Objective A Exercises

1. Since the number is greater than 10, move the decimal point 6 places to the left. The exponent on 10 is 6.
$2,370,000 = 2.37 \times 10^6$

3. Since the number is less than 10, move the decimal point 4 places to the right. The exponent on 10 is –4.
$0.00045 = 4.5 \times 10^{-4}$

5. Since the number is greater than 10, move the decimal point 5 places to the left. The exponent on 10 is 5.
$309,000 = 3.09 \times 10^5$

7. Since the number is less than 10, move the decimal point 7 places to the right. The exponent on 10 is –7.
$0.000000601 = 6.01 \times 10^{-7}$

9. Since the number is greater than 10, move the decimal point 10 places to the left. The exponent on 10 is 10.
$57,000,000,000 = 5.7 \times 10^{10}$

11. Since the number is less than 10, move the decimal point 8 places to the right. The exponent on 10 is –8.
$0.000000017 = 1.7 \times 10^{-8}$

13. The exponent on 10 is positive. Move the decimal point 5 places to the right.
$7.1 \times 10^5 = 710,000$

15. The exponent on 10 is negative. Move the decimal point 5 places to the left.
$4.3 \times 10^{-5} = 0.000043$

17. The exponent on 10 is positive. Move the decimal point 8 places to the right.
$6.71 \times 10^8 = 671,000,000$

19. The exponent on 10 is negative. Move the decimal point 6 places to the left.
$7.13 \times 10^{-6} = 0.00000713$

21. The exponent on 10 is positive. Move the decimal point 12 places to the right.
$5 \times 10^{12} = 5,000,000,000,000$

23. The exponent on 10 is negative. Move the decimal point 3 places to the left.
$8.01 \times 10^{-3} = 0.00801$

25. The number is greater than 10. Move the decimal point 10 places to the left. The exponent on 10 is 10.
$16,000,000,000 \text{ mi} = 1.6 \times 10^{10} \text{ mi}$

27. The monetary cost is 3.1 trillion. The exponent on 10 is 12.
$\$3.1 \times 10^{12}$

29. The number is less than 10. Move the decimal point 6 places to the right. The exponent on 10 is –6.
$0.0000037 \text{ m} = 3.7 \times 10^{-6} \text{ m}$

Objective B Exercises

31. $8 \div 4 + 2 = 2 + 2 = 4$

33. $4 + (-7) + 3 = -3 + 3 = 0$

35. $4^2 - 4 = 16 - 4 = 16 + (-4) = 12$

37. $2 \times (3 - 5) - 2 = 2 \times [3 + (-5)] + (-2)$
$= 2 \times (-2) + (-2)$
$= -4 + (-2) = -6$

39. $4 - (-3)^2 = 4 - 9 = 4 + (-9) = -5$

41. $4 - (-3) - 5 = 4 + 3 + (-5)$
$= 7 + (-5) = 2$

43. $4 - (-2)^2 + (-3) = 4 - 4 + (-3)$
$= 4 + (-4) + (-3)$
$= 0 + (-3) = -3$

45. $3^2 - 4 \times 2 = 9 - 4 \times 2$
$= 9 - 8 = 9 + (-8) = 1$

47. $3 \times (6 - 2) \div 6 = 3 \times [6 + (-2)] \div 6$
$= 3 \times 4 \div 6 = 12 \div 6 = 2$

49. $2^2 - (-3)^2 + 2 = 4 - 9 + 2$
$= 4 + (-9) + 2$
$= -5 + 2 = -3$

51. $6 - 2 \times (1 - 5) = 6 - 2 \times [1 + (-5)]$
$= 6 - 2 \times (-4)$
$= 6 - (-8) = 6 + 8 = 14$

53. $(-2)^2 - (-3)^2 + 1 = 4 - 9 + 1$
$= 4 + (-9) + 1$
$= -5 + 1 = -4$

55. $6 - (-3) \times (-3)^2 = 6 - (-3) \times 9$
$= 6 - (-27) = 6 + 27 = 33$

57. $4 \times 2 - 3 \times 7 = 8 - 3 \times 7$
$= 8 - 21 = 8 + (-21) = -13$

59. $(-2)^2 - 5 \times 3 - 1 = 4 - 5 \times 3 - 1$
$= 4 - 15 - 1$
$= 4 + (-15) + (-1)$
$= -11 + (-1) = -12$

61. $7 \times 6 - 5 \times 6 + 3 \times 2 - 2 + 1$
$= 42 - 5 \times 6 + 3 \times 2 - 2 + 1$
$= 42 - 30 + 3 \times 2 - 2 + 1$
$= 42 - 30 + 6 - 2 + 1$
$= 42 + (-30) + 6 + (-2) + 1$
$= 12 + 6 + (-2) + 1$
$= 18 + (-2) + 1$
$= 16 + 1 = 17$

63. $-4 \times 3 \times (-2) + 12 \times (3 - 4) + (-12)$
$= -4 \times 3 \times (-2) + 12 \times [3 + (-4)] + (-12)$
$= -4 \times 3 \times (-2) + 12 \times (-1) + (-12)$
$= -12 \times (-2) + 12 \times (-1) + (-12)$
$= 24 + 12 \times (-1) + (-12)$
$= 24 + (-12) + (-12)$
$= 12 + (-12) = 0$

65. $-12 \times (6 - 8) + 1^2 \times 3^2 \times 2 - 6 \times 2$
$= -12 \times [6 + (-8)] + 1^2 \times 3^2 \times 2 - 6 \times 2$
$= -12 \times (-2) + 1^2 \times 3^2 \times 2 - 6 \times 2$
$= -12 \times (-2) + 1 \times 9 \times 2 - 6 \times 2$
$= 24 + 1 \times 9 \times 2 - 6 \times 2$
$= 24 + 18 - 6 \times 2$
$= 24 + 18 - 12$
$= 24 + 18 + (-12)$
$= 42 + (-12) = 30$

67. $10 \times 9 - (8 + 7) \div 5 + 6 - 7 + 8$
$= 10 \times 9 - 15 \div 5 + 6 - 7 + 8$
$= 90 - 15 \div 5 + 6 - 7 + 8$
$= 90 - 3 + 6 - 7 + 8$
$= 90 + (-3) + 6 + (-7) + 8$
$= 87 + 6 + (-7) + 8$
$= 93 + (-7) + 8$
$= 86 + 8 = 94$

69. $3^2 \times (4 - 7) \div 9 + 6 - 3 - 4 \times 2$
$= 3^2 \times [4 + (-7)] \div 9 + 6 - 3 - 4 \times 2$
$= 3^2 \times (-3) \div 9 + 6 - 3 - 4 \times 2$
$= 9 \times (-3) \div 9 + 6 - 3 - 4 \times 2$
$= -27 \div 9 + 6 - 3 - 4 \times 2$
$= -3 + 6 - 3 - 4 \times 2$
$= -3 + 6 - 3 - 8$
$= -3 + 6 + (-3) + (-8)$
$= 3 + (-3) + (-8)$
$= 0 + (-8) = -8$

71. $(-3)^2 \times (5 - 7)^2 - (-9) \div 3$
$= (-3)^2 \times [5 + (-7)]^2 - (-9) \div 3$
$= (-3)^2 \times (-2)^2 - (-9) \div 3$
$= 9 \times 4 - (-9) \div 3$
$= 36 - (-9 \div 3)$
$= 36 - (-3)$
$= 36 + 3 = 39$

73. $4 - 6(2 - 5)^3 \div (17 - 8)$
$= 4 - 6[2 + (-5)]^3 \div (17 - 8)$
$= 4 - 6(-3)^3 \div (17 - 8)$
$= 4 - 6(-27) \div (17 - 8)$
$= 4 - 6(-27) \div 9$
$= 4 - (-162) \div 9$
$= 4 - (-18) = 4 + 18 = 22$

75. $(1.2)^2 - 4.1 \times 0.3 = 1.44 - 4.1 \times 0.3$
$= 1.44 - 1.23$
$= 1.44 + (-1.23) = 0.21$

77. $1.6 - (-1.6)^2 = 1.6 - 2.56$
$= 1.6 + (-2.56) = -0.96$

79. $(4.1 - 3.9) - 0.7^2 = [4.1 + (-3.9)] - 0.7^2$
$= 0.2 - 0.7^2$
$= 0.2 - 0.49$
$= 0.2 + (-0.49) = -0.29$

81. $(-0.4)^2 \times 1.5 - 2 = 0.16 \times 1.5 - 2$
$= 0.24 - 2$
$= 0.24 + (-2) = -1.76$

83. $4.2 - (-3.9) - 6 = 4.2 + 3.9 + (-6)$
$= 8.1 + (-6) = 2.1$

85. $\left(\dfrac{3}{4}\right)^2 - \dfrac{3}{8} = \dfrac{9}{16} - \dfrac{3}{8}$
$= \dfrac{9}{16} - \dfrac{6}{16} = \dfrac{3}{16}$

87. $\dfrac{5}{16} - \dfrac{3}{8} + \dfrac{1}{2} = \dfrac{5}{16} - \dfrac{6}{16} + \dfrac{1}{2}$
$= \dfrac{5}{16} + \left(-\dfrac{6}{16}\right) + \dfrac{1}{2}$
$= -\dfrac{1}{16} + \dfrac{1}{2} = -\dfrac{1}{16} + \dfrac{8}{16} = \dfrac{7}{16}$

89. $\dfrac{1}{2} \times \dfrac{1}{4} \times \dfrac{1}{2} - \dfrac{3}{8} = \dfrac{1}{8} \times \dfrac{1}{2} - \dfrac{3}{8}$
$= \dfrac{1}{16} - \dfrac{3}{8}$
$= \dfrac{1}{16} + \left(-\dfrac{3}{8}\right) = \dfrac{1}{16} + \left(-\dfrac{6}{16}\right) = -\dfrac{5}{16}$

91. $\dfrac{1}{2} - \left(\dfrac{3}{4} - \dfrac{3}{8}\right) \div \dfrac{1}{3} = \dfrac{1}{2} - \left(\dfrac{6}{8} - \dfrac{3}{8}\right) \div \dfrac{1}{3}$
$= \dfrac{1}{2} - \dfrac{3}{8} \div \dfrac{1}{3} = \dfrac{1}{2} - \dfrac{3}{8} \times \dfrac{3}{1}$
$= \dfrac{1}{2} - \dfrac{9}{8} = \dfrac{4}{8} + \left(-\dfrac{9}{8}\right) = -\dfrac{5}{8}$

Applying the Concepts

93. a. $3.45 \times 10^{-14} > 3.45 \times 10^{-15}$

b. $5.23 \times 10^{18} > 5.23 \times 10^{17}$

c. $3.12 \times 10^{12} > 3.12 \times 10^{11}$

95. a. $1^3 + 2^3 + 3^3 + 4^3 = 1 + 8 + 27 + 64 = 100$

b. $(-1)^3 + (-2)^3 + (-3)^3 + (-4)^3$
$= -1 + (-8) + (-27) + (-64)$
$= -100$

c. $1^3 + 2^3 + 3^3 + 4^3 + 5^3$
$= 1 + 8 + 27 + 64 + 125$
$= 225$

d. $(-1)^3 + (-2)^3 + (-3)^3 + (-4)^3 + (-5)^3 = -225$

97. Because the first statement is false (it was either Becky or Diana), neither Becky nor Diana could have done it. Therefore, it was either Abdul or Carl. Because the second statement is false, (it was neither Becky nor Carl), either Becky or Carl did it. Because the first statement ensured that it was Abdul or Carl, it must have been Carl.

99. Mass of the sun: 1.99×10^{30} kg
Mass of a neutron: 1.67×10^{-27} kg

CHAPTER REVIEW

1. -22

2. $-8 - (-2) - (-10) - 3 = -8 + 2 + 10 - 3$
$= -6 + 10 - 3 = 4 - 3 = 1$

3. $\dfrac{5}{8} - \dfrac{5}{6} = \dfrac{15}{24} - \dfrac{20}{24}$
$= \dfrac{15}{24} + \dfrac{(-20)}{24} = \dfrac{15 + (-20)}{24} = \dfrac{-5}{24} = -\dfrac{5}{24}$

4. $-0.33 + 1.98 - 1.44 = -0.33 + 1.98 + (-1.44)$
$= 1.65 + (-1.44) = 0.21$

5. $\left(-\dfrac{2}{3}\right)\left(\dfrac{6}{11}\right)\left(-\dfrac{22}{25}\right) = \dfrac{2 \cdot 6 \cdot 22}{3 \cdot 11 \cdot 25} = \dfrac{264}{825} = \dfrac{8}{25}$

6. $-0.08 \times 16 = -(0.08 \times 16) = -1.28$
$\begin{array}{r} 16 \\ \times\ .08 \\ \hline 1.28 \end{array}$

7. $12 - 6 \div 3 = 12 - 2 = 12 + (-2) = 10$

8. $\left(\dfrac{2}{3}\right)^2 - \dfrac{5}{6} = \left(\dfrac{2}{3} \cdot \dfrac{2}{3}\right) - \dfrac{5}{6}$
$= \dfrac{4}{9} - \dfrac{5}{6} = \dfrac{8}{18} - \dfrac{15}{18}$
$= \dfrac{8}{18} + \left(-\dfrac{15}{18}\right) = \dfrac{8 + (-15)}{18} = -\dfrac{7}{18}$

9. 4

10. $0 > -3$

11. $-|-6| = -6$

12. $-18 \div (-3) = 18 \div 3 = 6$

13. $-\dfrac{3}{8} + \dfrac{5}{12} + \dfrac{2}{3} = \dfrac{-9}{24} + \dfrac{10}{24} + \dfrac{16}{24}$
$= \dfrac{-9 + 10 + 16}{24} = \dfrac{17}{24}$

14. $\dfrac{1}{3} \times \left(-\dfrac{3}{4}\right) = -\left(\dfrac{1}{3} \times \dfrac{3}{4}\right) = -\dfrac{3}{12} = -\dfrac{1}{4}$

15. $-\dfrac{7}{12} \div \left(-\dfrac{14}{39}\right) = -\dfrac{7}{12} \times \left(-\dfrac{39}{14}\right)$
$= \dfrac{7 \cdot 39}{12 \cdot 14} = \dfrac{273}{168} = \dfrac{13}{8} = 1\dfrac{5}{8}$

16. $16 \div 4(8 - 2) = 16 \div 4[8 + (-2)]$
$= 16 \div 4(6) = 4(6) = 24$

17. $-22 + 14 + (-18) = -8 + (-18) = -26$

18. $3^2 - 9 + 2 = 9 - 9 + 2$
$= 9 + (-9) + 2$
$= 0 + 2 = 2$

19. The number is less than 10. Move the decimal point 5 places to the right. The exponent on 10 is -5.
$$0.0000397 = 3.97 \times 10^{-5}$$

20. $-1.464 \div 18.3 = -(1.464 \div 18.3) = -0.08$

$$
\begin{array}{r}
0.08 \\
18.3{\overline{)1.464}} \\
-1464 \\
\hline
0
\end{array}
$$

21. $-\dfrac{5}{12} + \dfrac{7}{9} - \dfrac{1}{3} = \dfrac{-15}{36} + \dfrac{28}{36} - \dfrac{12}{36}$

$$= \dfrac{-15}{36} + \dfrac{28}{36} + \dfrac{(-12)}{36}$$

$$= \dfrac{-15 + 28 + (-12)}{36} = \dfrac{1}{36}$$

22. $\dfrac{6}{34} \times \dfrac{17}{40} = \dfrac{6 \cdot 17}{34 \cdot 40} = \dfrac{102}{1360} = \dfrac{3}{40}$

23. $1.2 \times (-0.035) = -(1.2 \times 0.035) = -0.042$

$$
\begin{array}{r}
0.035 \\
\times \ 1.2 \\
\hline
70 \\
35 \\
\hline
0.042
\end{array}
$$

24. $-\dfrac{1}{2} + \dfrac{3}{8} \div \dfrac{9}{20} = -\dfrac{1}{2} + \dfrac{3}{8} \times \dfrac{20}{9}$

$$= -\dfrac{1}{2} + \dfrac{3 \cdot 20}{8 \cdot 9} = -\dfrac{1}{2} + \dfrac{60}{72}$$

$$= -\dfrac{1}{2} + \dfrac{5}{6} = -\dfrac{3}{6} + \dfrac{5}{6} = \dfrac{2}{6} = \dfrac{1}{3}$$

25. $|-5| = 5$

26. $-2 > -40$

27. $2 \times (-13) = -(2 \times 13) = -26$

28. $-0.4 \times 5 - (-3.33) = -2 - (-3.33)$
$$= -2 + 3.33 = 1.33$$

29. $\dfrac{5}{12} + \left(-\dfrac{2}{3}\right) = \dfrac{5}{12} + \dfrac{(-8)}{12}$

$$= \dfrac{5 + (-8)}{12} = \dfrac{-3}{12} = \dfrac{-1}{4} = -\dfrac{1}{4}$$

30. $-33.4 + 9.8 - (-16.2) = -33.4 + 9.8 + 16.2$
$$= -23.6 + 16.2 = -7.4$$

31. $\left(-\dfrac{3}{8}\right) \div \left(-\dfrac{4}{5}\right) = -\dfrac{3}{8} \times \left(-\dfrac{5}{4}\right) = \dfrac{3 \cdot 5}{8 \cdot 4} = \dfrac{15}{32}$

32. The exponent on 10 is positive. Move the decimal point 5 places to the right.
$$2.4 \times 10^5 = 240,000$$

33. Strategy To find the temperature, add the increase (18) to the original temperature (-22).

    Solution $-22 + 18 = -4$
The temperature is $-4°$.

34. Strategy To find the student's score:
• Multiply the number of questions answered correctly (38) by 3.
Multiply the number of questions left blank (8) by -1.
Multiply the number of questions answered incorrectly (4) by -2.
• Add the three products.

    Solution $38 \times 3 = 114$
$8 \times -1 = -8$
$4 \times -2 = -8$
$114 + (-8) + (-8) = 106 + (-8) = 98$
The student's score was 98.

35. Strategy To find the difference between the boiling point and the melting point of mercury, subtract the melting point (-38.87) from the boiling point (356.58).

    Solution $356.58 - (-38.87) = 395.45$
The difference between the melting and boiling point is $395.45°C$.

CHAPTER TEST

1. $-5 - (-8) = -5 + 8 = 3$

2. $-|-2| = -2$

3. $-\dfrac{2}{5} + \dfrac{7}{15} = \dfrac{-6}{15} + \dfrac{7}{15} = \dfrac{-6 + 7}{15} = \dfrac{1}{15}$

4. $0.032 \times (-1.9) = -(0.032 \times 1.9) = -0.0608$

$$
\begin{array}{r}
0.032 \\
\times \ 1.9 \\
\hline
288 \\
32 \\
\hline
0.0608
\end{array}
$$

5. $-8 > -10$

6. $1.22 + (-3.1) = -1.88$

7. $4 \times (4 - 7) \div (-2) - 4 \times 8$
$= 4 \times [4 + (-7)] \div (-2) - 4 \times 8$
$= 4 \times (-3) \div (-2) - 4 \times 8$
$= -12 \div (-2) - 4 \times 8$
$= 6 - 4 \times 8$
$= 6 - 32$
$= 6 + (-32) = -26$

8. $-5 \times (-6) \times 3 = 5 \times 6 \times 3 = 30 \times 3 = 90$

9. $-1.004 - 3.01 = -1.004 + (-3.01) = -4.014$

10. $-72 \div 8 = -(72 \div 8) = -9$

11. $-2 + 3 + (-8) = 1 + (-8) = -7$

12. $-\dfrac{3}{8} + \dfrac{2}{3} = \dfrac{-9}{24} + \dfrac{16}{24} = \dfrac{-9+16}{24} = \dfrac{7}{24}$

13. The number is greater than 10. Move the decimal point 10 places to the left. The exponent on 10 is 10.
$87,600,000,000 = 8.76 \times 10^{10}$

14. $-4 \times 12 = -(4 \times 12) = -48$

15. $\dfrac{0}{-17} = 0$

16. $16 - 4 - (-5) - 7 = 16 + (-4) + 5 + (-7)$
$\qquad = 12 + 5 + (-7)$
$\qquad = 17 + (-7) = 10$

17. $-\dfrac{2}{3} \div \dfrac{5}{6} = -\dfrac{2}{3} \times \dfrac{6}{5} = \left(-\dfrac{2 \cdot 6}{3 \cdot 5}\right) = -\dfrac{12}{15} = -\dfrac{4}{5}$

18. $0 > -4$

19. $16 + (-10) + (-20) = 6 + (-20) = -14$

20. $(-2)^2 - (-3)^2 \div (1-4)^2 \times 2 - 6$
$= (-2)^2 - (-3)^2 \div [1 + (-4)]^2 \times 2 - 6$
$= (-2)^2 - (-3)^2 \div (-3)^2 \times 2 - 6$
$= 4 - 9 \div 9 \times 2 - 6$
$= 4 - 1 \times 2 - 6$
$= 4 - 2 - 6$
$= 4 + (-2) + (-6)$
$= 2 + (-6) = -4$

21. $-\dfrac{2}{5} - \left(\dfrac{-7}{10}\right) = \dfrac{-4}{10} - \left(-\dfrac{7}{10}\right)$
$\qquad = \dfrac{-4}{10} + \dfrac{7}{10} = \dfrac{-4+7}{10} = \dfrac{3}{10}$

22. The exponent on 10 is negative. Move the decimal point 8 places to the left.
$9.601 \times 10^{-8} = 0.00000009601$

23. $-15.64 \div (-4.6) = (15.64 \div 4.6) = 3.4$

$$
\begin{array}{r}
3.4 \\
4.6\overline{)15.64} \\
-13\,8 \\
\hline
184 \\
-1\,84 \\
\hline
0
\end{array}
$$

24. $-\dfrac{1}{2} + \dfrac{1}{3} + \dfrac{1}{4} = \dfrac{-6}{12} + \dfrac{4}{12} + \dfrac{3}{12}$
$\qquad = \dfrac{-6+4+3}{12} = \dfrac{1}{12}$

25. $\dfrac{3}{8} \times \left(-\dfrac{5}{6}\right) \times \left(-\dfrac{4}{15}\right) = \dfrac{3}{8} \times \dfrac{5}{6} \times \dfrac{4}{15}$
$\qquad = \dfrac{3 \cdot 5 \cdot 4}{8 \cdot 6 \cdot 15} = \dfrac{60}{720} = \dfrac{1}{12}$

26. $2.113 - (-1.1) = 2.113 + 1.1 = 3.213$

27. Strategy   To find the temperature, add the increase (11) to the previous temperature (–4).

   Solution $-4 + 11 = 7$
   The temperature is 7°C.

28. Strategy   To find the melting point of oxygen, multiply the melting point of radon (–71) by 3.

   Solution $-71 \times 3 = -213$
   The melting point of oxygen is –213°C.

29. Strategy   To find the amount the temperature fell, subtract the temperature at midnight (–29.4) from the temperature at noon (17.22).

   Solution $17.22 - (-29.4) = 46.62$
   The temperature fell 46.62°C.

30. Strategy   To find the average daily low temperature:
   • Add the three temperature readings.
   • Divide by 3.

   Solution $-7 + 9 + (-8) = 2 + (-8) = -6$
   $-6 \div 3 = -2$
   The average low temperature was –2°.

CUMULATIVE REVIEW

1. $16 - 4 \cdot (3-2)^2 \cdot 4 = 16 - 4 \cdot (1)^2 \cdot 4$
$\qquad = 16 - 4 \cdot (1) \cdot 4 = 16 - 16 = 0$

2. $8\dfrac{1}{2} = 8\dfrac{7}{14} = 7\dfrac{21}{14}$
$-3\dfrac{4}{7} = 3\dfrac{8}{14} = 3\dfrac{8}{14}$
$\overline{\phantom{-3\dfrac{4}{7}}\,4\dfrac{13}{14}}$

3. $3\dfrac{7}{8} \div 1\dfrac{1}{2} = \dfrac{31}{8} \div \dfrac{3}{2}$
$\qquad = \dfrac{31}{8} \times \dfrac{2}{3}$
$\qquad = \dfrac{62}{24} = \dfrac{31}{12} = 2\dfrac{7}{12}$

4. $\dfrac{3}{8} \div \left(\dfrac{3}{8} - \dfrac{1}{4}\right) \div \dfrac{7}{3} = \dfrac{3}{8} \div \left(\dfrac{3}{8} - \dfrac{2}{8}\right) \div \dfrac{7}{3}$
$\qquad = \dfrac{3}{8} \div \left(\dfrac{1}{8}\right) \div \dfrac{7}{3} = \dfrac{3}{8} \times \dfrac{8}{1} \div \dfrac{7}{3}$
$\qquad = 3 \div \dfrac{7}{3} = 3 \times \dfrac{3}{7} = \dfrac{3 \cdot 3}{7} = \dfrac{9}{7} = 1\dfrac{2}{7}$

5. $\begin{array}{r} 2.90700 \\ -1.09761 \\ \hline 1.80939 \end{array}$

6.
$$\frac{7}{12} = \frac{n}{32}$$
$$7 \cdot 32 = 12 \times n$$
$$224 = 12 \times n$$
$$224 \div 12 = n$$
$$18.67 = n$$

7. $160\% \times n = 22$
$$1.6 \times n = 22$$
$$n = 22 \div 1.6$$
$$n = 13.75$$

8. 7 qt = 1 gal 3 qt

9. 6692 ml = 6 L 692 ml = 6 L + 0.692 L = 6.692 L

10. $4.2 \text{ ft} = 4.2 \cancel{\text{ ft}} \times \dfrac{1 \text{ m}}{3.28 \cancel{\text{ ft}}} = \dfrac{4.2}{3.28} \text{ m} \approx 1.28 \text{ m}$

11. $n = 0.32 \times 180$
$$n = 57.6$$

12. $3\dfrac{2}{5} \times 100\% = \dfrac{1700}{5}\% = 340\%$

13. $-8 + 5 = -3$

14. $3\dfrac{1}{4} + \left(-6\dfrac{5}{8}\right) = \dfrac{13}{4} + \left(\dfrac{-53}{8}\right) = \dfrac{26}{8} + \dfrac{(-53)}{8}$
$$= \dfrac{26 + (-53)}{8} = \dfrac{-27}{8} = -3\dfrac{3}{8}$$

15. $-6\dfrac{1}{8} - 4\dfrac{5}{12} = \dfrac{-49}{8} - \dfrac{53}{12}$
$$= \dfrac{-147}{24} - \dfrac{106}{24} = \dfrac{-147}{24} + \dfrac{(-106)}{24}$$
$$= \dfrac{-147 + (-106)}{24} = \dfrac{-253}{24} = -10\dfrac{13}{24}$$

16. $-12 - (-7) - 3(-8) = -12 + 7 + 24 = -5 + 24 = 19$

17. $-3.2 \times -1.09 = 3.2 \times 1.09 = 3.488$
$$\begin{array}{r} 1.09 \\ \times\ 3.2 \\ \hline 218 \\ 327 \\ \hline 3.488 \end{array}$$

18. $-6 \times 7 \times \left(-\dfrac{3}{4}\right) = 6 \times 7 \times \dfrac{3}{4}$
$$= \dfrac{6 \cdot 7 \cdot 3}{4} = \dfrac{126}{4} = 31\dfrac{1}{2}$$

19. $42 \div (-6) = -(42 \div 6) = -7$

20. $-2\dfrac{1}{7} \div \left(-3\dfrac{3}{5}\right) = -\dfrac{15}{7} \div \left(-\dfrac{18}{5}\right)$
$$= -\dfrac{15}{7} \times \left(-\dfrac{5}{18}\right) = \dfrac{75}{126} = \dfrac{25}{42}$$

21. $3 \times (3 - 7) \div 6 - 2 = 3 \times [3 + (-7)] \div 6 - 2$
$$= 3 \times (-4) \div 6 - 2$$
$$= -12 \div 6 - 2$$
$$= -2 + (-2) = -4$$

22. $4 - (-2)^2 \div (1 - 2)^2 \times 3 + 4$
$$= 4 - (-2)^2 \div [1 + (-2)]^2 \times 3 + 4$$
$$= 4 - (-2)^2 \div (-1)^2 \times 3 + 4$$
$$= 4 - 4 \div 1 \times 3 + 4$$
$$= 4 - 4 \times 3 + 4$$
$$= 4 - 12 + 4$$
$$= 4 + (-12) + 4$$
$$= -8 + 4 = -4$$

23. Strategy   To find the length of the remaining board, subtract the length cut $\left(5\dfrac{2}{3} \text{ ft}\right)$ from the original length (8 ft).

Solution
$$8 \text{ ft} = 7\dfrac{3}{3} \text{ ft}$$
$$-5\dfrac{2}{3} \text{ ft} = 5\dfrac{2}{3} \text{ ft}$$
$$\overline{2\dfrac{1}{3} \text{ ft}}$$

The length remaining is $2\dfrac{1}{3}$ ft.

24. Strategy   To find Nimisha's new balance, subtract the amounts of the checks written and add the amount of the deposit.

Solution
$$\begin{array}{r} \$763.56 \\ -\ 135.88 \\ \hline 627.68 \\ -\ 47.81 \\ \hline 579.87 \\ +\ 223.44 \\ \hline \$803.31 \end{array}$$

Nimisha's new balance is $803.31.

25. Strategy   To find the percent:
• Subtract the sale price ($120) from the original price ($165) to find the amount of the decrease.
• Solve the basic percent equation for percent. The base is $165 and the amount is the amount of the decrease.

Solution $\$165 - \$120 = \$45$
Percent × base = amount
$$n \times 165 = 45$$
$$n = 45 \div 165$$
$$n = 0.273 = 27.3\%$$
The percent decrease is 27.3%.

26. Strategy To find how many gallons of coffee must be prepared:
• Multiply the number of guests (80) by the amount of coffee each guest is expected to drink to find the number of cups of coffee to prepare.
• Convert cups to gallons.

Solution $80 \times 2 \text{ c} = 160 \text{ c}$

$$160 \text{ c} = 160 \text{ c} \times \frac{1 \text{ pt}}{2 \text{ c}} \times \frac{1 \text{ qt}}{2 \text{ pt}} \times \frac{1 \text{ gal}}{4 \text{ qt}}$$

$$= \frac{160}{16} \text{ gal} = 10 \text{ gal}$$

The amount of coffee that should be prepared is 10 gal.

27. Strategy To find the dividend per share:
• Multiply the dividend ($1.50) by the increase (12%) to find the amount of the increase.
• Add the amount of the increase to the dividend ($1.50).

Solution $1.50 \times 12\% = 0.18$

$$\begin{array}{r} 1.50 \\ + 0.18 \\ \hline \$1.68 \end{array}$$

The dividend per share after the increase was $1.68.

28. Strategy To find the median:
• Arrange the hourly wages in descending order.
• Pick the middle number.

Solution $8.32; $9.73; $10.40; $12.10; $14.25
The median hourly pay is $10.40.

29. Strategy To find the number of voters:
• Multiply the ratio $\left(\frac{5}{8}\right)$ by the number of registered voters (960,000).

Solution $\frac{5}{8} \times 960,000 = 600,000$

600,000 people would vote.

30. Strategy To find the mean high temperature, add the daily high temperatures ($-19°$, $-7°$, $1°$, and $9°$) and divide that sum by the number of temperatures (4).

Solution $(-19°) + (-7°) + (1°) + (9°)$
$= -26° + 1° + 9°$
$= -25° + 9°$
$= -16° = $ sum of temperatures
$-16° \div 4 = -4°$
The mean high temperature is $-4°$.

Chapter 11: Introduction to Algebra

PREP TEST and GO FIGURE

1. -7

2. -20

3. 0

4. 1

5. 1

6. $\left(\dfrac{3}{5}\right)^3 \cdot \left(\dfrac{5}{9}\right)^2 = \dfrac{\cancel{3}}{\cancel{5}} \cdot \dfrac{\cancel{3}}{\cancel{5}} \cdot \dfrac{\cancel{3}}{5} \cdot \dfrac{\cancel{5}}{\cancel{9}} \cdot \dfrac{\cancel{5}}{\cancel{9}} = \dfrac{1}{15}$

7. $\dfrac{2}{3} + \left(\dfrac{3}{4}\right)^2 \cdot \dfrac{2}{9} = \dfrac{2}{3} + \dfrac{9}{16} \cdot \dfrac{2}{9}$

$= \dfrac{2}{3} + \dfrac{18}{144}$

$= \dfrac{96}{144} + \dfrac{18}{144}$

$= \dfrac{114}{144}$

$= \dfrac{19}{24}$

8. $-8 \div (-2)^2 + 6 = -8 \div 4 + 6$

$= -2 + 6$

$= 4$

9. $4 + 5(2-7)^2 \div (-8+3)$

$= 4 + 5(-5)^2 \div (-5)$

$= 4 + 5(25) \div (-5)$

$= 4 + 125 \div (-5)$

$= 4 + (-25) = -21$

Go Figure

Replacing the known values,

```
  271
   51
+ 3HE
 S71
```

In the first column,
$1 + 1 + E = 11.$
$2 + E = 11$
$E = 9$
In the second column,
$1 + 7 + 5 + H = 17.$
$13 + H = 17$
$H = 4$
So then, the third column,
$1 + 2 + 3 = S$
$S = 6$

SECTION 11.1

Objective A Exercises

1. $5a - 3b = 5(-3) - 3(6)$

$= -15 - 18$

$= -15 + (-18)$

$= -33$

3. $2a + 3c = 2(-3) + 3(-2)$

$= -6 + (-6) = -12$

5. $-c^2 = -(-2)^2 = -4$

7. $b - a^2 = 6 - (-3)^2 = 6 - 9 = 6 + (-9) = -3$

9. $ab - c^2 = (-3)(6) - (-2)^2$

$= -18 - 4$

$= -18 + (-4) = -22$

11. $2ab - c^2 = 2(-3)6 - (-2)^2$

$= -36 - 4$

$= -36 + (-4) = -40$

13. $a - (b \div a) = -3 - [6 \div (-3)]$

$= -3 - (-2)$

$= -3 + 2 = -1$

15. $2ac - (b \div a) = 2(-3)(-2) - [6 \div (-3)]$

$= 12 - (-2)$

$= 12 + 2 = 14$

17. $b^2 - c^2 = (6)^2 - (-2)^2$

$= 36 - 4$

$= 36 + (-4) = 32$

19. $b^2 \div (ac) = (6)^2 \div (-3)(-2)$

$= 36 \div 6 = 6$

21. $c^2 - (b \div c) = (-2)^2 - [6 \div (-2)]$

$= 4 - (-3)$

$= 4 + 3 = 7$

23. $a^2 + b^2 + c^2 = (-3)^2 + 6^2 + (-2)^2$

$= 9 + 36 + 4$

$= 45 + 4 = 49$

25. $ac + bc + ab = (-3)(-2) + 6(-2) + (-3)(6)$

$= 6 + (-12) + (-18)$

$= -6 + (-18) = -24$

27. $a^2 + b^2 - ab = (-3)^2 + 6^2 - (-3)(6)$

$= 9 + 36 - (-3)(6)$

$= 9 + 36 - (-18)$

$= 9 + 36 + 18$

$= 45 + 18 = 63$

29. $2b - (3c + a^2) = 2(6) - [3(-2) + (-3)^2]$

$= 2(6) - [3(-2) + 9]$

$= 2(6) - (-6 + 9)$

$= 2(6) - 3$

$= 12 - 3$

$= 12 + (-3) = 9$

31. $\dfrac{1}{3}a + \left(\dfrac{1}{2}b - \dfrac{2}{3}a\right) = \dfrac{1}{3}(-3) + \left[\dfrac{1}{2} \cdot 6 - \dfrac{2}{3}(-3)\right]$

$= \dfrac{1}{3}(-3) + [3 - (-2)]$

$= \dfrac{1}{3}(-3) + (3 + 2) = -1 + 5 = 4$

33. $\dfrac{1}{6}b+\dfrac{1}{3}(c+a)=\dfrac{1}{6}\cdot 6+\dfrac{1}{3}[-2+(-3)]$

$=\dfrac{1}{6}\cdot 6+\dfrac{1}{3}(-5)$

$=1+\left(-\dfrac{5}{3}\right)=\dfrac{3}{3}+\left(-\dfrac{5}{3}\right)=-\dfrac{2}{3}$

35. $a+(b-c)=-\dfrac{1}{2}+\left(\dfrac{3}{4}-\dfrac{1}{4}\right)$

$=-\dfrac{1}{2}+\dfrac{2}{4}=-\dfrac{2}{4}+\dfrac{2}{4}=0$

37.

$2a-b^2\div c=2\left(-\dfrac{1}{2}\right)-\left(\dfrac{3}{4}\right)^2\div\dfrac{1}{4}$

$=2\left(-\dfrac{1}{2}\right)-\dfrac{9}{16}\div\dfrac{1}{4}$

$=-1-\dfrac{9}{16}\div\dfrac{1}{4}$

$=-1-\dfrac{9}{16}\times\dfrac{4}{1}$

$=-1-\dfrac{9}{4}$

$=-1+\left(-\dfrac{9}{4}\right)=-3\dfrac{1}{4}$

39. $a^2-b^2=(3.72)^2-(-2.31)^2$
$=13.8384-5.3361=8.5023$

41. $3ac-(c\div a)=3(3.72)(-1.74)-(-1.74\div 3.72)$
≈ -18.950658

43. $2x^2,\ 3x,\ \underline{-4}$

45. $3a^2,\ -4a,\ \underline{8}$

47. $\underline{3}x^2,\ \underline{-4}x$

49. $\underline{1}y^2,\ \underline{6}a$

Objective B Exercises

51. $16z$

53. $12m-3m=12m+(-3)m=9m$

55. $12at$

57. $3yt$

59. Unlike terms

61. $3t^2-5t^2=3t^2+(-5)t^2=-2t^2$

63. $6c-5+7c=6c+7c-5=13c-5$

65. $2t+3t-7t=2t+3t+(-7)t=5t+(-7)t=-2t$

67. $7y^2-2-4y^2=7y^2+(-2)+(-4)y^2$
$=7y^2+(-4)y^2+(-2)$
$=3y^2-2$

69. $6w-8u+8w=6w+(-8)u+8w$
$=6w+8w+(-8)u$
$=14w-8u$

71. $10-11xy-12xy=10+(-11)xy+(-12)xy$
$=10+(-23)xy$
$=10-23xy=-23xy+10$

73. $3v^2-6v^2-8v^2=3v^2+(-6)v^2+(-8)v^2$
$=-3v^2+(-8)v^2$
$=-11v^2$

75. $-10ab-3a+2ab=-10ab+2ab-3a$
$=-8ab-3a$

77. $-3y^2-y+7y^2=-3y^2+7y^2-y$
$=4y^2-y$

79. $2a-3b^2-5a+b^2=2a+(-3)b^2+(-5)a+b^2$
$=2a+(-5)a+(-3)b^2+b^2$
$=-3a-2b^2$

81. $3x^2-7x+4x^2-x=3x^2+(-7)x+4x^2+(-1)x$
$=3x^2+4x^2+(-7)x+(-1)x$
$=7x^2-8x$

83. $6s-t-9s+7t=6s+(-1)t+(-9)s+7t$
$=6s+(-9)s+(-1)t+7t$
$=-3s+6t$

85. $4m+8n-7m+2n=4m+8n+(-7)m+2n$
$=4m+(-7)m+8n+2n$
$=-3m+10n$

87. $-5ab+7ac+10ab-3ac$
$=-5ab+7ac+10ab+(-3)ac$
$=-5ab+10ab+7ac+(-3)ac$
$=5ab+4ac$

89. $\dfrac{4}{9}a^2-\dfrac{1}{5}b^2+\dfrac{2}{9}a^2+\dfrac{4}{5}b^2$

$=\dfrac{4}{9}a^2+\left(-\dfrac{1}{5}\right)b^2+\dfrac{2}{9}a^2+\dfrac{4}{5}b^2$

$=\dfrac{4}{9}a^2+\dfrac{2}{9}a^2+\left(-\dfrac{1}{5}\right)b^2+\dfrac{4}{5}b^2$

$=\dfrac{6}{9}a^2+\dfrac{3}{5}b^2$

$=\dfrac{2}{3}a^2+\dfrac{3}{5}b^2$

91. $6.994x$

93. $1.56m-3.77n$

Objective C Exercises

95. $5(x+4)=5x+5\cdot 4=5x+20$

97. $(y-3)4=[y+(-3)]4$
$=y\cdot 4+(-3)4$
$=4y+(-12)$
$=4y-12$

99. $-2(a+4)=-2(a)+(-2)(4)$
$=-2a+(-8)$
$=-2a-8$

101. $3(5x+10)=3(5x)+3(10)=15x+30$

103.
$$5(3c - 5) = 5[3c + (-5)]$$
$$= 5(3c) + 5(-5)$$
$$= 15c + (-25)$$
$$= 15c - 25$$

105.
$$-3(y-6) = -3[y + (-6)]$$
$$= -3y + (-3)(-6)$$
$$= -3y + 18$$

107. $5x + 2(x + 7) = 5x + 2x + 2(7) = 7x + 14$

109.
$$8y - 4(y + 2) = 8y + (-4)(y + 2)$$
$$= 8y + (-4)y + (-4)(2)$$
$$= 8y + (-4y) + (-8)$$
$$= 4y - 8$$

111.
$$9x - 4(x - 6) = 9x + (-4)[x + (-6)]$$
$$= 9x + (-4)(x) + (-4)(-6)$$
$$= 9x + (-4)x + 24$$
$$= 5x + 24$$

113.
$$-2y + 3(y - 2) = -2y + 3[y + (-2)]$$
$$= -2y + 3y + 3(-2)$$
$$= -2y + 3y + (-6)$$
$$= y - 6$$

115.
$$4n + 2(n + 1) - 5 = 4n + 2(n + 1) + (-5)$$
$$= 4n + 2n + 2(1) + (-5)$$
$$= 4n + 2n + 2 + (-5)$$
$$= 6n - 3$$

117.
$$9y - 3(y - 4) + 8 = 9y + (-3)[y + (-4)] + 8$$
$$= 9y + (-3)(y) + (-3)(-4) + 8$$
$$= 9y + (-3)y + 12 + 8$$
$$= 6y + 20$$

119.
$$3x + 2(x + 2) + 5x = 3x + 2x + 2(2) + 5x$$
$$= 3x + 2x + 4 + 5x$$
$$= 3x + 2x + 5x + 4$$
$$= 5x + 5x + 4$$
$$= 10x + 4$$

121.
$$-7t + 2(t - 3) - t = -7t + 2[t + (-3)] + (-1)t$$
$$= -7t + 2t + 2(-3) + (-1)t$$
$$= -7t + 2t + (-6) + (-1)t$$
$$= -7t + 2t + (-1)t + (-6)$$
$$= -5t + (-1)t + (-6)$$
$$= -6t - 6$$

123.
$$z - 2(1 - z) - 2z = z + (-2)[1 + (-z)] + (-2)z$$
$$= z + (-2)(1) + (-2)(-z) + (-2)z$$
$$= z + (-2) + 2z + (-2)z$$
$$= z + 2z + (-2)z + (-2)$$
$$= 3z + (-2)z + (-2)$$
$$= z - 2$$

125.
$$3(y - 2) - 2(y - 6) = 3[y + (-2)] + (-2)[y + (-6)]$$
$$= 3y + 3(-2) + (-2)y + (-2)(-6)$$
$$= 3y + (-6) + (-2)y + 12$$
$$= 3y + (-2)y + (-6) + 12$$
$$= y + 6$$

127.
$$2(t - 3) + 7(t + 3) = 2[t + (-3)] + 7(t + 3)$$
$$= 2t + 2(-3) + 7t + 7(3)$$
$$= 2t + (-6) + 7t + 21$$
$$= 2t + 7t + (-6) + 21$$
$$= 9t + 15$$

129.
$$3t - 6(t - 4) + 8t = 3t + (-6)[t + (-4)] + 8t$$
$$= 3t + (-6)(t) + (-6)(-4) + 8t$$
$$= 3t + (-6)t + 24 + 8t$$
$$= 3t + (-6)t + 8t + 24$$
$$= -3t + 8t + 24$$
$$= 5t + 24$$

Applying the Concepts

131a. $3 + 2x$ | 1 | 1 | 1 | x | x |

$4x + 6$ | x | x | x | x | 1 | 1 | 1 | 1 | 1 | 1 |

$3x + 2$ | x | x | x | 1 | 1 |

$2x + 4$ | x | x | 1 | 1 | 1 | 1 |

131b. $3x + 3$ | x | x | x | 1 | 1 | 1 |

133. **a.** Answers will vary. Combining like terms is like sorting things out: apples with apples, oranges with oranges.

  b. 4 feet measures distance from one point to the other, whereas 4 square feet measures an area of 2 ft × 2 ft. Therefore, 4 feet and 4 square feet are not to be combined.

SECTION 11.2

Objective A Exercises

1.
$$\frac{2x + 9 = 3}{}$$
$2(-3) + 9$	3
$-6 + 9$	3
	$3 = 3$

Yes, −3 is a solution.

3.
$$\frac{4 - 2x = 8}{}$$
$4 - 2(2)$	8
$4 - 4$	8
	$0 \neq 8$

No, 2 is not a solution.

5.
$$\frac{3x - 2 = x + 4}{}$$
$3(3) - 2$	$3 + 4$
$9 - 2$	7
	$7 = 7$

Yes, 3 is a solution.

7.
$$\frac{x^2 - 5x + 1 = 10 - 5x}{}$$
$3^2 - 5(3) + 1$	$10 - 5(3)$
$9 - 15 + 1$	$10 - 15$
	$-5 = -5$

Yes, 3 is a solution.

9.
$$2x(x-1)=3-x$$

$2(-1)(-1-1)$	$3-(-1)$
$-2(-2)$	$3+1$
$4=4$	

Yes, -1 is a solution.

11.
$$x(x-2)=x^2-4$$

$2(2-2)$	2^2-4
$2(0)$	$4-4$
$0=0$	

Yes, 2 is a solution.

13.
$$3x+6=4$$

$3\left(-\frac{2}{3}\right)+6$	4
$-2+6$	4
$4=4$	

Yes, $-\dfrac{2}{3}$ is a solution.

15.
$$2x-3=1-14x$$

$2\left(\frac{1}{4}\right)-3$	$1-14\left(\frac{1}{4}\right)$
$\frac{2}{4}-3$	$1-\frac{14}{4}$
$\frac{1}{2}-3$	$1-\frac{7}{2}$
$-2\frac{1}{2}=-2\frac{1}{2}$	

Yes, $\dfrac{1}{4}$ is a solution.

17.
$$3x(x-2)=x-4$$

$3\left(\frac{3}{4}\right)\left(\frac{3}{4}-2\right)$	$\frac{3}{4}-4$
$\frac{9}{4}\left(-\frac{5}{4}\right)$	$\frac{3}{4}-\frac{16}{4}$
$-\frac{45}{16}\neq -\frac{13}{4}$	

No, $\dfrac{3}{4}$ is not a solution.

19.
$$x^2-3x=-0.8776-x$$

$(1.32)^2-3(1.32)$	$-0.8776-1.32$
$1.7424-3.96$	-2.1976
$-2.2176\neq -2.1976$	

No, 1.32 is not a solution.

21.
$$x^2+3x=x(x+3)$$

$(1.05)^2+3(1.05)$	$1.05(1.05+3)$
$1.1025+3.15$	4.2525
$4.2525=4.2525$	

Yes, 1.05 is a solution.

Objective B Exercises

23.
$$x+7=5$$
$$x+7-7=5-7$$
$$x+0=-2$$
$$x=-2$$

25.
$$z-4=10$$
$$z-4+4=10+4$$
$$z+0=14$$
$$z=14$$

27.
$$6+x=8$$
$$6-6+x=8-6$$
$$0+x=2$$
$$x=2$$

29.
$$w+9=5$$
$$w+9-9=5-9$$
$$w+0=-4$$
$$w=-4$$

31.
$$m-4=-9$$
$$m-4+4=-9+4$$
$$m+0=-5$$
$$m=-5$$

33.
$$t-3=-3$$
$$t-3+3=-3+3$$
$$t+0=0$$
$$t=0$$

35.
$$x-3=-1$$
$$x-3+3=-1+3$$
$$x+0=2$$
$$x=2$$

37.
$$3+y=0$$
$$3-3+y=0-3$$
$$0+y=0+(-3)$$
$$0+y=-3$$
$$y=-3$$

39.
$$y-7=3$$
$$y-7+7=3+7$$
$$y+0=10$$
$$y=10$$

41.
$$t-3=-8$$
$$t-3+3=-8+3$$
$$t+0=-5$$
$$t=-5$$

43.
$$z+6=-6$$
$$z+6-6=-6-6$$
$$z+0=-6+(-6)$$
$$z+0=-12$$
$$z=-12$$

45.
$$x+2=-5$$
$$x+2-2=-5-2$$
$$x+0=-5+(-2)$$
$$x=-7$$

47.
$$x-\frac{5}{6}=-\frac{1}{6}$$
$$x-\frac{5}{6}+\frac{5}{6}=-\frac{1}{6}+\frac{5}{6}$$
$$x+0=\frac{4}{6}$$
$$x=\frac{4}{6}=\frac{2}{3}$$

49.
$$\frac{2}{5} + x = -\frac{3}{5}$$
$$\frac{2}{5} - \frac{2}{5} + x = -\frac{3}{5} - \frac{2}{5}$$
$$0 + x = -\frac{3}{5} + \left(-\frac{2}{5}\right)$$
$$x = -\frac{5}{5}$$
$$x = -1$$

51.
$$\frac{1}{3} + x = \frac{2}{3}$$
$$\frac{1}{3} - \frac{1}{3} + x = \frac{2}{3} - \frac{1}{3}$$
$$0 + x = \frac{2}{3} + \left(-\frac{1}{3}\right)$$
$$x = \frac{1}{3}$$

53.
$$y + \frac{3}{8} = \frac{1}{4}$$
$$y + \frac{3}{8} - \frac{3}{8} = \frac{1}{4} - \frac{3}{8}$$
$$y + 0 = \frac{1}{4} + \left(-\frac{3}{8}\right)$$
$$y = -\frac{1}{8}$$

55.
$$t + \frac{1}{4} = -\frac{1}{2}$$
$$t + \frac{1}{4} - \frac{1}{4} = -\frac{1}{2} - \frac{1}{4}$$
$$t + 0 = -\frac{1}{2} + \left(-\frac{1}{4}\right)$$
$$t = -\frac{3}{4}$$

57.
$$y + \frac{2}{3} = -\frac{5}{12}$$
$$y + \frac{2}{3} - \frac{2}{3} = -\frac{5}{12} - \frac{2}{3}$$
$$y + 0 = -\frac{5}{12} + \left(-\frac{2}{3}\right)$$
$$y = -\frac{13}{12}$$
$$y = -1\frac{1}{12}$$

Objective C Exercises

59.
$$5x = 30$$
$$\frac{5x}{5} = \frac{30}{5}$$
$$1x = 6$$
$$x = 6$$

61.
$$3z = -27$$
$$\frac{3z}{3} = \frac{-27}{3}$$
$$1z = -9$$
$$z = -9$$

63.
$$-4t = 20$$
$$\frac{-4t}{-4} = \frac{20}{-4}$$
$$1t = -5$$
$$t = -5$$

65.
$$-2y = -28$$
$$\frac{-2y}{-2} = \frac{-28}{-2}$$
$$1y = 14$$
$$y = 14$$

67.
$$24 = 3y$$
$$\frac{24}{3} = \frac{3y}{3}$$
$$8 = 1y$$
$$8 = y$$

69.
$$-21 = 7y$$
$$\frac{-21}{7} = \frac{7y}{7}$$
$$-3 = 1y$$
$$-3 = y$$

71.
$$\frac{y}{2} = 10$$
$$\frac{1}{2}y = 10$$
$$2\left(\frac{1}{2}y\right) = 2(10)$$
$$1y = 20$$
$$y = 20$$

73.
$$\frac{y}{7} = -3$$
$$\frac{1}{7}y = -3$$
$$7\left(\frac{1}{7}y\right) = 7(-3)$$
$$1y = -21$$
$$y = -21$$

75.
$$\frac{-y}{3} = 5$$
$$-\frac{1}{3}y = 5$$
$$-3\left(-\frac{1}{3}y\right) = -3(5)$$
$$1y = -15$$
$$y = -15$$

77.
$$\frac{5}{8}x = 10$$
$$\frac{8}{5}\left(\frac{5}{8}x\right) = \frac{8}{5}(10)$$
$$1x = 16$$
$$x = 16$$

79.
$$\frac{2}{7}x = -12$$
$$\frac{7}{2}\left(\frac{2}{7}x\right) = \frac{7}{2}(-12)$$
$$1x = -42$$
$$x = -42$$

81.
$$-\frac{1}{5}y = -3$$
$$-5\left(-\frac{1}{5}y\right) = -5(-3)$$
$$1y = 15$$
$$y = 15$$

83.
$$\frac{5}{12}y = -16$$
$$\frac{12}{5}\left(\frac{5}{12}y\right) = \frac{12}{5}(-16)$$
$$1y = -\frac{192}{5}$$
$$y = -38\frac{2}{5}$$

85.
$$-8 = -\frac{5}{6}x$$
$$\left(-\frac{6}{5}\right)(-8) = \left(-\frac{6}{5}\right)\left(-\frac{5}{6}\right)x$$
$$\frac{48}{5} = 1x$$
$$9\frac{3}{5} = x$$

87.
$$-9 = \frac{5}{6}t$$
$$\frac{6}{5}(-9) = \left(\frac{6}{5}\right)\left(\frac{5}{6}\right)t$$
$$-\frac{54}{5} = 1t$$
$$-10\frac{4}{5} = t$$

89.
$$\frac{3}{7}y = \frac{5}{6}$$
$$\frac{7}{3}\left(\frac{3}{7}\right)y = \frac{7}{3}\left(\frac{5}{6}\right)$$
$$1y = \frac{35}{18}$$
$$y = 1\frac{17}{18}$$

91.
$$3a - 6a = 8$$
$$-3a = 8$$
$$\frac{-3a}{-3} = \frac{8}{-3}$$
$$1a = -\frac{8}{3}$$
$$a = -2\frac{2}{3}$$

93.
$$\frac{1}{3}b - \frac{2}{3}b = -1$$
$$-\frac{1}{3}b = -1$$
$$(-3)\left(-\frac{1}{3}b\right) = (-3)(-1)$$
$$1b = 3$$
$$b = 3$$

Objective D Application Problems

95. Strategy   To find the value of the original investment, replace the variables A and I by the given values and solve for P.

Solution
$$A = P + I$$
$$26,440 = P + 2830$$
$$26,440 - 2830 = P + 2830 - 2830$$
$$26,440 + (-2830) = P + 0$$
$$23,610 = P$$

The original investment was $23,610.

97. Strategy   To find the increase in the value of the investment, replace the variables A and P by the given values and solve for I.

Solution
$$A = P + I$$
$$8690 = 7500 + I$$
$$8690 - 7500 = 7500 - 7500 + I$$
$$8690 + (-7500) = 0 + I$$
$$1190 = I$$

The value of the fund increased by $1190.

99. Strategy   To find the number of gallons of gasoline used, replace the variables D and M in the formula by the given values and solve for G.

Solution
$$D = M \cdot G$$
$$592 = 32 \cdot G$$
$$\frac{592}{32} = \frac{32G}{32}$$
$$18.5 = G$$

18.5 gallons of gasoline was used.

101. Strategy   To find the number of miles per gallon, replace the variables D and G in the formula by the given values and solve for M.

Solution
$$D = M \cdot G$$
$$410 = M \cdot 12$$
$$\frac{410}{12} = \frac{12M}{12}$$
$$34.2 \approx M$$

The car gets 34.2 mi/gal.

103. Strategy   To find the markup, replace the variables S and C in the formula by the given values and solve for M.

Solution
$$S = C + M$$
$$39.80 = 23.50 + M$$
$$39.80 - 23.50 = 23.50 - 23.50 + M$$
$$39.80 + (-23.50) = 0 + M$$
$$16.30 = M$$

The markup on each shirt is $16.30.

105. Strategy   To find the cost of a compact disc, replace the variables S and R in the formula by the given values and solve for C.

Solution
$$S = C + RC$$
$$18.85 = C + 0.30C$$
$$18.85 = 1.30C$$
$$\frac{18.85}{1.30} = \frac{1.30C}{1.30}$$
$$14.50 = C$$

The compact disc costs $14.50.

Applying the Concepts

107.
$$\frac{3}{4}x = 6$$
$$\left(\frac{4}{3}\right)\left(\frac{3}{4}x\right) = \left(\frac{4}{3}\right)6 \quad \text{Multiplication Property of}$$
$$\text{Equations.}$$
$$x = 8$$

109. Answers will vary. For example, $x + 5 = 1$.

111. Students should rephrase the Addition Property of Equations (the same number or variable expression can be added to each side of an equation without changing the solution of the equation) and the Multiplication Property of Equations (each side of an equation can be multiplied by the same nonzero number without changing the solution of the equation).

SECTION 11.3

Objective A Exercises

1.
$$3x + 5 = 14$$
$$3x + 5 - 5 = 14 - 5$$
$$3x = 9$$
$$\frac{3x}{3} = \frac{9}{3}$$
$$x = 3$$

3.
$$2n - 3 = 7$$
$$2n - 3 + 3 = 7 + 3$$
$$2n = 10$$
$$\frac{2n}{2} = \frac{10}{2}$$
$$n = 5$$

5.
$$5w + 8 = 3$$
$$5w + 8 - 8 = 3 - 8$$
$$5w = -5$$
$$\frac{5w}{5} = \frac{-5}{5}$$
$$w = -1$$

7.
$$3z - 4 = -16$$
$$3z - 4 + 4 = -16 + 4$$
$$3z = -12$$
$$\frac{3z}{3} = \frac{-12}{3}$$
$$z = -4$$

9.
$$5 + 2x = 7$$
$$5 - 5 + 2x = 7 - 5$$
$$2x = 2$$
$$\frac{2x}{2} = \frac{2}{2}$$
$$x = 1$$

11.
$$6 - x = 3$$
$$6 + (-1)x = 3$$
$$6 - 6 + (-1)x = 3 - 6$$
$$(-1)x = -3$$
$$(-1)(-1)x = (-1)(-3)$$
$$x = 3$$

13.
$$3 - 4x = 11$$
$$3 - 3 - 4x = 11 - 3$$
$$-4x = 8$$
$$\frac{-4x}{-4} = \frac{8}{-4}$$
$$x = -2$$

15.
$$5 - 4x = 17$$
$$5 - 5 - 4x = 17 - 5$$
$$-4x = 12$$
$$\frac{-4x}{-4} = \frac{12}{-4}$$
$$x = -3$$

17.
$$3x + 6 = 0$$
$$3x + 6 - 6 = 0 - 6$$
$$3x = -6$$
$$\frac{3x}{3} = \frac{-6}{3}$$
$$x = -2$$

19.
$$-3x - 4 = -1$$
$$-3x - 4 + 4 = -1 + 4$$
$$-3x = 3$$
$$\frac{-3x}{-3} = \frac{3}{-3}$$
$$x = -1$$

21.
$$12x - 30 = 6$$
$$12x - 30 + 30 = 6 + 30$$
$$12x = 36$$
$$\frac{12x}{12} = \frac{36}{12}$$
$$x = 3$$

23.
$$3x + 7 = 4$$
$$3x + 7 - 7 = 4 - 7$$
$$3x = -3$$
$$\frac{3x}{3} = \frac{-3}{3}$$
$$x = -1$$

25.
$$-2x + 11 = -3$$
$$-2x + 11 - 11 = -3 - 11$$
$$-2x = -14$$
$$\frac{-2x}{-2} = \frac{-14}{-2}$$
$$x = 7$$

27.
$$14 - 5x = 4$$
$$14 - 14 - 5x = 4 - 14$$
$$-5x = -10$$
$$\frac{-5x}{-5} = \frac{-10}{-5}$$
$$x = 2$$

29.
$$-8x + 7 = -9$$
$$-8x + 7 - 7 = -9 - 7$$
$$-8x = -16$$
$$\frac{-8x}{-8} = \frac{-16}{-8}$$
$$x = 2$$

31.
$$9x + 13 = 13$$
$$9x + 13 - 13 = 13 - 13$$
$$9x = 0$$
$$\frac{9x}{9} = \frac{0}{9}$$
$$x = 0$$

33.
$$7x - 14 = 0$$
$$7x - 14 + 14 = 0 + 14$$
$$7x = 14$$
$$\frac{7x}{7} = \frac{14}{7}$$
$$x = 2$$

35.
$$4x - 4 = -4$$
$$4x - 4 + 4 = -4 + 4$$
$$4x = 0$$
$$\frac{4x}{4} = \frac{0}{4}$$
$$x = 0$$

37.
$$3x + 5 = 7$$
$$3x + 5 - 5 = 7 - 5$$
$$3x = 2$$
$$\frac{3x}{3} = \frac{2}{3}$$
$$x = \frac{2}{3}$$

39.
$$6x - 1 = 16$$
$$6x - 1 + 1 = 16 + 1$$
$$6x = 17$$
$$\frac{6x}{6} = \frac{17}{6}$$
$$x = 2\frac{5}{6}$$

41.
$$2x - 3 = -8$$
$$2x - 3 + 3 = -8 + 3$$
$$2x = -5$$
$$\frac{2x}{2} = \frac{-5}{2}$$
$$x = -2\frac{1}{2}$$

43.
$$-6x + 2 = -7$$
$$-6x + 2 - 2 = -7 - 2$$
$$-6x = -9$$
$$\frac{-6x}{-6} = \frac{-9}{-6}$$
$$x = 1\frac{1}{2}$$

45.
$$-2x - 3 = -7$$
$$-2x - 3 + 3 = -7 + 3$$
$$-2x = -4$$
$$\frac{-2x}{-2} = \frac{-4}{-2}$$
$$x = 2$$

47.
$$3x + 8 = 2$$
$$3x + 8 - 8 = 2 - 8$$
$$3x = -6$$
$$\frac{3x}{3} = \frac{-6}{3}$$
$$x = -2$$

49.
$$3x - 7 = 0$$
$$3x - 7 + 7 = 0 + 7$$
$$3x = 7$$
$$\frac{3x}{3} = \frac{7}{3}$$
$$x = 2\frac{1}{3}$$

51.
$$-2x + 9 = 12$$
$$-2x + 9 - 9 = 12 - 9$$
$$-2x = 3$$
$$\frac{-2x}{-2} = \frac{3}{-2}$$
$$x = -1\frac{1}{2}$$

53.
$$\frac{1}{2}x - 2 = 3$$
$$\frac{1}{2}x - 2 + 2 = 3 + 2$$
$$\frac{1}{2}x = 5$$
$$2\left(\frac{1}{2}x\right) = 5 \cdot 2$$
$$x = 10$$

55.
$$\frac{3}{5}w - 1 = 2$$
$$\frac{3}{5}w - 1 + 1 = 2 + 1$$
$$\frac{3}{5}w = 3$$
$$\frac{5}{3} \cdot \frac{3}{5}w = 3 \cdot \frac{5}{3}$$
$$w = 5$$

57.
$$\frac{2}{9}t - 3 = 5$$
$$\frac{2}{9}t - 3 + 3 = 5 + 3$$
$$\frac{2}{9}t = 8$$
$$\frac{9}{2} \cdot \frac{2}{9}t = 8 \cdot \frac{9}{2}$$
$$t = 36$$

59.
$$\frac{y}{3} - 6 = -8$$
$$\frac{y}{3} - 6 + 6 = -8 + 6$$
$$\frac{y}{3} = -2$$
$$3 \cdot \frac{y}{3} = 3(-2)$$
$$y = -6$$

61.
$$\frac{x}{3} - 2 = -5$$
$$\frac{x}{3} - 2 + 2 = -5 + 2$$
$$\frac{x}{3} = -3$$
$$3 \cdot \frac{x}{3} = 3(-3)$$
$$x = -9$$

63.
$$\frac{5}{8}v + 6 = 3$$
$$\frac{5}{8}v + 6 - 6 = 3 - 6$$
$$\frac{5}{8}v = -3$$
$$\frac{8}{5} \cdot \frac{5}{8}v = \frac{8}{5} \cdot (-3)$$
$$v = -\frac{24}{5} = -4\frac{4}{5}$$

65.
$$\frac{4}{7}z + 10 = 5$$
$$\frac{4}{7}z + 10 - 10 = 5 - 10$$
$$\frac{4}{7}z = -5$$
$$\frac{7}{4} \cdot \frac{4}{7}z = \frac{7}{4} \cdot (-5)$$
$$z = -\frac{35}{4} = -8\frac{3}{4}$$

67.
$$\frac{2}{9}x - 3 = 5$$
$$\frac{2}{9}x - 3 + 3 = 5 + 3$$
$$\frac{2}{9}x = 8$$
$$\frac{9}{2} \cdot \frac{2}{9}x = \frac{9}{2} \cdot 8$$
$$x = 36$$

69.
$$\frac{3}{4}x - 5 = -4$$
$$\frac{3}{4}x - 5 + 5 = -4 + 5$$
$$\frac{3}{4}x = 1$$
$$\frac{4}{3} \cdot \frac{3}{4}x = \frac{4}{3} \cdot 1$$
$$x = \frac{4}{3} = 1\frac{1}{3}$$

71.
$$1.5x - 0.5 = 2.5$$
$$1.5x - 0.5 + 0.5 = 2.5 + 0.5$$
$$1.5x = 3$$
$$\frac{1.5x}{1.5} = \frac{3}{1.5}$$
$$x = 2$$

73.
$$0.8t + 1.1 = 4.3$$
$$0.8t + 1.1 - 1.1 = 4.3 - 1.1$$
$$0.8t = 3.2$$
$$\frac{0.8t}{0.8} = \frac{3.2}{0.8}$$
$$t = 4$$

75.
$$0.4x - 2.3 = 1.3$$
$$0.4x - 2.3 + 2.3 = 1.3 + 2.3$$
$$0.4x = 3.6$$
$$\frac{0.4x}{0.4} = \frac{3.6}{0.4}$$
$$x = 9$$

77.
$$3.5y - 3.5 = 10.5$$
$$3.5y - 3.5 + 3.5 = 10.5 + 3.5$$
$$3.5y = 14$$
$$\frac{3.5y}{3.5} = \frac{14}{3.5}$$
$$y = 4$$

79.
$$0.32x + 4.2 = 3.2$$
$$0.32x + 4.2 - 4.2 = 3.2 - 4.2$$
$$0.32x = -1$$
$$\frac{0.32x}{0.32} = \frac{-1}{0.32}$$
$$x = -3.125$$

81.
$$6m + 2m - 3 = 5$$
$$8m - 3 = 5$$
$$8m - 3 + 3 = 5 + 3$$
$$8m = 8$$
$$\frac{8m}{8} = \frac{8}{8}$$
$$m = 1$$

83.
$$3y - 8y - 9 = 6$$
$$-5y - 9 = 6$$
$$-5y - 9 + 9 = 6 + 9$$
$$-5y = 15$$
$$\frac{-5y}{-5} = \frac{15}{-5}$$
$$y = -3$$

85.
$$-2y + y - 3 = 6$$
$$-y - 3 = 6$$
$$-y - 3 + 3 = 6 + 3$$
$$-y = 9$$
$$(-1)(-y) = (-1)9$$
$$y = -9$$

87.
$$0.032x - 0.0194 = 0.139$$
$$0.032x - 0.0194 + 0.0194 = 0.139 + 0.0194$$
$$0.032x = 0.1584$$
$$\frac{0.032x}{0.032} = \frac{0.1584}{0.032}$$
$$x = 4.95$$

89.
$$6.09x + 17.33 = 16.805$$
$$6.09x + 17.33 - 17.33 = 16.805 - 17.33$$
$$6.09x = -0.525$$
$$\frac{6.09x}{6.09} = \frac{-0.525}{6.09}$$
$$x \approx -0.0862069$$

Objective B Application Problems

91. Strategy To find the Celsius temperature, replace the variable F in the formula by the given value and solve for C.

Solution
$$F = \frac{9}{5}C + 32$$
$$-40 = \frac{9}{5}C + 32$$
$$-40 - 32 = \frac{9}{5}C + 32 - 32$$
$$-72 = \frac{9}{5}C$$
$$\frac{5}{9}(-72) = \frac{5}{9} \cdot \frac{9}{5}C$$
$$-40 = C$$

The Celsius temperature is $-40°$.

93. Strategy To find the time required, replace the variables V and V_0 in the formula by the given values and solve for t.

Solution
$$V = V_0 + 32t$$
$$472 = 8 + 32t$$
$$472 - 8 = 8 - 8 + 32t$$
$$464 = 32t$$
$$\frac{464}{32} = \frac{32t}{32}$$
$$14.5 = t$$
The time is 14.5 s.

95. Strategy To find the number of units made, replace the variables T, U, and F in the formula by the given values and solve for N.

Solution
$$T = U \cdot N + F$$
$$25,000 = 8 \cdot N + 5000$$
$$25,000 - 5000 = 8N + 5000 - 5000$$
$$20,000 = 8N$$
$$\frac{20,000}{8} = \frac{8N}{8}$$
$$2500 = N$$
The number of units made was 2500.

97. Strategy To find the monthly income, replace the variables T, R, and B in the formula by the given values and solve for I.

Solution
$$T = I \cdot R + B$$
$$476 = 0.22 \cdot I + 80$$
$$476 - 80 = 0.22I + 80 - 80$$
$$396 = 0.22I$$
$$\frac{396}{0.22} = \frac{0.22I}{0.22}$$
$$1800 = I$$
The mechanic's monthly income is $1800.

99. Strategy To find the total sales, replace the variables M, R, and B in the formula by the given values and solve for S.

Solution
$$M = S \cdot R + B$$
$$3480 = 0.09S + 600$$
$$3480 - 600 = 0.09S + 600 - 600$$
$$2880 = 0.09S$$
$$\frac{2880}{0.09} = \frac{0.09S}{0.09}$$
$$32,000 = S$$
The total sales were $32,000.

101. Strategy To find the commission rate, replace the variables M, S, and B in the formula by the given values and solve for R.

Solution
$$M = S \cdot R + B$$
$$2640 = 42{,}000\,R + 750$$
$$2640 - 750 = 42{,}000\,R + 750 - 750$$
$$1890 = 42{,}000\,R$$
$$\frac{1890}{42{,}000} = \frac{42{,}000\,R}{42{,}000}$$
$$0.045 = R$$
$$4.5\% = R$$

Miguel's commission rate was 4.5%.

Applying the Concepts

103.
$$\frac{2}{3}x - 4 = 10$$
$$\underline{+\qquad 4 \qquad +4}$$ Add 4 to both sides (Addition Property).
$$\frac{2}{3}x = 14$$
$$\left(\frac{3}{2}\right)\frac{2}{3}x = \left(\frac{3}{2}\right)14$$ Multiply both sides by $\frac{3}{2}$ (Multiplication Property).
$$x = 21$$

105. No, the sentence "Solve $3x + 4(x-3)$" does not make sense because
$$3x + 4(x-3)$$
is an expression, and you cannot solve an expression. You can solve an equation.

SECTION 11.4

1.
$$6x + 3 = 2x + 5$$
$$6x - 2x + 3 = 2x - 2x + 5$$
$$4x + 3 = 5$$
$$4x + 3 - 3 = 5 - 3$$
$$4x = 2$$
$$\frac{4x}{4} = \frac{2}{4}$$
$$x = \frac{1}{2}$$

3.
$$3x + 3 = 2x + 2$$
$$3x - 2x + 3 = 2x - 2x + 2$$
$$x + 3 = 2$$
$$x + 3 - 3 = 2 - 3$$
$$x = -1$$

5.
$$5x + 4 = x - 12$$
$$5x - x + 4 = x - x - 12$$
$$4x + 4 = -12$$
$$4x + 4 - 4 = -12 - 4$$
$$4x = -16$$
$$\frac{4x}{4} = \frac{-16}{4}$$
$$x = -4$$

7.
$$7x - 2 = 3x - 6$$
$$7x - 3x - 2 = 3x - 3x - 6$$
$$4x - 2 = -6$$
$$4x - 2 + 2 = -6 + 2$$
$$4x = -4$$
$$\frac{4x}{4} = \frac{-4}{4}$$
$$x = -1$$

9.
$$9x - 4 = 5x - 20$$
$$9x - 5x - 4 = 5x - 5x - 20$$
$$4x - 4 = -20$$
$$4x - 4 + 4 = -20 + 4$$
$$4x = -16$$
$$\frac{4x}{4} = \frac{-16}{4}$$
$$x = -4$$

11.
$$2x + 1 = 16 - 3x$$
$$2x + 3x + 1 = 16 - 3x + 3x$$
$$5x + 1 = 16$$
$$5x + 1 - 1 = 16 - 1$$
$$5x = 15$$
$$\frac{5x}{5} = \frac{15}{5}$$
$$x = 3$$

13.
$$5x - 2 = -10 - 3x$$
$$5x + 3x - 2 = -10 - 3x + 3x$$
$$8x - 2 = -10$$
$$8x - 2 + 2 = -10 + 2$$
$$8x = -8$$
$$\frac{8x}{8} = \frac{-8}{8}$$
$$x = -1$$

15.
$$2x + 7 = 4x + 3$$
$$2x - 4x + 7 = 4x - 4x + 3$$
$$-2x + 7 = 3$$
$$-2x + 7 - 7 = 3 - 7$$
$$-2x = -4$$
$$\frac{-2x}{-2} = \frac{-4}{-2}$$
$$x = 2$$

17.
$$x + 4 = 6x - 11$$
$$x - 6x + 4 = 6x - 6x - 11$$
$$-5x + 4 = -11$$
$$-5x + 4 - 4 = -11 - 4$$
$$-5x = -15$$
$$\frac{-5x}{-5} = \frac{-15}{-5}$$
$$x = 3$$

19.
$$3x - 7 = x - 7$$
$$3x - x - 7 = x - x - 7$$
$$2x - 7 = -7$$
$$2x - 7 + 7 = -7 + 7$$
$$2x = 0$$
$$\frac{2x}{2} = \frac{0}{2}$$
$$x = 0$$

21.
$$3 - 4x = 5 - 3x$$
$$3 - 4x + 3x = 5 - 3x + 3x$$
$$3 - x = 5$$
$$3 - 3 - x = 5 - 3$$
$$-x = 2$$
$$(-1)(-x) = (-1)2$$
$$x = -2$$

23.
$$7 + 3x = 9 + 5x$$
$$7 + 3x - 5x = 9 + 5x - 5x$$
$$7 - 2x = 9$$
$$7 - 7 - 2x = 9 - 7$$
$$-2x = 2$$
$$\frac{-2x}{-2} = \frac{2}{-2}$$
$$x = -1$$

25.
$$5 + 2x = 7 + 5x$$
$$5 + 2x - 5x = 7 + 5x - 5x$$
$$5 - 3x = 7$$
$$5 - 5 - 3x = 7 - 5$$
$$-3x = 2$$
$$\frac{-3x}{-3} = \frac{2}{-3}$$
$$x = -\frac{2}{3}$$

27.
$$8 - 5x = 4 - 6x$$
$$8 - 5x + 6x = 4 - 6x + 6x$$
$$8 + x = 4$$
$$8 - 8 + x = 4 - 8$$
$$x = -4$$

29.
$$6x + 1 = 3x + 2$$
$$6x - 3x + 1 = 3x - 3x + 2$$
$$3x + 1 = 2$$
$$3x + 1 - 1 = 2 - 1$$
$$3x = 1$$
$$\frac{3x}{3} = \frac{1}{3}$$
$$x = \frac{1}{3}$$

31.
$$5x + 8 = x + 5$$
$$5x - x + 8 = x - x + 5$$
$$4x + 8 = 5$$
$$4x + 8 - 8 = 5 - 8$$
$$4x = -3$$
$$\frac{4x}{4} = \frac{-3}{4}$$
$$x = -\frac{3}{4}$$

33.
$$2x - 3 = 6x - 4$$
$$2x - 6x - 3 = 6x - 6x - 4$$
$$-4x - 3 = -4$$
$$-4x - 3 + 3 = -4 + 3$$
$$-4x = -1$$
$$\frac{-4x}{-4} = \frac{-1}{-4}$$
$$x = \frac{1}{4}$$

35.
$$6 - 3x = 6 - 5x$$
$$6 - 3x + 5x = 6 - 5x + 5x$$
$$6 + 2x = 6$$
$$6 - 6 + 2x = 6 - 6$$
$$2x = 0$$
$$\frac{2x}{2} = \frac{0}{2}$$
$$x = 0$$

37.
$$6x - 2 = 2x - 9$$
$$6x - 2x - 2 = 2x - 2x - 9$$
$$4x - 2 = -9$$
$$4x - 2 + 2 = -9 + 2$$
$$4x = -7$$
$$\frac{4x}{4} = \frac{-7}{4}$$
$$x = -\frac{7}{4}$$
$$x = -1\frac{3}{4}$$

39.
$$6x - 3 = -5x + 8$$
$$6x + 5x - 3 = -5x + 5x + 8$$
$$11x - 3 = 8$$
$$11x - 3 + 3 = 8 + 3$$
$$11x = 11$$
$$\frac{11x}{11} = \frac{11}{11}$$
$$x = 1$$

41.
$$-6x - 2 = -8x - 4$$
$$-6x + 8x - 2 = -8x + 8x - 4$$
$$2x - 2 = -4$$
$$2x - 2 + 2 = -4 + 2$$
$$2x = -2$$
$$\frac{2x}{2} = \frac{-2}{2}$$
$$x = -1$$

43.
$$-3 - 4x = 7 - 2x$$
$$-3 - 4x + 2x = 7 - 2x + 2x$$
$$-3 - 2x = 7$$
$$-3 + 3 - 2x = 7 + 3$$
$$-2x = 10$$
$$\frac{-2x}{-2} = \frac{10}{-2}$$
$$x = -5$$

45.
$$3 - 7x = -2 + 5x$$
$$3 - 7x - 5x = -2 + 5x - 5x$$
$$3 - 12x = -2$$
$$3 - 3 - 12x = -2 - 3$$
$$-12x = -5$$
$$\frac{-12x}{-12} = \frac{-5}{-12}$$
$$x = \frac{5}{12}$$

47.
$$5x + 8 = 4 - 2x$$
$$5x + 2x + 8 = 4 - 2x + 2x$$
$$7x + 8 = 4$$
$$7x + 8 - 8 = 4 - 8$$
$$7x = -4$$
$$\frac{7x}{7} = \frac{-4}{7}$$
$$x = -\frac{4}{7}$$

49.
$$12x - 9 = 3x + 12$$
$$12x - 3x - 9 = 3x - 3x + 12$$
$$9x - 9 = 12$$
$$9x - 9 + 9 = 12 + 9$$
$$9x = 21$$
$$\frac{9x}{9} = \frac{21}{9}$$
$$x = \frac{7}{3}$$
$$x = 2\frac{1}{3}$$

51.
$$\frac{5}{7}x - 3 = \frac{2}{7}x + 6$$
$$\frac{5}{7}x - \frac{2}{7}x - 3 = \frac{2}{7}x - \frac{2}{7}x + 6$$
$$\frac{3}{7}x - 3 = 6$$
$$\frac{3}{7}x - 3 + 3 = 6 + 3$$
$$\frac{3}{7}x = 9$$
$$\frac{7}{3} \cdot \frac{3}{7}x = \frac{7}{3} \cdot 9$$
$$x = 21$$

53.
$$\frac{3}{7}x + 5 = \frac{5}{7}x - 1$$
$$\frac{3}{7}x - \frac{5}{7}x + 5 = \frac{5}{7}x - \frac{5}{7}x - 1$$
$$\frac{-2}{7}x + 5 = -1$$
$$\frac{-2}{7}x + 5 - 5 = -1 - 5$$
$$\frac{-2}{7}x = -6$$
$$\left(\frac{-7}{2}\right)\left(\frac{-2}{7}x\right) = \left(\frac{-7}{2}\right)(-6)$$
$$x = 21$$

Objective B Exercises

55.
$$6x + 2(x - 1) = 14$$
$$6x + 2x - 2 = 14$$
$$8x - 2 = 14$$
$$8x - 2 + 2 = 14 + 2$$
$$8x = 16$$
$$\frac{8x}{8} = \frac{16}{8}$$
$$x = 2$$

57.
$$-3 + 4(x + 3) = 5$$
$$-3 + 4x + 12 = 5$$
$$4x + 9 = 5$$
$$4x + 9 - 9 = 5 - 9$$
$$4x = -4$$
$$\frac{4x}{4} = \frac{-4}{4}$$
$$x = -1$$

59.
$$6 - 2(x + 4) = 6$$
$$6 - 2x - 8 = 6$$
$$-2x - 2 = 6$$
$$-2x - 2 + 2 = 6 + 2$$
$$-2x = 8$$
$$\frac{-2x}{-2} = \frac{8}{-2}$$
$$x = -4$$

61.
$$5 + 7(x + 3) = 20$$
$$5 + 7x + 21 = 20$$
$$7x + 26 = 20$$
$$7x + 26 - 26 = 20 - 26$$
$$7x = -6$$
$$\frac{7x}{7} = \frac{-6}{7}$$
$$x = -\frac{6}{7}$$

63.
$$2x + 3(x - 5) = 10$$
$$2x + 3x - 15 = 10$$
$$5x - 15 = 10$$
$$5x - 15 + 15 = 10 + 15$$
$$5x = 25$$
$$\frac{5x}{5} = \frac{25}{5}$$
$$x = 5$$

65.
$$3(x - 4) + 2x = 3$$
$$3x - 12 + 2x = 3$$
$$5x - 12 = 3$$
$$5x - 12 + 12 = 3 + 12$$
$$5x = 15$$
$$\frac{5x}{5} = \frac{15}{5}$$
$$x = 3$$

67.
$$2x - 3(x - 4) = 12$$
$$2x - 3x + 12 = 12$$
$$-x + 12 = 12$$
$$-x + 12 - 12 = 12 - 12$$
$$-x = 0$$
$$(-1)(-x) = (-1)0$$
$$x = 0$$

69.
$$2x + 3(x + 4) = 7$$
$$2x + 3x + 12 = 7$$
$$5x + 12 = 7$$
$$5x + 12 - 12 = 7 - 12$$
$$5x = -5$$
$$\frac{5x}{5} = \frac{-5}{5}$$
$$x = -1$$

71.
$$3(x-2)+5=5$$
$$3x-6+5=5$$
$$3x-1=5$$
$$3x-1+1=5+1$$
$$3x=6$$
$$\frac{3x}{3}=\frac{6}{3}$$
$$x=2$$

73.
$$3x+7(x-2)=5$$
$$3x+7x-14=5$$
$$10x-14=5$$
$$10x-14+14=5+14$$
$$10x=19$$
$$\frac{10x}{10}=\frac{19}{10}$$
$$x=1\frac{9}{10}$$

75.
$$4x-2(x+9)=8$$
$$4x-2x-18=8$$
$$2x-18=8$$
$$2x-18+18=8+18$$
$$2x=26$$
$$\frac{2x}{2}=\frac{26}{2}$$
$$x=13$$

77.
$$3x+5(x-2)=10$$
$$3x+5x-10=10$$
$$8x-10=10$$
$$8x-10+10=10+10$$
$$8x=20$$
$$\frac{8x}{8}=\frac{20}{8}$$
$$x=\frac{5}{2}=2\frac{1}{2}$$

79.
$$3x+4(x+2)=2(x+9)$$
$$3x+4x+8=2x+18$$
$$7x+8=2x+18$$
$$7x-2x+8=2x-2x+18$$
$$5x+8=18$$
$$5x+8-8=18-8$$
$$5x=10$$
$$\frac{5x}{5}=\frac{10}{5}$$
$$x=2$$

81.
$$2x-3(x-4)=2(x+6)$$
$$2x-3x+12=2x+12$$
$$-x+12-12=2x+12-12$$
$$-x=2x$$
$$-x-2x=2x-2x$$
$$-3x=0$$
$$\frac{-3x}{-3}=\frac{0}{-3}$$
$$x=0$$

83.
$$7-2(x-3)=3(x-1)$$
$$7-2x+6=3x-3$$
$$-2x+13=3x-3$$
$$-2x+13-13=3x-3-13$$
$$-2x=3x-16$$
$$-2x-3x=3x-3x-16$$
$$-5x=-16$$
$$\frac{-5x}{-5}=\frac{-16}{-5}$$
$$x=3\frac{1}{5}$$

85.
$$6x-2(x-3)=11(x-2)$$
$$6x-2x+6=11x-22$$
$$4x+6=11x-22$$
$$4x-11x+6=11x-11x-22$$
$$-7x+6=-22$$
$$-7x+6-6=-22-6$$
$$-7x=-28$$
$$\frac{-7x}{-7}=\frac{-28}{-7}$$
$$x=4$$

87.
$$6x-3(x+1)=5(x+2)$$
$$6x-3x-3=5x+10$$
$$3x-3=5x+10$$
$$3x-5x-3=5x-5x+10$$
$$-2x-3=10$$
$$-2x-3+3=10+3$$
$$-2x=13$$
$$\frac{-2x}{-2}=\frac{13}{-2}$$
$$x=-6\frac{1}{2}$$

89.
$$7-(x+1)=3(x+3)$$
$$7-x-1=3x+9$$
$$-x+6=3x+9$$
$$-x-3x+6=3x-3x+9$$
$$-4x+6=9$$
$$-4x+6-6=9-6$$
$$-4x=3$$
$$\frac{-4x}{-4}=\frac{3}{-4}$$
$$x=-\frac{3}{4}$$

91.
$$2x-3(x+4)=2(x-5)$$
$$2x-3x-12=2x-10$$
$$-x-12=2x-10$$
$$-x-2x-12=2x-2x-10$$
$$-3x-12=-10$$
$$-3x-12+12=-10+12$$
$$-3x=2$$
$$\frac{-3x}{-3}=\frac{2}{-3}$$
$$x=-\frac{2}{3}$$

93.
$$x + 5(x - 4) = 3(x - 8) - 5$$
$$x + 5x - 20 = 3x - 24 - 5$$
$$6x - 20 = 3x - 29$$
$$6x - 3x - 20 = 3x - 3x - 29$$
$$3x - 20 = -29$$
$$3x - 20 + 20 = -29 + 20$$
$$3x = -9$$
$$\frac{3x}{3} = \frac{-9}{3}$$
$$x = -3$$

95.
$$9x - 3(x - 4) = 13 + 2(x - 3)$$
$$9x - 3x + 12 = 13 + 2x - 6$$
$$6x + 12 = 2x + 7$$
$$6x - 2x + 12 = 2x - 2x + 7$$
$$4x + 12 = 7$$
$$4x + 12 - 12 = 7 - 12$$
$$4x = -5$$
$$\frac{4x}{4} = \frac{-5}{4}$$
$$x = -1\frac{1}{4}$$

97.
$$3(x - 4) + 3x = 7 - 2(x - 1)$$
$$3x - 12 + 3x = 7 - 2x + 2$$
$$6x - 12 = -2x + 9$$
$$6x + 2x - 12 = -2x + 2x + 9$$
$$8x - 12 = 9$$
$$8x - 12 + 12 = 9 + 12$$
$$8x = 21$$
$$\frac{8x}{8} = \frac{21}{8}$$
$$x = 2\frac{5}{8}$$

99.
$$3.67x - 5.3(x - 1.932) = 6.99$$
$$3.67x - 5.3x + 10.2396 = 6.99$$
$$-1.63x + 10.2396 = 6.99$$
$$-1.63x + 10.2396 - 10.2396 = 6.99 - 10.2396$$
$$-1.63x = -3.2496$$
$$\frac{-1.63x}{-1.63} = \frac{-3.2496}{-1.63}$$
$$x \approx 1.9936196$$

101.
$$8.45(x - 10) = 3(x - 3.854)$$
$$8.45x - 84.5 = 3x - 11.562$$
$$8.45x - 3x - 84.5 = 3x - 3x - 11.562$$
$$5.45x - 84.5 = -11.562$$
$$5.45x - 84.5 + 84.5 = -11.562 + 84.5$$
$$5.45x = 72.938$$
$$\frac{5.45x}{5.45} = \frac{72.938}{5.45}$$
$$x \approx 13.3831193$$

Applying the Concepts

103.
$$2x - 2 = 4x + 6$$
$$-8 = 2x$$
$$-4 = x$$
Then $3x^2 = 3(-4)^2 = 48$.

105. Students should explain that the solution of the original equation is $x = 0$. Therefore, the fourth line, where each side of the equation is divided by x, involved division by zero, which is not defined.

SECTION 11.5

Objective A Exercises

1. $y - 9$

3. $z + 3$

5. $\frac{2}{3}n + n$

7. $\frac{m}{m - 3}$

9. $9(x + 4)$

11. $n - (-5)n$

13. $c\left(\frac{1}{4}c\right)$

15. $m^2 + 2m^2$

17. $2(t + 6)$

19. $\frac{x}{9 + x}$

21. $3(b + 6)$

Objective B Exercises

23. The *square* of a number
The unknown number: x
x^2

25. A number *divided* by twenty
The unknown number: x
$\frac{x}{20}$

27. Four *times* some number
The unknown number: x
$4x$

29. Three-fourths *of* a number
The unknown number: x
$\frac{3}{4}x$

31. Four *increased* by some number
The unknown number: x
$4 + x$

33. The *difference between* five *times* a number and the number
The unknown number: x
Five times the number: $5x$
$5x - x$

35. The *product* of a number and two *more than* the number
The unknown number: x
Two more than the number: $x + 2$
$x(x + 2)$

37. Seven *times* the *total* of a number and eight
The unknown number: x
The total of the number and eight: $x + 8$
$7(x + 8)$

39. The *square* of a number *plus* the *product* of three and the number
The unknown number: x
The square of the number: x^2
The product of three and the number: $3x$
$x^2 + 3x$

41. The *sum* of three *more than* a number and one-half *of* the number
The unknown number: x
Three more than the number: $x + 3$
One-half of the number: $\dfrac{1}{2}x$

$(x + 3) + \dfrac{1}{2}x$

43. The *quotient* of three times a number and the number
The unknown number: x
Three times the number: $3x$
$\dfrac{3x}{x}$

Applying the Concepts

45. $2x + 3$: The sum of twice a number and 3.
$2(x + 3)$: Twice the sum of a number and 3.

47. Students will provide different explanations of how variables are used. Look for the idea that a variable is used to represent a number that is unknown or a number that can change, or vary.

49. $\dfrac{1}{2}x$: number of carbon atoms

$\dfrac{1}{2}x$: number of oxygen atoms

SECTION 11.6

Objective A Exercises

1. The unknown number: x
$x + 7 = 12$
$x + 7 - 7 = 12 - 7$
$x = 5$
The number is 5.

3. The unknown number: x
$3x = 18$
$\dfrac{3x}{3} = \dfrac{18}{3}$
$x = 6$

5. The unknown number: x
$x + 5 = 3$
$x + 5 - 5 = 3 - 5$
$x = -2$
The number is -2.

7. The unknown number: x
$6x = 14$
$\dfrac{6x}{6} = \dfrac{14}{6}$
$x = \dfrac{7}{3} = 2\dfrac{1}{3}$
The number is $2\dfrac{1}{3}$.

9. The unknown number: x
$\dfrac{5}{6}x = 15$
$\dfrac{6}{5} \cdot \dfrac{5}{6}x = \dfrac{6}{5} \cdot 15$
$x = 18$
The number is 18.

11. The unknown number: x
$3x + 4 = 8$
$3x + 4 - 4 = 8 - 4$
$3x = 4$
$\dfrac{3x}{3} = \dfrac{4}{3}$
$x = \dfrac{4}{3} = 1\dfrac{1}{3}$
The number is $1\dfrac{1}{3}$.

13. The unknown number: x
$\dfrac{1}{4}x - 7 = 9$
$\dfrac{1}{4}x - 7 + 7 = 9 + 7$
$\dfrac{1}{4}x = 16$
$4 \cdot \dfrac{1}{4}x = 4 \cdot 16$
$x = 64$
The number is 64.

15. The unknown number: x
$\dfrac{x}{9} = 14$
$9 \cdot \dfrac{x}{9} = 9 \cdot 14$
$x = 126$
The number is 126.

17. The unknown number: x

$$\frac{x}{4} - 6 = -2$$

$$\frac{x}{4} - 6 + 6 = -2 + 6$$

$$\frac{x}{4} = 4$$

$$4 \cdot \frac{x}{4} = 4 \cdot 4$$

$$x = 16$$

The number is 16.

19. The unknown number: x

$$7 - 2x = 13$$

$$7 - 7 - 2x = 13 - 7$$

$$-2x = 6$$

$$\frac{-2x}{-2} = \frac{6}{-2}$$

$$x = -3$$

The number is –3.

21. The unknown number: x

$$9 - \frac{x}{2} = 5$$

$$9 - 9 - \frac{x}{2} = 5 - 9$$

$$-\frac{x}{2} = -4$$

$$(-2)\left(-\frac{x}{2}\right) = (-2)(-4)$$

$$x = 8$$

The number is 8.

23. The unknown number: x

$$\frac{3}{5}x + 8 = 2$$

$$\frac{3}{5}x + 8 - 8 = 2 - 8$$

$$\frac{3}{5}x = -6$$

$$\frac{5}{3} \cdot \frac{3}{5}x = \frac{5}{3} \cdot (-6)$$

$$x = -10$$

The number is –10.

25. The unknown number: x

$$\frac{x}{4.186} - 7.92 = 12.529$$

$$\frac{x}{4.186} - 7.92 + 7.92 = 12.529 + 7.92$$

$$\frac{x}{4.186} = 20.449$$

$$4.186 \cdot \frac{x}{4.186} = 4.186(20.449)$$

$$x = 85.599514$$

The number is 85.599514.

Objective B Application Problems

27. Strategy   To find the price of a pair of shoes at Target, write and solve an equation using P to represent the price at the competitor's store.

Solution
$$72.50 = P - 4.25$$
$$72.50 + 4.25 = P - 4.25 + 4.25$$
$$76.75 = P$$
The price at Target is $76.75.

29. Strategy   To find the value of the SUV last year, write and solve an equation using V to represent the value of the SUV last year.

Solution
$$\frac{4}{5}V = 16,000$$
$$\frac{5}{4} \cdot \frac{4}{5}V = \frac{5}{4} \cdot 16,000$$
$$V = 20,000$$
The value of the SUV last year was $20,000.

31. Strategy   To find the total quick-serve restaurant market in 2000, write and solve an equation using M to represent the quick-serve market.

Solution
$$7.3\% \times M = 19$$
$$0.073M = 19$$
$$\frac{0.073M}{0.073} = \frac{19}{0.073}$$
$$M \approx 260$$
The total quick-serve restaurant market in 2000 was approximately $260 billion.

33. Strategy   To find the family's monthly income, write and solve an equation using S to represent the family's income.

Solution
$$860 = \frac{1}{4}S$$
$$4 \cdot 860 = 4 \cdot \frac{1}{4}S$$
$$3440 = S$$
The family's monthly salary is $3440.

35. Strategy   To find the monthly output a year ago, write and solve an equation using M to represent the monthly output a year ago.

Solution
$$400 = 0.08M$$
$$\frac{400}{0.08} = \frac{0.08M}{0.08}$$
$$5000 = M$$
The monthly output a year ago was 5000 computers.

37. Strategy   To find the recommended daily intake of sodium:
- Write and solve an equation using x to represent the daily intake of sodium.
- Convert the milligrams to grams.

Solution
$$8\% \cdot x = 200$$
$$0.08x = 200$$
$$\frac{0.08x}{0.08} = \frac{200}{0.08}$$
$$x = 2,500$$
$$2,500 \text{ mg} = 2.5 \text{ g}$$
The recommended daily intake of sodium is 2.5 g.

39. Strategy   To find how long it took the plumber to install a water softener, write and solve an equation using T to represent the time it took to install a water softener.

Solution
$$385 = 310 + 25T$$
$$385 - 310 = 310 - 310 + 25T$$
$$75 = 25T$$
$$\frac{75}{25} = \frac{25T}{25}$$
$$3 = T$$
It took 3 h to install a water softener.

41. Strategy   To find the number of plants and animals at risk, write and solve an equation using P to represent the number of plants and animals.

Solution
$$10.7\% \cdot P = 1184$$
$$0.107P = 1184$$
$$\frac{0.107P}{0.107} = \frac{1184}{0.107}$$
$$P \approx 11,065$$
About 11,065 plants and animals are at risk of extinction in the world.

43. Strategy   To find the percent spent in New York, write and solve the basic percent equation using P to represent the percent. The base is 295 and the amount is 9.8.

Solution
$$\text{Percent} \times \text{base} = \text{amount}$$
$$P \times 295 = 9.8$$
$$295P = 9.8$$
$$\frac{295P}{295} = \frac{9.8}{295}$$
$$P \approx 0.0332$$
The percent is 3.3%.

45. Strategy   To find the original rate of water flow, write and solve an equation using R to represent the original rate.

Solution
$$2 = \frac{3}{5}R - 1$$
$$2 + 1 = \frac{3}{5}R - 1 + 1$$
$$3 = \frac{3}{5}R$$
$$\frac{5}{3} \cdot 3 = \frac{5}{3} \cdot \frac{3}{5}R$$
The original water flow rate was 5 gal/min.

47. Strategy   To find the total sales for the month, write and solve an equation using T to represent the total sales.

Solution
$$600 + 0.0825T = 4109.55$$
$$600 - 600 + 0.0825T = 4109.55 - 600$$
$$0.0825T = 3509.55$$
$$\frac{0.0825T}{0.0825} = \frac{3509.55}{0.0825}$$
$$T = 42,540$$
The total sales for the month were $42,540.

49. Strategy   After adding subsidy and employee contribution to find total cost, convert the appropriate fraction to percent.

Solution
$$0.71 + 2.59 = 3.30$$
$$\frac{0.71}{3.30} \approx 0.215$$
21.5% of lunchroom meals are subsidized.

Applying the Concepts

51. The problem states that a 4-quart mixture of fruit juice is made from apple juice and cranberry juice. There are 6 more quarts of apple juice than of cranberry juice. If we let x = the number of quarts of cranberry juice, then $x + 6$ = the number of quarts apple juice. The total number of quarts is 4. Therefore, we can write the equation $x + (x + 6) = 4$.
$$x + (x + 6) = 4$$
$$2x = -2$$
$$x = -1$$
Since x = the number of quarts of cranberry juice, there are -1 quarts of cranberry juice in the mixture. We cannot add -1 quarts to a mixture. The solution is not reasonable.
We see from the original problem that the answer will not be reasonable. If the total number of quarts in the mixture is 4, we cannot have more than 6 quarts of apple juice in the mixture.

CHAPTER REVIEW

1. $-2(a-b) = -2[a+(-b)]$
$\qquad = -2(a)+(-2)(-b)$
$\qquad = -2a+2b$

2.
$$3x-2 = -8$$

$3(-2)-2$	-8
$-6-2$	-8
$-6+(-2)$	-8
$-8 = -8$	

Yes, -2 is a solution.

3. $\quad x-3 = -7$
$\quad x-3+3 = -7+3$
$\qquad\qquad x = -4$

4. $\quad -2x+5 = -9$
$-2x+5-5 = -9-5$
$\qquad -2x = -14$
$\qquad \dfrac{-2x}{-2} = \dfrac{-14}{-2}$
$\qquad\qquad x = 7$

5. $a^2-3b = (2)^2-3(-3)$
$\qquad\qquad = 4+9 = 13$

6. $\quad -3x = 27$
$\quad \dfrac{-3x}{-3} = \dfrac{27}{-3}$
$\qquad\quad x = -9$

7. $\quad \dfrac{2}{3}x+3 = -9$
$\quad \dfrac{2}{3}x+3-3 = -9-3$
$\qquad\quad \dfrac{2}{3}x = -12$
$\qquad \dfrac{3}{2}\cdot\dfrac{2}{3}x = \dfrac{3}{2}(-12)$
$\qquad\qquad x = -18$

8. $3x-2(3x-2) = 3x+(-2)[3x+(-2)]$
$\qquad\qquad = 3x+(-2)(3x)+(-2)(-2)$
$\qquad\qquad = 3x+(-6x)+4$
$\qquad\qquad = -3x+4$

9. $\quad 6x-9 = -3x+36$
$6x+3x-9 = -3x+3x+36$
$\qquad 9x-9 = 36$
$9x-9+9 = 36+9$
$\qquad\quad 9x = 45$
$\qquad\quad \dfrac{9x}{9} = \dfrac{45}{9}$
$\qquad\qquad x = 5$

10. $\quad x+3 = -2$
$\quad x+3-3 = -2-3$
$\qquad\qquad x = -5$

11.
$$3x-5 = -10$$

$3(5)-5$	-10
$15-5$	-10
$15+(-5)$	-10
$10 \neq -10$	

No, 5 is not a solution.

12. $a^2-(b\div c) = (-2)^2-[8\div(-4)]$
$\qquad\qquad\qquad = 4-(-2) = 6$

13. $3(x-2)+2 = 11$
$\quad 3x-6+2 = 11$
$\qquad 3x-4 = 11$
$3x-4+4 = 11+4$
$\qquad\quad 3x = 15$
$\qquad\quad \dfrac{3x}{3} = \dfrac{15}{3}$
$\qquad\qquad x = 5$

14. $\qquad 35-3x = 5$
$35-35-3x = 5-35$
$\qquad\quad -3x = -30$
$\qquad\quad \dfrac{-3x}{-3} = \dfrac{-30}{-3}$
$\qquad\qquad x = 10$

15. $6bc-7bc+2bc-5bc$
$= 6bc+(-7)bc+2b+(-5)bc$
$= (-1)bc+2bc+(-5)bc$
$= 1bc+(-5)bc$
$= -4bc$

16. $\qquad 7-3x = 2-5x$
$7-3x+5x = 2-5x+5x$
$\qquad 7+2x = 2$
$7-7+2x = 2-7$
$\qquad\quad 2x = -5$
$\qquad\quad \dfrac{2x}{2} = \dfrac{-5}{2}$
$\qquad\qquad x = \dfrac{-5}{2} = -2\dfrac{1}{2}$

17. $\qquad -\dfrac{3}{8}x = -\dfrac{15}{32}$
$-\dfrac{8}{3}\left(-\dfrac{3}{8}x\right) = \left(-\dfrac{8}{3}\right)\left(-\dfrac{15}{32}\right)$
$\qquad\qquad x = \dfrac{5}{4} = 1\dfrac{1}{4}$

18. $\dfrac{1}{2}x^2 - \dfrac{1}{3}x^2 + \dfrac{1}{5}x^2 + 2x^2$

$= \dfrac{1}{2}x^2 + \left(-\dfrac{1}{3}\right)x^2 + \dfrac{1}{5}x^2 + 2x^2$

$= \dfrac{3}{6}x^2 + \left(-\dfrac{2}{6}\right)x^2 + \dfrac{1}{5}x^2 + 2x^2$

$= \dfrac{1}{6}x^2 + \dfrac{1}{5}x^2 + 2x^2$

$= \dfrac{5}{30}x^2 + \dfrac{6}{30}x^2 + 2x^2$

$= \dfrac{11}{30}x^2 + 2x^2$

$= \dfrac{11}{30}x^2 + \dfrac{60}{30}x^2$

$= \dfrac{71}{30}x^2$

19. $5x - 3(1 - 2x) = 4(2x - 1)$

$5x - 3 + 6x = 8x - 4$

$11x - 3 = 8x - 4$

$11x - 8x - 3 = 8x - 8x - 4$

$3x - 3 = -4$

$3x - 3 + 3 = -4 + 3$

$3x = -1$

$\dfrac{3x}{3} = \dfrac{-1}{3}$

$x = -\dfrac{1}{3}$

20. $\dfrac{5}{6}x - 4 = 5$

$\dfrac{5}{6}x - 4 + 4 = 5 + 4$

$\dfrac{5}{6}x = 9$

$\dfrac{6}{5} \cdot \dfrac{5}{6}x = \dfrac{6}{5} \cdot 9$

$x = \dfrac{54}{5} = 10\dfrac{4}{5}$

21. Strategy To find the number of miles per gallon of gas, replace D and G in the formula by the given values and solve for M.

Solution

$D = M \cdot G$

$621 = M \cdot 27$

$\dfrac{621}{27} = \dfrac{27M}{27}$

$23 = M$

The number of miles per gallon of gas is 23.

22. Strategy To find the Celsius temperature, replace the variable F in the formula by the given value and solve for C.

Solution

$F = \dfrac{9}{5}C + 32$

$100 = \dfrac{9}{5}C + 32$

$100 - 32 = \dfrac{9}{5}C + 32 - 32$

$68 = \dfrac{9}{5}C$

$\dfrac{5}{9}(68) = \dfrac{5}{9} \cdot \dfrac{9}{5}C$

$37.8 \approx C$

The Celsius temperature is 37.8°.

23. The *total* of n and the *quotient* of n and 5

The unknown number: n

The quotient of n and 5: $\dfrac{n}{5}$

$n + \dfrac{n}{5}$

24. The *sum* of five more than a number and one-third *of* the number

The unknown number: n

Five more than a number: $n + 5$

$(n + 5) + \dfrac{1}{3}n$

25. The unknown number: x

$9 - 2x = 5$

$9 - 9 - 2x = 5 - 9$

$-2x = -4$

$\dfrac{-2x}{-2} = \dfrac{-4}{-2}$

$x = 2$

The number is 2.

26. The unknown number: p

$5p = 50$

$\dfrac{5p}{5} = \dfrac{50}{5}$

$p = 10$

The number is 10.

27. Strategy To find the regular price, write and solve an equation using R to represent the regular price.

Solution

$228 = 60\% \cdot R$

$228 = 0.60R$

$\dfrac{228}{0.60} = \dfrac{0.60R}{0.60}$

$380 = R$

The regular price of the CD player is $380.

28. Strategy Let x represent last year's crop. Then $0.12x$ is the increase in last year's crop. Last year's crop plus the increase is this year's crop (28,336 bushels).

Solution $0.12x + x = 26{,}336$
$1.12x = 28{,}336$
$x = 25{,}300$
Last year's crop was 25,300 bushels.

CHAPTER TEST

1. $\dfrac{x}{5} - 12 = 7$

$\dfrac{x}{5} - 12 + 12 = 7 + 12$

$\dfrac{x}{5} = 19$

$5 \cdot \dfrac{x}{5} = 5 \cdot 19$

$x = 95$

2. $x - 12 = 14$
$x - 12 + 12 = 14 + 12$
$x = 26$

3. $3y - 2x - 7y - 9x = 3y + (-2x) + (-7y) + (-9x)$
$\qquad = 3y + (-7y) + (-2x) + (-9x)$
$\qquad = -4y + (-2x) + (-9x)$
$\qquad = -4y + (-11x)$
$\qquad = -4y - 11x = -11x - 4y$

4. $8 - 3x = 2x - 8$
$8 - 3x + 3x = 2x + 3x - 8$
$8 = 5x - 8$
$8 + 8 = 5x - 8 + 8$
$16 = 5x$
$\dfrac{16}{5} = \dfrac{5x}{5}$
$\dfrac{16}{5} = x$
$3\dfrac{1}{5} = x$

5. $3x - 12 = -18$
$3x - 12 + 12 = -18 + 12$
$3x = -6$
$\dfrac{3x}{3} = \dfrac{-6}{3}$
$x = -2$

6. $c^2 - (2a + b^2) = (-2)^2 - [2(3) + (-6)^2]$
$\qquad = 4 - (6 + 36)$
$\qquad = 4 - (42)$
$\qquad = 4 + (-42) = -38$

7. $x^2 + 3x - 7 = 3x - 2$

$3^2 + 3(3) - 7$	$3(3) - 2$
$9 + 9 - 7$	$9 - 2$
$18 - 7$	7
$11 \ne 7$	

No, 3 is not a solution.

8. $9 - 8ab - 6ab = 9 - 14ab = -14ab + 9$

9. $-5x = 14$
$\dfrac{-5x}{-5} = \dfrac{14}{-5}$
$x = -2\dfrac{4}{5}$

10. $3y + 5(y - 3) + 8 = 3y + 5[y + (-3)] + 8$
$\qquad = 3y + 5y + 5(-3) + 8$
$\qquad = 8y + (-15) + 8$
$\qquad = 8y + (-7)$
$\qquad = 8y - 7$

11. $3x - 4(x - 2) = 8$
$3x - 4x + 8 = 8$
$-x + 8 = 8$
$-x + 8 - 8 = 8 - 8$
$-x = 0$
$(-1)(-x) = (-1)0$
$x = 0$

12. $5 = 3 - 4x$
$5 - 3 = 3 - 3 - 4x$
$2 = -4x$
$\dfrac{2}{-4} = \dfrac{-4x}{-4}$
$-\dfrac{1}{2} = x$

13. $\dfrac{x^2}{y} - \dfrac{y^2}{x} = \dfrac{3^2}{-2} - \dfrac{(-2)^2}{3}$
$\qquad = \dfrac{9}{-2} - \dfrac{4}{3}$
$\qquad = \dfrac{-27}{6} - \dfrac{8}{6}$
$\qquad = \dfrac{-35}{6} = -5\dfrac{5}{6}$

14. $\dfrac{5}{8}x = -10$
$\dfrac{8}{5} \cdot \dfrac{5}{8}x = \dfrac{8}{5}(-10)$
$x = -16$

15. $y - 4y + 3 = 12$
$-3y + 3 = 12$
$-3y + 3 - 3 = 12 - 3$
$-3y = 9$
$\dfrac{-3y}{-3} = \dfrac{9}{-3}$
$y = -3$

16. $2x + 4(x - 3) = 5x - 1$
$2x + 4x - 12 = 5x - 1$
$6x - 12 = 5x - 1$
$6x - 5x - 12 = 5x - 5x - 1$
$x - 12 = -1$
$x - 12 + 12 = -1 + 12$
$x = 11$

17. Strategy   To find the monthly payment, replace the variables L and N in the formula by the given values and solve for P.

Solution
$$L = P \cdot N$$
$$6600 = P \cdot 48$$
$$\frac{6600}{48} = \frac{48P}{48}$$
$$137.50 = P$$
The monthly payment is $137.50.

18. Strategy   To find the number of clocks made during a month, replace the variables T, U, and F in the formula by the given values and solve for N.

Solution
$$T = U \cdot N + F$$
$$65,000 = 15N + 5000$$
$$65,000 - 5000 = 15N + 5000 - 5000$$
$$60,000 = 15N$$
$$\frac{60,000}{15} = \frac{15N}{15}$$
$$4000 = N$$
The number of clocks made during a month was 4000.

19. Strategy   To find the time, replace the variables V and V_0 in the formula by the given values and solve for t.

Solution
$$V = V_0 + 32t$$
$$392 = 24 + 32t$$
$$392 - 24 = 24 - 24 + 32t$$
$$368 = 32t$$
$$\frac{368}{32} = \frac{32t}{32}$$
$$11.5 = t$$
The object will fall for 11.5 s.

20. The *sum* of x and one-third *of* x
The unknown number: x
One-third of x: $\frac{1}{3}x$

$$x + \frac{1}{3}x$$

21. Five *times* the *sum* of a number and three
The unknown number: x
The sum of a number and three: $x + 3$
$5(x + 3)$

22. The unknown number: x
$$2x - 3 = 7$$
$$2x - 3 + 3 = 7 + 3$$
$$2x = 10$$
$$\frac{2x}{2} = \frac{10}{2}$$
$$x = 5$$
The number is 5.

23. The unknown number: w
$$5 + 3w = w - 2$$
$$5 + 3w - w = w - w - 2$$
$$5 + 2w = -2$$
$$5 - 5 + 2w = -2 - 5$$
$$2w = -7$$
$$\frac{2w}{2} = \frac{-7}{2}$$
$$w = -3\frac{1}{2}$$
The number is $-3\frac{1}{2}$.

24. Strategy   To find Santos's total sales for the month, write and solve an equation using T to represent the total sales.

Solution
$$3600 = 1200 + 0.06T$$
$$3600 - 1200 = 1200 - 1200 + 0.06T$$
$$2400 = 0.06T$$
$$\frac{2400}{0.06} = \frac{0.06T}{0.06}$$
$$40,000 = T$$
Santos's total sales for the month were $40,000.

25. Strategy   To find the number of hours worked, write and solve an equation using h to represent the number of hours worked.

Solution
$$152 + 42h = 278$$
$$152 - 152 + 42h = 278 - 152$$
$$42h = 126$$
$$h = 3$$
The mechanic worked for 3 hours.

CUMULATIVE REVIEW

1. $6^2 - (18 - 6) \div 4 + 8 = 36 - (12) \div 4 + 8$
$$= 36 - 3 + 8$$
$$= 33 + 8 = 41$$

2.
$$3\frac{1}{6} = 3\frac{5}{30} = 2\frac{35}{30}$$
$$-1\frac{7}{15} = 1\frac{14}{30} = 1\frac{14}{30}$$
$$\overline{\qquad\qquad 1\frac{21}{30} = 1\frac{7}{10}}$$

3. $\left(\frac{3}{8} - \frac{1}{4}\right) \div \frac{3}{4} + \frac{4}{9} = \left(\frac{3}{8} - \frac{2}{8}\right) \div \frac{3}{4} + \frac{4}{9}$
$$= \frac{1}{8} \div \frac{3}{4} + \frac{4}{9}$$
$$= \frac{1}{8} \times \frac{4}{3} + \frac{4}{9}$$
$$= \frac{4}{24} + \frac{4}{9}$$
$$= \frac{1}{6} + \frac{4}{9}$$
$$= \frac{3}{18} + \frac{8}{18} = \frac{11}{18}$$

4.
$$\begin{array}{r} 9.67 \\ \times\ \ 0.0049 \\ \hline 8703 \\ 3868\ \ \\ \hline 0.047383 \end{array}$$

5. $\dfrac{\$84}{20\ \text{h}} = \$4.20/\text{h}$

6.
$$\dfrac{2}{3} = \dfrac{n}{40}$$
$$2 \times 40 = 3 \cdot n$$
$$80 = 3 \cdot n$$
$$80 \div 3 = n$$
$$26.67 \approx n$$

7. $5\dfrac{1}{3}\% = \dfrac{16}{3} \times \dfrac{1}{100} = \dfrac{16}{300} = \dfrac{4}{75}$

8. Percent × base = amount
$$n \times 30 = 42$$
$$n = 42 \div 30$$
$$n = 1.40 = 140\%$$

9. Percent × base = amount
$$125\% \times n = 8$$
$$1.25 \times n = 8$$
$$n = 8 \div 1.25 = 6.4$$

10.
$$\begin{array}{r} 3\ \text{ft}\ 9\ \text{in.} \\ \times\ \ \ \ \ \ \ 5 \\ \hline 15\ \text{ft}\ 45\ \text{in.} = 18\ \text{ft}\ 9\ \text{in.} \end{array}$$

11. $1\dfrac{3}{8}\ \text{lb} = \dfrac{11}{8}\ \cancel{\text{lb}} \times \dfrac{16\ \text{oz}}{1\ \cancel{\text{lb}}} = \dfrac{11}{\underset{1}{\cancel{8}}} \cdot \dfrac{\overset{2}{\cancel{16}}\ \text{oz}}{1} = 22\ \text{oz}$

12. $282\ \text{mg} = 0.282\ \text{g}$

13. $-2 + 5 + (-8) + 4 = 3 + (-8) + 4$
$$= -5 + 4 = -1$$

14. $13 - (-6) = 13 + 6 = 19$

15. $(-2)^2 - (-8) \div (3-5)^2 = (-2)^2 - (-8) \div (-2)^2$
$$= 4 - (-8) \div 4$$
$$= 4 - (-2) = 4 + 2 = 6$$

16. $3ab - 2ac = 3(-2)(6) - 2(-2)(-3)$
$$= -36 - 12 = -36 + (-12) = -48$$

17. $3z - 2x + 5z - 8x = 3z + (-2x) + 5z + (-8x)$
$$= 3z + 5z + (-2x) + (-8x)$$
$$= 8z + (-10x)$$
$$= 8z - 10x = -10x + 8z$$

18. $6y - 3(y-5) + 8 = 6y + (-3)[y + (-5)] + 8$
$$= 6y + (-3)y + (-3)(-5) + 8$$
$$= 6y + (-3y) + 15 + 8$$
$$= 3y + 23$$

19.
$$2x - 5 = -7$$
$$2x - 5 + 5 = -7 + 5$$
$$2x = -2$$
$$2x = -2$$
$$\dfrac{2x}{2} = \dfrac{-2}{2}$$
$$x = -1$$

20.
$$7x - 3(x-5) = -10$$
$$7x - 3x + 15 = -10$$
$$4x + 15 = -10$$
$$4x + 15 - 15 = -10 - 15$$
$$4x = -25$$
$$\dfrac{4x}{4} = \dfrac{-25}{4}$$
$$x = -6\dfrac{1}{4}$$

21.
$$-\dfrac{2}{3}x = 5$$
$$\left(-\dfrac{3}{2}\right)\left(-\dfrac{2}{3}\right)x = -\dfrac{3}{2} \cdot 5$$
$$x = -\dfrac{15}{2} = -7\dfrac{1}{2}$$

22.
$$\dfrac{x}{3} - 5 = -12$$
$$\dfrac{x}{3} - 5 + 5 = -12 + 5$$
$$\dfrac{x}{3} = -7$$
$$3 \cdot \dfrac{x}{3} = 3(-7)$$
$$x = -21$$

23. Strategy   To find the percent of the students who received an A grade, solve the basic percent equation for percent.

Solution   Percent · base = amount
$$n \cdot 34 = 6$$
$$n = 6 \div 34$$
$$n \approx 0.176 = 17.6\%$$
The percent is 17.6%.

24. Strategy   To find the price:
- Find the amount of the markup by solving the basic percent equation for amount. The base is $28.50 and the percent is 40%.
- Add the amount of the markup to the cost.

Solution
$$0.40 \times 28.50 = n$$
$$11.40 = n$$
$$\begin{array}{r} \$28.50 \\ +11.40 \\ \hline \$39.90 \end{array}$$

The price of the piece of pottery is $39.90.

25a. Strategy   To find the discount subtract the sale price ($369) from the regular price ($450).

Solution $450 - 369 = 81$
The discount is $81.

25b. Strategy   To find the discount rate, write and solve the basic percent equation for percent. The base is the discount and the amount is the regular price

     Solution

$$\text{Percent} \times \text{base} = \text{amount}$$
$$n \times 450 = 81$$
$$n = 81 \div 450$$
$$n = 0.18$$

The discount rate is 18%.

26. Strategy   To find the simple interest due, multiply the principal and rate and time (in years).

     Solution

$$\text{Interest} = 80,000 \times 11\% \times \frac{4}{12}$$
$$= 80,000 \times 0.11 \times \frac{4}{12}$$
$$= 2933.3\overline{3}$$

The simple interest due on the loan is $2933.33.

27. The *sum* of three *times* a number and four
The unknown number: n
Three times the number: $3n$
$3n + 4$

28. Strategy   To calculate the probability:
    • Count the number of possible outcomes.
    • Count the number of favorable outcomes.
    • Use the probability formula.

     Solution   There are 16 possible outcomes.
There are 2 favorable outcomes: (3, 4), (4, 3).

$$\text{Probability} = \frac{2}{16} = \frac{1}{8}$$

The probability is $\frac{1}{8}$ that the sum of the upward faces on the two dice is 7.

29. Strategy   To find the total sales, replace the variables M, R, and B in the formula with the given values and solve for S.

     Solution

$$M = S \cdot R + B$$
$$3400 = S \cdot 0.08 + 800$$
$$3400 - 800 = S \cdot 0.08 + 800 - 800$$
$$2600 = S \cdot 0.08$$
$$\frac{2600}{0.08} = \frac{S \cdot 0.08}{0.08}$$
$$32,500 = S$$

The total sales were $32,500.

30. The unknown number: x
$$8x - 3 = 3 + 5x$$
$$8x - 5x - 3 = 3 + 5x - 5x$$
$$3x - 3 = 3$$
$$3x - 3 + 3 = 3 + 3$$
$$3x = 6$$
$$\frac{3x}{3} = \frac{6}{3}$$
$$x = 2$$
The number is 2.

Chapter 12: Geometry

PREP TEST and GO FIGURE

1. $$x + 47 = 90$$
    $$x + 47 - 47 = 90 - 47$$
    $$x = 43$$

2. $$32 + 97 + x = 180$$
    $$129 + x = 180$$
    $$129 - 129 + x = 180 - 129$$
    $$x = 51$$

3. $$2(18) + 2(10) = 36 + 20 = 56$$

4. $$abc$$
    $$= (2)(3.14)(9)$$
    $$= (6.28)(9)$$
    $$= 56.52$$

5. $$xyz^3$$
    $$= \left(\frac{4}{3}\right)(3.14)(3)^3$$
    $$= 113.04$$

6. $$\frac{5}{12} = \frac{6}{x}$$
    $$5x = 12 \times 6$$
    $$\frac{5x}{5} = \frac{36}{5}$$
    $$x = 14.4$$

Go Figure

The first figure is a diamond (D) inside a square (S) inside a triangle (T) inside a circle (C), or DSTC.
The second figure is STCD.
The third figure is TCDS.
The next figure would be CDST: a circle inside a diamond inside a square inside a triangle.

SECTION 12.1

Objective A Exercises

1. $0°$ and $90°$

3. $180°$

5. $$EG = EF + FG$$
    $$EG = 20 + 10 = 30$$

7. $$RS = QS - QR$$
    $$RS = 28 - 7 = 21$$

9. $$BC = AD - AB - CD$$
    $$BC = 35 - 12 - 9 = 14$$

11. $90° - 31° = 59°$
    $59°$ is the complement of $31°$.

13. $180° - 72° = 108°$
    $108°$ is the supplement of $72°$.

15. $90° - 13° = 77°$
    $77°$ is the complement of $13°$.

17. $180° - 127° = 53°$
    $53°$ is the supplement of $127°$.

19. $\angle AOB = 32° + 45° = 77°$

21. $\angle a = 160° - 42° = 118°$

23. $\angle a = 180° - 47° = 133°$

25. $\angle MON = \angle LON - \angle LOM$
    $\angle MON = 139° - 53° = 86°$

Objective B Exercises

27. $180°$

29. Rectangle or square

31. Cube

33. Parallelogram, rectangle, or square

35. Cylinder

37. $13° + 65° = 78°$; $180° - 78° = 102°$.
    The third angle is $102°$.

39. $90° - 45° = 45°$. The other angles of the triangle are $90°$ and $45°$.

41. $62° + 104° = 166°$; $180° - 166° = 14°$. The third angle is $14°$.

43. $90° - 25° = 65°$. The other angles of the triangle are $90°$ and $65°$.

45. Radius $= \frac{1}{2} \cdot$ diameter $= \frac{1}{2} \cdot 16$ in. $= 8$ in.

47. Diameter $= 2 \cdot$ radius $= 2 \cdot 2\frac{1}{3}$ ft $= 4\frac{2}{3}$ ft

49. Diameter $= 2 \cdot$ radius $= 2 \cdot 3.5$ cm $= 7$ cm

51. Radius $= \frac{1}{2} \cdot$ diameter $= \frac{1}{2}(4$ ft 8 in.$) = 2$ ft 4 in.

Objective C Exercises

53. $\angle b = 74°$. $\angle a$ and $\angle b$ are supplementary; therefore $\angle a = 180° - 74° = 106°$.

55. $\angle a = 112°$. $\angle a$ and $\angle b$ are supplementary; therefore $\angle b = 180° - 112° = 68°$.

57. $\angle a = 38°$. $\angle a$ and $\angle b$ are supplementary; therefore $\angle b = 180° - 38° = 142°$.

59. $\angle a = 180° - 122° = 58°$. $\angle a$ and $\angle b$ are alternate interior angles; therefore $\angle b = 58°$.

61. $\angle b = 180° - 28° = 152°$. $\angle a$ and $\angle b$ are corresponding; therefore $\angle a = 152°$.

63. $\angle b = 180° - 130° = 50°$; $\angle a = 130°$.

Applying the Concepts

65. **a.** $1°$

  b. $90°$

67. $\angle AOC$ and $\angle BOC$ are supplementary angles. Therefore, $\angle AOC + \angle BOC = 180°$. Because $\angle AOC = \angle BOC$, by substitution $\angle AOC + \angle AOC = 180°$. Therefore, $2\angle AOC = 180°$ and $\angle AOC = 90°$. Therefore, AB is perpendicular to CD.

SECTION 12.2

Objective A Exercises

1. Perimeter = side 1 + side 2 + side 3
  = 12 in. + 20 in. + 24 in. = 56 in.

3. Perimeter = $4 \cdot$ side
  = $4(5$ ft$) = 20$ ft

5. Perimeter = 2(width) + 2(length)
  = 2(14 cm) + 2(32 cm)
  = 28 cm + 64 cm = 92 cm

7. Circumference = $\pi \cdot$ diameter = $15\pi = 47.1$ cm

9. Perimeter = side 1 + side 2 + side 3
  = 2 ft 4 in. + 3 ft + 4 ft 6 in.
  = 9 ft 10 in.

11. Circumference = $2 \cdot \pi \cdot$ radius
  = $2 \cdot 3.14 \cdot 8$ cm = 50.24 cm

13. Perimeter = $4 \cdot$ side = $4 \cdot 60$ m = 240 m

15. Perimeter = sum of sides
  = 22 cm + 47 cm + 29 cm + 42 cm
    + 17 cm
  = 157 cm

Objective B Exercises

17. Perimeter = sum of sides
  = 19 cm + 20 cm + 8 cm + 5 cm + 27 cm + 42 cm
  = 121 cm

19. Perimeter = length of three sides
    $+ \frac{1}{2}$ circumference of circle
  = 8 m + (2 · 15) m + $\frac{1}{2}$ · (3.14 · 8 m)
  = 8 m + 30 m + 12.56 m = 50.56 m

21. Perimeter = length of two sides
    $+ \frac{1}{2}$ circumference of circle
  = $2 \cdot 1$ ft + $\frac{1}{2}$(3.14·1 ft)
  = 2 ft + 1.57 ft
  = 3.57 ft

23. Perimeter = sum of six sides of figure
  = 22.75 m + 25.73 m + 15.94 m
    + 18.3 m + 21.61 m + 34.97 m
  = 139.3 m

Objective C Application Problems

25. Strategy To find the amount of fencing, use the formula for the perimeter of a rectangle.

  Solution Perimeter = $2 \cdot 18 + 2 \cdot 12$
    = 36 + 24 = 60
  The amount of fencing needed is 60 ft.

27. Strategy To find the amount of binding, find the perimeter of a rectangle.

  Solution Perimeter = $2 \cdot 3.5 + 2 \cdot 8.5$
    = 7 + 17 = 24
  The amount of binding needed is 24 ft.

29. Strategy To find the cost of the fence:
  • Find the length of fence that is not along the road.
  • Multiply the length of fence not along the road by $5.85.
  • Multiply the length along the road by $6.20.
  • Add the two products to find the total cost.

  Solution Length = 800 + 1250 + 800 = 2850
  $2850 \times \$5.85 = \$16,672.50$
  $1250 \times \$6.20 = \$7,750$
  $\$16,672.50 + \$7,750 = \$24,422.50$
  The total cost of the fence is $24,422.50.

31. Strategy To find the distance the bike travels:
  • Convert diameter (24 in.) to feet.
  • Use the formula for circumference to find the distance traveled in 1 revolution.
  • Multiply the distance traveled in 1 revolution by the number of revolutions (5).

  Solution 24 in. = $24 \text{ in.} \times \frac{1 \text{ ft}}{12 \text{ in.}} = \frac{24}{12}$ ft = 2 ft
  Circumference = $\pi \cdot$ diameter
    = $3.14 \cdot 2 = 6.28$ ft
  6.28 ft · 5 = 31.4 ft
  The bicycle travels 31.4 ft.

33a. Strategy   To estimate whether the perimeter is more or less than 70 m, assume that the figure is a rectangle with length 25 m and width 10 m.

Solution   Perimeter $= 2 \cdot 25 + 2 \cdot 10$
$= 50 + 20 = 70$
The perimeter of the rink is larger than the perimeter of the rectangle. The perimeter of the rink is larger than 70 m.

33b. Strategy   To find the perimeter of the roller rink, find the perimeter of the composite figure.

Solution   Perimeter
$=$ sum of length of two sides
$+ 2$ times $\frac{1}{2}$ circumference of
a circle
$\approx 2 \cdot 25 + 2\frac{1}{2}(3.14 \cdot 10)$
$= 50 + 31.4 = 81.4$
The perimeter of the rink is 81.4 m.

35. Strategy   To find the length of weather stripping, find the perimeter of the composite figure.

Solution   Perimeter
$=$ sum of three sides
$+ \frac{1}{2}$ circumference of circle
$\approx (3 \text{ ft}) + (2 \cdot 6 \text{ ft } 6 \text{ in.}) + \frac{1}{2}(3.14 \cdot 3 \text{ ft})$
$= 3 \text{ ft} + 13 \text{ ft} + 4.71 \text{ ft} = 20.71 \text{ ft}$
Approximately 20.71 ft of weather stripping are installed.

37. Strategy   To find the circumference of Earth, use the formula for the circumference of a circle.

Solution   Circumference $= 2 \cdot \pi \cdot \text{radius}$
$\approx 2 \cdot 3.14 \cdot 6356 \text{ km}$
$= 39,915.68 \text{ km}$
The circumference of Earth is approximately 39,915.68 km.

Applying the Concepts

39. Length $= 8$
Width $= 3$
Perimeter $= 2(8) + 2(3) = 16 + 6 = 22$ units

41. The perimeter of one large triangle is 3 cm. (1 on each side) as in Figure A.

The perimeter of three medium triangles is $3\left(\frac{3}{2}\right)$

$\left(\frac{1}{2} \text{ on each side}\right)$ as in Figure B.

The perimeter of nine small triangles is $9\left(\frac{3}{4}\right)$

$\left(\frac{1}{4} \text{ on each side}\right)$ as in Figure C.

The perimeter of all shaded triangles
$= 3 + 3\left(\frac{3}{2}\right) + 9\left(\frac{3}{4}\right)$
$= 3 + \frac{9}{2} + \frac{27}{4}$
$= 13\frac{5}{4} = 14\frac{1}{4}$ cm

43. The ranger could measure the circumference of the trunk of the tree and then solve the equation $C = \pi d$ for d.

SECTION 12.3

Objective A Exercises

1. Area $=$ length $\cdot$ width $= 24 \text{ ft} \cdot 6 \text{ ft} = 144 \text{ ft}^2$

3. Area $= (\text{side})^2 = (9 \text{ in.})^2 = 81 \text{ in}^2$

5. Area $= \pi(\text{radius})^2$
$\approx 3.14(4 \text{ ft})^2 = 50.24 \text{ ft}^2$

7. Area $= \frac{1}{2} \cdot \text{base} \cdot \text{height}$
$= \frac{1}{2} \cdot (10 \text{ in.})(4 \text{ in.}) = 20 \text{ in}^2$

9. Area $= \frac{1}{2} \cdot \text{base} \cdot \text{height}$
$= \frac{1}{2} \cdot 3 \text{ cm} \cdot 1.42 \text{ cm} = 2.13 \text{ cm}^2$

11. Area $= (\text{side})^2 = 4 \text{ ft} \cdot 4 \text{ ft} = 16 \text{ ft}^2$

13. Area $=$ length $\cdot$ width $= 43 \text{ in.} \cdot 19 \text{ in.} = 817 \text{ in}^2$

15. Area $= \pi(\text{radius})^2 \approx \frac{22}{7} \cdot 7 \text{ in.} \cdot 7 \text{ in.} = 154 \text{ in}^2$

Objective B Exercises

17. Area $=$ area of rectangle $-$ area of triangle
$= (\text{length} \cdot \text{width}) - \left(\frac{1}{2} \cdot \text{base} \cdot \text{height}\right)$
$= (8 \text{ cm} \cdot 4 \text{ cm}) - \left(\frac{1}{2} \cdot 4 \text{ cm} \cdot 3 \text{ cm}\right)$
$= 32 \text{ cm}^2 - 6 \text{ cm}^2$
$= 26 \text{ cm}^2$

19. Area = area of rectangle− area of triangle

$$= (\text{length} \cdot \text{width}) - \left(\frac{1}{2} \cdot \text{base} \cdot \text{height}\right)$$

$$= (80 \text{ cm} \cdot 30 \text{ cm}) - \left(\frac{1}{2} \cdot 30 \text{ cm} \cdot 12 \text{ cm}\right)$$

$$= 2400 \text{ cm}^2 - 180 \text{ cm}^2$$

$$= 2220 \text{ cm}^2$$

21. Area = area of circle $- \frac{1}{4}$ area of circle

$$= \pi(\text{radius})^2 - \frac{1}{4} \cdot \pi(\text{radius})^2$$

$$\approx 3.14 \cdot 8 \text{ in.} \cdot 8 \text{ in.} - \frac{1}{4} \cdot 3.14 \cdot 8 \text{ in.} \cdot 8 \text{ in.}$$

$$= 200.96 \text{ in}^2 - 50.24 \text{ in}^2$$

$$= 150.72 \text{ in}^2$$

23. Area = area of rectangle $- \frac{1}{2}$ area of circle

$$= \text{length} \cdot \text{width} - \frac{1}{2} \cdot \pi(\text{radius})^2$$

$$\approx 4.38 \text{ ft} \cdot 3.74 \text{ ft} - \frac{1}{2} \cdot 3.14 \cdot 2.19 \text{ ft} \cdot 2.19 \text{ ft}$$

$$= 16.3812 \text{ft}^2 - 7.529877 \text{ ft}^2$$

$$= 8.851323 \text{ ft}^2$$

Objective C Application Problems

25. Strategy To find area of the playing field, find the area of a rectangle with length 100 yd and width 75 yd.

Solution Area = length · width
= 100 yd · 75 yd
= 7500 yd^2

The area of the playing field is 7500 yd^2.

27. Strategy To find the area of the field, find the area of a circle with a radius of 50 ft.

Solution Area = $\pi(\text{radius})^2 \approx 3.14 \cdot 50$ ft · 50 ft
= 7850 ft^2

The area watered by the irrigation system is approximately 7850 ft^2.

29. Strategy To find the amount of stain:
• Find the area of a rectangle that measures 10 ft 3 8 ft.
• Divide the area by the area that one quart of stain will cover (50 ft^2).

Solution Area = length · width
= 10 ft · 8 ft
= 80 ft^2
80 ft^2 ÷ 50 ft^2 = 1.6
It will take 1.6 quarts of stain. You should buy 2 quarts.

31. Strategy To find the area of the driveway, subtract the area of the small rectangle from the area of the large rectangle.

Solution Area = area of large rectangle
− area of small rectangle
= (length · width) − (length · width)
= (75 ft · 30 ft) − (50 ft · 20 ft)
= 2250 ft^2 − 1000 ft^2
= 1250 ft^2

The area of the driveway is 1250 ft^2.

33. Strategy To find the cost of the wallpaper:
• Use the formula for the area of a rectangle to find the areas of the two walls.
• Add the areas of the two walls.
• Divide the total area by the area in one roll (40 ft^2) to find the total number of rolls.
• Multiply the number of rolls by $18.50.

Solution Area$_1$ = 9 ft · 8 ft = 72 ft^2
Area$_2$ = 11 ft · 8 ft = 88 ft^2
Total area = 72 ft^2 + 88 ft^2 = 160 ft^2
160 ft^2 ÷ 40 ft^2 = 4
4 · $18.50 = $74
The cost to wallpaper the two walls is $74.

35. Strategy To find the amount budgeted:
• Use the formula for the area of a square to find the area of the park.
• Divide the area by 1,200 ft^2 to find the number of bags of seed.
• Multiply the number of bags of seed by $5.75.

Solution Area = side · side
= 60 ft · 60 ft = 3,600 ft^2
3,600 ft^2 ÷ 1,200 ft^2 = 3
3 · $5.75 = $17.25
$17.25 should be budgeted for grass seed.

37. Strategy   To find the total area of the park:
 • Find the area of the first rectangle.
 • Find the area of the second rectangle.
 • Find the area of the semicircle.
 • Add all the areas together.

Solution   Area of first rectangle $= 12.7 \times 2.5 = 31.75$
Area of second rectangle $= 17.5 \times 4.3 = 75.25$

Area of semicircle $= 0.5 \times 3.14 \times (3.4)^2 = 18.1492$
Total area $= 31.75 + 75.25 + 18.1492 = 125.1492$
The total area of the park is 125.1492 mi^2.

39a. Strategy   To find the increase in area:
 • Find the area of the original circle.
 • Find the area of the increased circle.
 • Subtract the area of the original circle from the area of the larger circle.

Solution   Area of original circle $\approx (64)(3.14) = 200.96 \text{ in}^2$
Area of increased circle $\approx (100)(3.14) = 314 \text{ in}^2$
$314 - 200.96 = 113.04 \text{ in}^2$
The amount of the increase is about 113.04 in^2.

39b. Strategy   To find the increase in area:
 • Find the area of the original circle.
 • Find the area of the increased circle.
 • Subtract the area of the original circle from the area of the larger circle.

Solution   Area of original circle $\approx (25)(3.14) = 78.5 \text{ cm}^2$
Area of increased circle $\approx (100)(3.14) = 314 \text{ cm}^2$
$314 - 78.5 = 235.5 \text{ cm}^2$
The amount of the increase is about 235.5 cm^2.

Applying the Concepts

41. a. Area $=$ length $\cdot$ width
$= 2 \cdot$ length $\cdot 2 \cdot$ width
$= 4 \cdot$ length $\cdot$ width
If the length and width are doubled, the area is increased 4 times.

b. Area $= \pi \cdot (\text{radius})^2$
$= \pi \cdot (2 \cdot \text{radius})^2$
$= \pi \cdot 4 \cdot (\text{radius})^2$
$= 4\pi \cdot (\text{radius})^2$
If the radius is doubled, the area is quadrupled.

c. If the diameter is doubled, the radius is doubled. From (b) if the radius is doubled, the area is quadrupled.

43. a. Sometimes true.

b. Sometimes true.

c. Always true.

SECTION 12.4

Objective A Exercises

1. Volume = length · width · height
    = 12 cm · 4 cm · 3 cm = 144 cm^3

3. Volume = (side)^3 = (8 in.)^3 = 512 in^3

5. Volume = $\frac{4}{3}\pi$(radius)^3 ≈ $\frac{4}{3}$(3.14)(8 in.)^3
    ≈ 2143.57 in^3

7. Volume = π(radius)^2 · height
    ≈ 3.14(2 cm)^2 · 12 cm = 150.72 cm^3

9. Volume = length · width · height
    = 2 m · 0.8 m · 4 m = 6.4 m^3

11. Volume = $\frac{4}{3}$ · π(radius)^3
    ≈ $\frac{4}{3}$ · 3.14 · 11 mm · 11 mm · 11 mm
    ≈ 5572.45 mm^3

13. Radius = $\frac{1}{2}$ · diameter = $\frac{1}{2}$ · 12 ft = 6 ft
    Volume = π(radius)^2 · height
    ≈ 3.14 · 6 ft · 6 ft · 30 ft
    = 3391.2 ft^3

15. Volume = (side)^3
    = $3\frac{1}{2}$ ft · $3\frac{1}{2}$ ft · $3\frac{1}{2}$ ft = $42\frac{7}{8}$ ft^3

Objective B Exercises

17. Volume = $\frac{1}{2}$ volume of cylinder + volume of rectangular solid
    = $\left[\frac{1}{2} \cdot \pi(\text{radius})^2 \cdot \text{height}\right]$ + (length · width · height)
    ≈ $\left(\frac{1}{2} \cdot 3.14 \cdot 3 \text{ in.} \cdot 3 \text{ in.} \cdot 2 \text{ in.}\right)$ + (6 in. · 9 in. · 1 in.) = 28.26 in^3 + 54 in^3 = 82.26 in^3

19. Volume = volume of rectangular solid − volume of cylinder = (length · width · height) − [π(radius)^2 · height]
    ≈ (1.20 m · 2 m · 0.80 m) − (3.14 · 0.20 m · 0.20 m · 2 m) = 1.92 m^3 − 0.2512 m^3 = 1.6688 m^3

21. Volume = volume of cylinder + volume of cylinder = [π(radius)^2 · height] + [π(radius)^2 · height]
    ≈ (3.14 · 3 in. · 3 in. · 2 in.) + (3.14 · 1 in. · 1 in. · 4 in.)
    = 56.52 in^3 + 12.56 in^3 = 69.08 in^3

Objective C Application Problems

23. Strategy   To find the volume of the tank, use the formula for the volume of a rectangular solid.

    Solution   Volume = length · width · height
    = 9 m · 3 m · 1.5 m
    = 40.5 m^3
    The volume of the water in the tank is 40.5 m^3.

25. Strategy   To find the volume of the balloon, use the formula for the volume of a sphere.

    Solution   Volume = $\frac{4}{3}$ · π(radius)^3
    ≈ $\frac{4}{3}$ · 3.14 · 16 ft · 16 ft · 16 ft
    ≈ 17,148.59 ft^3
    The volume is approximately 17,148.59 ft^3.

27. Strategy   To find the amount of oil:
    • Use the formula for the volume of a cylinder to find the volume of the tank.
    • Multiply the volume by $\frac{2}{3}$ to find the amount of oil.

    Solution   Volume = π(radius)^2 · height
    ≈ 3.14 · 3 m · 3 m · 4 m
    = 113.04 m^3
    Amount of oil = $\frac{2}{3}$ · 113.04 m^3
    = 75.36 m^3
    The amount of oil is approximately 75.36 m^3.

29. Strategy To find the volume of the auditorium, find the volume of the rectangular solid and add half the volume of the cylinder.

Solution Volume = length · width · height

$$+\frac{1}{2}[\pi(\text{radius})^2 \cdot \text{height}]$$

$$\approx 125 \text{ ft} \cdot 94 \text{ ft} \cdot 32 \text{ ft}$$

$$+\frac{1}{2}[3.14(47 \text{ ft})^2 \cdot 125 \text{ ft}]$$

$$= 376,000 \text{ ft}^3 + 433,516.25 \text{ ft}^3$$

$$= 809,516.25 \text{ ft}^3$$

The volume of the auditorium is approximately $809,516.25 \text{ ft}^3$.

31. Strategy To find the volume of the bushing, subtract the volume of the half-cylinder from the volume of the rectangular solid.

Solution Volume = length · width · height

$$-\frac{1}{2}[\pi(\text{radius})^2 \cdot \text{height}]$$

$$\approx (12 \text{ in.} \cdot 8 \text{ in.} \cdot 3 \text{ in.})$$

$$-\frac{1}{2}(3.14 \cdot 2 \text{ in.} \cdot 2 \text{ in.} \cdot 12 \text{ in.})$$

$$= 288 \text{ in}^3 - 75.36 \text{ in}^3$$

$$= 212.64 \text{ in}^3$$

The volume of the bushing is approximately 212.64 in^3.

33. Strategy To find the number of gallons in the fish tank:
• Use the formula for the volume of a rectangular solid.
• Convert the volume to gallons.

Solution Volume = length · width · height

$$= 12 \text{ in.} \cdot 8 \text{ in.} \cdot 9 \text{ in.}$$

$$= 864 \text{ in}^3$$

$$864 \text{ in}^3 = 864 \text{ in}^3 \times \frac{1 \text{ gal}}{231 \text{ in}^3} \approx 3.7 \text{ gal}$$

The tank will hold 3.7 gal.

35. Strategy To find the cost of the floor:
• Find the volume. The volume is equal to the volume of a rectangular solid plus one half the volume of the cylinder. The radius is one half the length of the rectangular solid.
• Multiply the volume by $5.85.

Solution Volume = length · width · height

$$+\frac{1}{2} \cdot \pi(\text{radius})^2 \cdot \text{height}$$

$$\approx 50 \text{ ft} \cdot \frac{1}{2} \text{ ft} \cdot 25 \text{ ft} +$$

$$\frac{1}{2}\left(3.14 \cdot 25 \text{ ft} \cdot 25 \text{ ft} \cdot \frac{1}{2} \text{ ft}\right)$$

$$= 625 \text{ ft}^3 + 490.625 \text{ ft}^3$$

$$= 1115.625 \text{ ft}^3$$

Cost = $1115.625 \times \$5.85 \approx 6526.406$
The cost is approximately $6526.41.

Applying the Concepts

37. a. Volume = length · width · height. If the length and width are doubled,
   Volume = 2 · length · 2 · width · height
   = 4 · length · width · height
   The volume will be 4 times larger.

b. Volume = length · width · height. If the length, width and height are doubled,
   Volume = 2 · length · 2 · width · 2 · height
   = 8 · length · width · height
   The volume will be 8 times larger.

c. Volume = $(\text{side})^3$. If the side is doubled,
   Volume = $(2 \cdot \text{side})^3 = 8 \cdot (\text{side})^3$
   The volume will be 8 times larger.

39. The length of the rectangular solid is equal to half the circumference of the base of the cylinder.

$$L = \frac{1}{2}C$$

The width of the rectangular solid is equal to the radius of the base of the cylinder. $W = r$ The height of the rectangular solid is equal to the height of the cylinder. $h = h$ For the base of the rectangular solid:

$$L = \frac{1}{2}C \; ; W = r;$$

$$A = LW = \frac{1}{2}C(r) = \frac{1}{2}(\pi d)r = \frac{1}{2}\pi(2r)r = \pi r^2$$

The volume of the rectangular solid and of the cylinder is $V = LWH = (LW)h = \pi r^2 h$.

SECTION 12.5

Objective A Exercises

1. 2.646

3. 6.481

5. 12.845

7. 13.748

Objective B Exercises

9. Hypotenuse $= \sqrt{(\text{leg})^2 + (\text{leg})^2}$
$= \sqrt{(3 \text{ in.})^2 + (4 \text{ in.})^2}$
$= \sqrt{9 \text{ in}^2 + 16 \text{ in}^2}$
$= \sqrt{25 \text{ in}^2}$
$= 5 \text{ in.}$

11. Hypotenuse $= \sqrt{(\text{leg})^2 + (\text{leg})^2}$
$= \sqrt{(5 \text{ cm})^2 + (7 \text{ cm})^2}$
$= \sqrt{25 \text{ cm}^2 + 49 \text{ cm}^2}$
$= \sqrt{74 \text{ cm}^2}$
$\approx 8.602 \text{ cm}$

13. Leg $= \sqrt{(\text{hypotenuse})^2 - (\text{leg})^2}$
$= \sqrt{(15 \text{ ft})^2 - (10 \text{ ft})^2}$
$= \sqrt{225 \text{ ft}^2 - 100 \text{ ft}^2}$
$= \sqrt{125 \text{ ft}^2}$
$\approx 11.180 \text{ ft}$

15. Leg $= \sqrt{(\text{hypotenuse})^2 - (\text{leg})^2}$
$= \sqrt{(6 \text{ cm})^2 - (4 \text{ cm})^2}$
$= \sqrt{36 \text{ cm}^2 - 16 \text{ cm}^2}$
$= \sqrt{20 \text{ cm}^2}$
$\approx 4.472 \text{ cm}$

17. Hypotenuse $= \sqrt{(\text{leg})^2 + (\text{leg})^2}$
$= \sqrt{(9 \text{ yd})^2 + (9 \text{ yd})^2}$
$= \sqrt{81 \text{ yd}^2 + 81 \text{ yd}^2}$
$= \sqrt{162 \text{ yd}^2}$
$\approx 12.728 \text{ yd}$

19. The leg opposite the 30° angle is one-half the hypotenuse.
Leg $= \frac{1}{2}(\text{hypotenuse})$
$= \frac{1}{2}(12) = 6$
The leg is 6 ft.
Use the Pythagorean Theorem to find the other leg.
Leg $= \sqrt{(\text{hypotenuse})^2 - (\text{leg})^2}$
$= \sqrt{12^2 - 6^2}$
$= \sqrt{144 - 36}$
$= \sqrt{108}$
≈ 10.392
The lengths of the two legs are 6 ft and 10.392 ft.

21. This is a 45-45-90 triangle.
Hypotenuse $= \sqrt{2} \times (\text{leg})$
$= \sqrt{2} \times 15$
$\approx 21.213 \text{ cm}$

23. For a 30-60-90 triangle, the leg opposite the 30° angle is one-half the hypotenuse. Therefore the hypotenuse is twice the length of the leg.
Leg $= \frac{1}{2} \times (\text{hypotenuse})$
Hypotenuse $= 2 \times (\text{leg})$
$= 2 \times 4$
$= 8 \text{ m}$

25. This is a 45-45-90 triangle.
Hypotenuse $= \sqrt{2} \times (\text{leg})$
$= \sqrt{2} \times 8$
$\approx 11.314 \text{ yd}$

Objective C Application Problems

27. Strategy   To find the length of the ramp, use the Pythagorean Theorem. The ramp is the hypotenuse of a right triangle.

Solution   Hypotenuse $= \sqrt{(\text{leg})^2 + (\text{leg})^2}$
$= \sqrt{(9 \text{ ft})^2 + (3.5 \text{ ft})^2}$
$= \sqrt{81 \text{ ft}^2 + 12.25 \text{ ft}^2}$
$= \sqrt{93.25 \text{ ft}^2}$
$\approx 9.66 \text{ ft}$
The ramp is 9.66 ft long.

29. Strategy   • Traveling 18 mi east and then 12 mi north forms a right angle. The distance from the starting point is the hypotenuse of the triangle with legs 18 mi and 12 mi.
• Find the hypotenuse of the right triangle.

Solution   Hypotenuse $= \sqrt{(\text{leg})^2 + (\text{leg})^2}$
$= \sqrt{(18 \text{ mi})^2 + (12 \text{ mi})^2}$
$= \sqrt{324 \text{ mi}^2 + 144 \text{ mi}^2}$
$= \sqrt{468 \text{ mi}^2}$
$\approx 21.63 \text{ mi}$
The distance is 21.6 mi.

31. Strategy To find the distance between adjacent holes, use the Pythagorean Theorem. The distance between holes is the hypotenuse of a right triangle. The legs are each 3 in. long.

Solution Hypotenuse $= \sqrt{(\text{leg})^2 + (\text{leg})^2}$
$$= \sqrt{(3\text{ in.})^2 + (3\text{ in.})^2}$$
$$= \sqrt{9\text{ in}^2 + 9\text{ in}^2}$$
$$= \sqrt{18\text{ in}^2}$$
$$\approx 4.243\text{ in.}$$
The distance is 4.243 in.

Applying the Concepts

33. a. Always true

b. Always true

35. To find x:
$$x^2 = 4^2 + 3^2$$
$$x = \sqrt{16+9}$$
$$x = \sqrt{25}$$
$$x = 5$$
Total length of the pipe $= (2 + 5 + 1)$ m $= 8$ m

SECTION 12.6

Objective A Exercises

1. $\dfrac{5\text{ m}}{10\text{ m}} = \dfrac{1}{2}$

3. $\dfrac{9\text{ in.}}{12\text{ in.}} = \dfrac{3}{4}$

5. $\angle CAB = \angle DEF$
$AC = ED$ and
$AB = EF$
Therefore SAS applies and the triangles are congruent.

7. $AC = EF$
$AB = DE$ and
$BC = FD$
Therefore SSS applies and the triangles are congruent.

9. $\dfrac{AC}{DF} = \dfrac{AB}{DE}$
$\dfrac{5\text{ cm}}{9\text{ cm}} = \dfrac{4\text{ cm}}{DE}$
$5 \times DE = 4\text{ cm} \times 9$
$5 \times DE = 36\text{ cm}$
$DE = 36\text{ cm} \div 5$
$DE = 7.2\text{ cm}$

11. $\dfrac{AC}{DF} = \dfrac{\text{height of triangle } ABC}{\text{height of triangle } DEF}$
$\dfrac{3\text{ m}}{5\text{ m}} = \dfrac{2\text{ m}}{\text{height}}$
$3 \times \text{height} = 5 \times 2\text{ m}$
$3 \times \text{height} = 10\text{ m}$
Height $= 10\text{ m} \div 3$
Height $\approx 3.3\text{ m}$

Objective B Application Problems

13. Strategy To find the height of the building, solve a proportion.

Solution $\dfrac{\text{height}}{8\text{ m}} = \dfrac{8\text{ m}}{4\text{ m}}$
Height $\times 4 = 8\text{ m} \times 8$
Height $\times 4 = 64\text{ m}$
Height $= 64\text{ m} \div 4$
Height $= 16\text{ m}$
The height of the building is 16 m.

15. Strategy To find the perimeter:
• Solve a proportion to find the length of side BC.
• Add the three sides of triangle ABC.

Solution $\dfrac{AC}{DF} = \dfrac{BC}{EF}$
$\dfrac{4\text{ m}}{8\text{ m}} = \dfrac{BC}{6\text{ m}}$
$8 \times BC = 4 \times 6\text{ m}$
$8 \times BC = 24\text{ m}$
$BC = 24\text{ m} \div 8$
$BC = 3\text{ m}$
$3\text{ m} + 4\text{ m} + 5\text{ m} = 12\text{ m}$
The perimeter is 12 m.

17. Strategy To find the area:
• Solve a proportion to find the height of triangle ABC.
• Use the formula Area $= \dfrac{1}{2} \cdot$ base $\cdot$ height.

Solution $\dfrac{AB}{DE} = \dfrac{\text{height of triangle } ABC}{\text{height of triangle } DEF}$
$\dfrac{15\text{ cm}}{40\text{ cm}} = \dfrac{\text{height}}{20\text{ cm}}$
$40 \times \text{height} = 15 \times 20\text{ cm}$
$40 \times \text{height} = 300\text{ cm}$
Height $= 300\text{ cm} \div 40$
Height $= 7.5\text{ cm}$
Area $= \dfrac{1}{2} \cdot$ base $\cdot$ height
$= \dfrac{1}{2} \cdot 15\text{ cm} \cdot 7.5\text{ cm}$
$= 56.25\text{ cm}^2$
The area is 56.25 cm^2.

Applying the Concepts

19. a. Always true.

b. Sometimes true.

c. Always true.

21. $\triangle ABC$ and $\triangle DEC$ are similar triangles.

$\dfrac{8}{5}=\dfrac{6}{DE}$ $\dfrac{8}{5}=\dfrac{10}{EC}$

$8DE=30$ $8EC=50$

$DE=\dfrac{30}{8}=\dfrac{15}{4}$ $EC=\dfrac{50}{8}=\dfrac{25}{4}$

The perimeter of the trapezoid
$ABDE = AB + AD + DE + BE$

$= 6 + (8-5) + \dfrac{15}{4} + \left(10 - \dfrac{25}{4}\right)$

$= 6 + 3 + 3\dfrac{3}{4} + 3\dfrac{3}{4}$

$= 15\dfrac{6}{4} = 16\dfrac{2}{4} = 16\dfrac{1}{2}$

CHAPTER REVIEW

1. Radius $= \dfrac{1}{2}$ diameter $= \dfrac{1}{2}(1.5 \text{ m}) = 0.75 \text{ m}$

2. Circumference $= 2\pi \cdot$ radius
$\approx 2(3.14)(5 \text{ cm}) = 31.4 \text{ cm}$

3. Perimeter $= 2(\text{length}) + 2(\text{width})$
$= 2(8 \text{ ft}) + 2(5 \text{ ft})$
$= 16 \text{ ft} + 10 \text{ ft} = 26 \text{ ft}$

4. $BC = AD - AB - CD$
$= 24 - 15 - 6 = 3$

5. Volume $=$ length $\cdot$ width $\cdot$ height
$= 10 \text{ ft} \cdot 5 \text{ ft} \cdot 4 \text{ ft} = 200 \text{ ft}^3$

6. Hypotenuse $= \sqrt{(\text{leg})^2 + (\text{leg})^2}$
$= \sqrt{(10 \text{ cm})^2 + (24 \text{ cm})^2}$
$= \sqrt{100 \text{ cm}^2 + 576 \text{ cm}^2}$
$= \sqrt{676 \text{ cm}^2}$
$= 26 \text{ cm}$

7. $180° - 105° = 75°$
$75°$ is the supplement of $105°$.

8. $\sqrt{15} \approx 3.873$

9. $\dfrac{BC}{EF} = \dfrac{\text{height of triangle} ABC}{\text{height of triangle} DEF}$
$\dfrac{12 \text{ cm}}{24 \text{ cm}} = \dfrac{8 \text{ cm}}{h}$
$12 \times h = 24 \times 8 \text{ cm}$
$12 \times h = 192 \text{ cm}$
$h = 192 \text{ cm} \div 12 = 16 \text{ cm}$

10. Area $= \pi \cdot (\text{radius})^2$
$\approx 3.14 \cdot (4.5 \text{ cm})^2$
$= 63.585 \text{ cm}^2$

11. $b = 45°$; $a = 180° - 45° = 135°$

12. Area $=$ length $\times$ width
$= 11 \text{ m} \times 5 \text{ m} = 55 \text{ m}^2$

13. Volume $=$ volume of larger rectangular solid
$\quad -$ volume of smaller rectangular solid
$=$ length $\cdot$ width $\cdot$ height
$\quad -$ length $\cdot$ width $\cdot$ height
$= 8 \text{ in.} \cdot 7 \text{ in.} \cdot 6 \text{ in.} - 8 \text{ in.} \cdot 4 \text{ in.} \cdot 3 \text{ in.}$
$= 336 \text{ in}^3 - 96 \text{ in}^3 = 240 \text{ in}^3$

14. Area $=$ area of rectangle $+ \dfrac{1}{2}$ (area of circle)
$=$ length $\cdot$ width $+ \dfrac{1}{2}\pi(\text{radius})^2$
$\approx 8 \text{ in.} \cdot 4 \text{ in.} + \dfrac{1}{2}(3.14)(4 \text{ in.})^2$
$= 32 \text{ in}^2 + 25.12 \text{ in}^2$
$= 57.12 \text{ in}^2$

15. Volume $= \dfrac{4}{3} \cdot \pi(\text{radius})^3$
$\approx \dfrac{4}{3}(3.14)(4 \text{ ft})^3$
$\approx 267.9 \text{ ft}^3$

16. Strategy   To find the area:
• Solve a proportion to find the length of side DF (the base of the triangle DEF).
• Use the formula
Area $= \left(\dfrac{1}{2}\right) \cdot$ base $\cdot$ height .

Solution $\dfrac{AC}{DF} = \dfrac{\text{height of triangle} ABC}{\text{height of triangle} DEF}$
$\dfrac{8 \text{ m}}{DF} = \dfrac{5 \text{ m}}{9 \text{ m}}$
$8 \text{ m} \times 9 = 5 \times DF$
$72 \text{ m} = 5 \times DF$
$72 \text{ m} \div 5 = DF$
$14.4 \text{ m} = DF$
Area $= \dfrac{1}{2} \cdot$ base $\cdot$ height
$= \dfrac{1}{2}(14.4 \text{ m})(9 \text{ m}) = 64.8 \text{ m}^2$
The area is 64.8 m^2.

17. Perimeter $=$ length of two sides
$\quad + \dfrac{1}{2}$ circumference of circle
$\approx 2(16 \text{ in.}) + \dfrac{1}{2}(2 \cdot 3.14 \cdot 5 \text{ in.})$
$= 32 \text{ in.} + 15.7 \text{ in.} = 47.7 \text{ in.}$

18. $b = 80°$; $a = 180° - 80° = 100°$

19. Strategy To find how high on the building the ladder will reach, use the Pythagorean Theorem. The hypotenuse is 17 ft and one leg is 8 ft. The other leg is the height up the building.

Solution $\text{leg} = \sqrt{(\text{hypotenuse})^2 - (\text{leg})^2}$
$= \sqrt{(17\ \text{ft})^2 - (8\ \text{ft})^2}$
$= \sqrt{289\ \text{ft}^2 - 64\ \text{ft}^2}$
$= \sqrt{225\ \text{ft}^2}$
$= 15\ \text{ft}$
The ladder will reach 15 ft up the building.

20. $90° - 32° = 58°$
The other angles of the triangle are 90° and 58°.

21. Strategy To find how many feet the wheel travels if it makes 10 revolutions:
• Find how far the wheel travels when it makes 1 revolution by using the circumference formula.
• Convert the circumference to feet.
• Multiply the distance traveled in 1 revolution by 10.

Solution $\text{Circumference} = \pi \cdot \text{diameter}$
$= \pi \cdot 28\ \text{in.}$
$\approx 3.14 \cdot (28\ \text{in.})$
$= 87.92\ \text{in.}$
$87.92\ \text{in.} = 87.92\ \text{in.} \times \dfrac{1\ \text{ft}}{12\ \text{in.}}$
$= \dfrac{87.92}{12}\ \text{ft} \approx 7.33\ \text{ft}$
$10 \cdot 7.33\ \text{ft} = 73.3\ \text{ft}$
The wheel travels approximately 73.3 ft in 10 revolutions.

22. Strategy To find the area of the room in square yards:
• Use the area of the rectangle formula.
• Convert square feet to square yards.

Solution $\text{Area} = \text{length} \cdot \text{width}$
$= 18\ \text{ft} \cdot 14\ \text{ft} = 252\ \text{ft}^2$
$252\ \text{ft}^2 = 252\ \text{ft}^2 \times \dfrac{1\ \text{yd}^2}{9\ \text{ft}^2}$
$= \dfrac{252}{9}\ \text{yd}^2 = 28\ \text{yd}^2$
The area of the room is $28\ \text{yd}^2$.

23. Strategy To find the volume of the silo, use the formula for the volume of a cylinder.

Solution $\text{Volume} = \pi \cdot (\text{radius})^2 (\text{height})$
$\approx 3.14 (4.5\ \text{ft})^2 (18\ \text{ft})$
$= 1144.53\ \text{ft}^3$
The volume of the silo is approximately $1144.53\ \text{ft}^3$.

24. $\text{Area} = \dfrac{1}{2} \cdot \text{base} \cdot \text{height}$
$= \dfrac{1}{2} (8\ \text{m})(2.75\ \text{m})$
$= 11\ \text{m}^2$

25. Strategy • Traveling 20 mi west and then 21 mi south forms a right angle. The distance from the starting point is the hypotenuse of the triangle with legs 20 mi and 21 mi.
• Find the hypotenuse of the right triangle.

Solution $\text{hypotenuse} = \sqrt{(\text{leg})^2 + (\text{leg})^2}$
$= \sqrt{(20\ \text{mi})^2 + (21\ \text{mi})^2}$
$= \sqrt{400\ \text{mi}^2 + 441\ \text{mi}^2}$
$= \sqrt{841\ \text{mi}^2}$
$= 29$
The distance from the starting point is 29 mi.

CHAPTER TEST

1. $\text{Volume} = \pi(\text{radius})^2 \cdot \text{height}$
$\approx 3.14 \cdot (3\ \text{m})^2 \cdot 6\ \text{m}$
$= 169.56\ \text{m}^3$

2. $\text{Perimeter} = 2 \cdot \text{length} + 2 \cdot \text{width}$
$= 2(2\ \text{m}) + 2(1.4\ \text{m})$
$= 4\ \text{m} + 2.8\ \text{m}$
$= 6.8\ \text{m}$

3. Strategy To find the volume of the composite figure, subtract the volume of the smaller cylinder from the volume of the larger cylinder.

Solution $\text{Volume} = \text{volume of larger cylinder}$
$\quad\quad - \text{volume of smaller cylinder}$
$\doteq \pi(\text{radius})^2 \cdot \text{height}$
$\quad\quad - \pi \cdot (\text{radius})^2 \cdot \text{height}$
$\approx 3.14(6\ \text{cm})^2 \cdot 14\ \text{cm}$
$\quad\quad - 3.14 \cdot (2\ \text{cm})^2 \cdot 14\ \text{cm}$
$= 1582.56\ \text{cm}^3 - 175.84\ \text{cm}^3$
$= 1406.72\ \text{cm}^3$
The volume of the composite figure is approximately $1406.72\ \text{cm}^3$.

4. Strategy To find the missing length, use the Pythagorean Theorem. $AB = FE$ is the hypotenuse. The legs are 6 and 8 m.

Solution $\text{Hypotenuse} = \sqrt{(8\ \text{m})^2 + (6\ \text{m})^2}$
$= \sqrt{64\ \text{m}^2 + 36\ \text{m}^2}$
$= \sqrt{100\ \text{m}^2}$
$= 10\ \text{m}$
The length of FE is 10 m.

5. $90° - 32° = 58°$
   $58°$ is the complement of $32°$.

6. Area $= \pi(\text{radius})^2$
   $\approx \dfrac{22}{7}(1 \text{ m})^2$
   $= \dfrac{22 \text{ m}^2}{7} = 3\dfrac{1}{7} \text{ m}^2$

7. x and z are supplementary; therefore
   $z = 180° - 30° = 150°$. y and z are corresponding
   angles; therefore $y = z = 150°$.

8. Perimeter $=$ two lengths $+$ circumference of circle
   $= 2(4 \text{ ft}) + \pi \cdot \text{diameter}$
   $= 8 \text{ ft} + \pi\left(2\dfrac{1}{2} \text{ ft}\right)$
   $\approx 8 \text{ ft} + 3.14(2.5 \text{ ft})$
   $= 15.85 \text{ ft}$

9. $\sqrt{189} \approx 13.748$

10. Leg $= \sqrt{(\text{hypotenuse})^2 - (\text{leg})^2}$
    $= \sqrt{(12 \text{ ft})^2 - (7 \text{ ft})^2} = \sqrt{144 \text{ ft}^2 - 49 \text{ ft}^2}$
    $= \sqrt{95 \text{ ft}^2}$
    $\approx 9.747 \text{ ft}$

11. Area $=$ area of rectangle $-$ area of triangle
    $= \text{length} \cdot \text{width} - \dfrac{1}{2} \cdot \text{base} \cdot \text{height}$
    $= 3 \text{ ft} \cdot 4\dfrac{1}{2} \text{ ft} - \dfrac{1}{2}\left(4\dfrac{1}{2} \text{ ft}\right)\left(1\dfrac{1}{2} \text{ ft}\right)$
    $= 13.5 \text{ ft}^2 - 3.375 \text{ ft}^2 = 10.125 \text{ ft}^2$

12. Angles x and b are supplementary angles.
    $\angle x + \angle b = 180°$
    $45° + \angle b = 180°$
    $45° - 45° + \angle b = 180° - 45°$
    $\angle b = 135°$
    $\angle a = \angle x$ because $\angle a$ and $\angle x$ are alternate exterior
    angles. $\angle a = 45°$

13. $\dfrac{AB}{DE} = \dfrac{BC}{EF}$
    $\dfrac{\frac{3}{4} \text{ ft}}{2\frac{1}{2} \text{ ft}} = \dfrac{BC}{4 \text{ ft}}$
    $\dfrac{3}{4} \times 4 \text{ ft} = 2\dfrac{1}{2} \times BC$
    $3 \text{ ft} = 2\dfrac{1}{2} \times BC$
    $3 \text{ ft} \div 2\dfrac{1}{2} = BC$
    $3 \text{ ft} \times \dfrac{2}{5} = BC$
    $BC = \dfrac{6}{5} \text{ ft} = 1\dfrac{1}{5} \text{ ft}$

14. $90° - 40° = 50°$. The other two angles of the
    triangle are $50°$ and $90°$.

15. **Strategy** To find the width of the canal, solve a
    proportion.

    Solution $\dfrac{5 \text{ ft}}{\text{Width of canal}} = \dfrac{12 \text{ ft}}{60 \text{ ft}}$
    $5 \text{ ft} \times 60 = 12 \times \text{width of canal}$
    $300 \text{ ft} = 12 \times \text{width of canal}$
    $300 \text{ ft} \div 12 = \text{width of canal}$
    $25 \text{ ft} = \text{width of canal}$
    The width of the canal is 25 ft.

16. **Strategy** To find how much more pizza is
    contained in the larger pizza, subtract
    the area of the smaller pizza from the
    area of the larger pizza.

    Solution $A = \pi \cdot (\text{radius})^2$
    $\approx 3.14 \cdot (10 \text{ in.})^2 = 314 \text{ in}^2$
    $A = \pi \cdot (\text{radius})^2$
    $\approx 3.14 \cdot (8 \text{ in.})^2 = 200.96 \text{ in}^2$
    $314 \text{ in}^2 - 200.96 \text{ in}^2 = 113.04 \text{ in}^2$
    The amount of extra pizza is
    113.04 in^2.

17. **Strategy** To find the cost of the carpet:
    • Subtract the area of the smaller
    rectangle from the area of the larger
    rectangle.
    • Convert the area to square yards.
    • Multiply the area in square yards by
    the cost per square yard.

    Solution
    Area $=$ area of larger rectangle
    $-$ area of smaller rectangle
    $= \text{length} \cdot \text{width} - \text{length} \cdot \text{width}$
    $= 20 \text{ ft} \cdot 22 \text{ ft} - 6 \text{ ft} \cdot 11 \text{ ft}$
    $= 440 \text{ ft}^2 - 66 \text{ ft}^2 = 374 \text{ ft}^2$

    $374 \text{ ft}^2 = 374 \text{ ft}^2 \times \dfrac{1 \text{ yd}^2}{9 \text{ ft}^2} = 41.\overline{55} \text{ yd}^2$

    $41.\overline{55} \text{ yd}^2 \; 3 \; \$26.80 \approx \$1113.69$
    It will cost $1113.69 to carpet the area.

18. Area $= \pi \cdot (\text{radius})^2$
    $\approx 3.14 \cdot (5.75 \text{ ft})^2$
    $\approx 103.82 \text{ ft}^2$

19. **Strategy** To find the length of the rafter:
    • Use the Pythagorean Theorem to find
    the part of the rafter that covers the
    roof.
    • Find the total length of the rafter by
    adding the 2 ft overhang to the part
    that covers the roof.

    Solution Hypotenuse $= \sqrt{(5 \text{ ft})^2 + (12 \text{ ft})^2}$
    $= \sqrt{25 \text{ ft}^2 + 144 \text{ ft}^2}$
    $= \sqrt{169 \text{ ft}^2} = 13 \text{ ft}$
    $13 \text{ ft} + 2 \text{ ft} = 15 \text{ ft}$
    The length of the rafter is 15 ft.

20. Strategy   To find the volume of the interior of the box, subtract the thickness of the sides and bottom of the box from the exterior dimensions of the box, then use the formula for volume of a rectangular solid.

Solution   Length $= 1$ ft 2 in.$- 2\left(\dfrac{1}{2} \text{ in.}\right)$

$\qquad\qquad = 1$ ft 1 in.$= 13$ in.

Width $= 9$ in.$-2\left(\dfrac{1}{2} \text{ in.}\right) = 8$ in.

Height $= 8$ in.$-\dfrac{1}{2}$ in.$= 7.5$ in.

Volume $=$ length$\cdot$ width$\cdot$height

$\qquad\qquad = 13$ in.$\cdot 8$ in.$\cdot 7.5$ in.$= 780$ in^3

The volume of the interior of the toolbox is 780 in^3.

CUMULATIVE REVIEW

1.
$96 = 2\cdot2\cdot2\cdot2\cdot2 \mid \boxed{3}$
$144 = \boxed{2\cdot2\cdot2\cdot2} \mid 3\cdot3$
GCF $= 2\cdot2\cdot2\cdot2\cdot3 = 48$

2.
$3\dfrac{5}{12} = 3\dfrac{20}{48}$
$2\dfrac{9}{16} = 2\dfrac{27}{48}$
$+\ 1\dfrac{7}{8} = 1\dfrac{42}{48}$
$\overline{\qquad 6\dfrac{89}{48} = 7\dfrac{41}{48}}$

3. $4\dfrac{1}{3} \div 6\dfrac{2}{9} = \dfrac{13}{3} \div \dfrac{56}{9} = \dfrac{13}{3} \times \dfrac{9}{56}$

$\qquad = \dfrac{13\cdot\overset{1}{3}\cdot3}{\underset{1}{3}\cdot7\cdot2\cdot2\cdot2} = \dfrac{39}{56}$

4. $\left(\dfrac{2}{3}\right)^2 \div \left(\dfrac{1}{3}+\dfrac{1}{2}\right)-\dfrac{2}{5}$

$= \left(\dfrac{2}{3}\cdot\dfrac{2}{3}\right) \div \left(\dfrac{2}{6}+\dfrac{3}{6}\right)-\dfrac{2}{5}$

$= \dfrac{4}{9} \div \dfrac{5}{6}-\dfrac{2}{5}$

$= \dfrac{4}{9} \times \dfrac{6}{5}-\dfrac{2}{5}$

$= \dfrac{2\cdot2\cdot2\cdot\overset{1}{3}}{\underset{1}{3}\cdot3\cdot5}-\dfrac{2}{5}$

$= \dfrac{8}{15}-\dfrac{2}{5}$

$= \dfrac{8}{15}-\dfrac{6}{15} = \dfrac{2}{15}$

5. $-\dfrac{2}{3}-\left(-\dfrac{5}{8}\right) = -\dfrac{16}{24}+\dfrac{15}{64} = -\dfrac{1}{24}$

6. $\dfrac{\$348.80}{40 \text{ h}} = \$8.72/\text{h}$

7. $\dfrac{3}{8} = \dfrac{n}{100}$
$3\times100 = n\times8$
$300 = n\times8$
$300\div8 = n$
$37.5 = n$

8. $37\dfrac{1}{2}\% = \dfrac{75}{2}\% = \dfrac{75}{2}\times\dfrac{1}{100}$
$\qquad = \dfrac{75}{200} = \dfrac{3}{8}$

9. $2^2 -[(-2)^2-(-4)] = 4-[4+4]$
$\qquad\qquad\qquad\qquad = 4-8 = -4$

10. $36.4\%\times n = 30.94$
$0.364\times n = 30.94$
$n = 30.94 \div 0.364$
$n = 85$

11. $\dfrac{x}{3}+3 = 1$
$\dfrac{x}{3} = -2$
$x = -6$

12. $2(x-3)+2 = 5x-8$
$2x-6+2 = 5x-8$
$2x-4 = 5x-8$
$4 = 3x$
$\dfrac{4}{3} = x$

13. 32.5 km $= 32{,}500$ m

14. $\quad 32$ m $= 32.00$ m
$-\ 42$ cm $=\ \ 0.42$ m
$\overline{\qquad\qquad 31.58 \text{ m}}$

15. $\dfrac{2}{3}x = -10$
$x = \dfrac{3}{2}(-10)$
$x = -15$

16. $2x-4(x-3) = 8$
$2x-4x+12 = 8$
$-2x+12 = 8$
$-2x = -4$
$x = 2$

17. Strategy To find the monthly payment:
• Find the amount paid in payments by the subtracting the down payment ($1000) from the price ($26,488).
• Divide the amount paid in payments by the number of payments (36).

Solution $26,488 − $1000 = $25,488

$$36\overline{)25,488}\ \ \frac{708}{}$$

The monthly payment is $708.

18. Strategy To find the sales tax, solve a proportion.

Solution $\dfrac{\$175}{\$6.75} = \dfrac{\$1220}{n}$

$175 \times n = 6.75 \times 1220$

$175 \times n = 8235$

$n = 8235 \div 175 \approx 47.06$

The sales tax on the stereo system is $47.06.

19. Strategy To find the operator's original wage, solve the basic percent equation for the base. The percent is 110% and the amount is $16.06.

Solution $110\% \times n = 16.06$

$1.10 \times n = 16.06$

$n = 16.06 \div 1.10 = 14.60$

The original wage was $14.60.

20. Strategy To find the sale price:
• Find the amount of the markdown by solving the basic percent equation for amount. The base is $120 and the percent is 55%.
• Subtract the amount of the markdown from the original price ($120).

Solution
$55\% \times 120 = n$
$0.55 \times 120 = 66$

$\begin{array}{r} \$120 \\ -\ \ 66 \\ \hline \$54 \end{array}$

The sale price of the dress is $54.

21. Strategy To find the value of the investment, multiply the amount invested by the compound interest factor.

Solution $25,000 3 7.38703 = $184,675.75

The value of the investment after 20 years would be $184,675.75.

22. Strategy To find the weight of the package:
• Find the weight of the package in ounces by multiplying the weight of one tile (6 oz) by the number of tiles in the package (144).
• Convert the weight in ounces to pounds.

Solution 6 oz 3 144 = 864 oz

$864 \text{ oz} = 864 \ \cancel{oz} \times \dfrac{1 \text{ lb}}{16 \ \cancel{oz}} = 54 \text{ lb}$

The weight of the package is 54 lb.

23. Strategy To find the distance between the rivets:
• Divide the length of the plates (5.4 m) by the number of spaces (24).
• Convert the meters to centimeters.

Solution $24\overline{)5.400}\ \ \overset{0.225}{}$

0.225 m = 22.5 cm

The distance between the rivets is 22.5 cm.

24. Let x = the number.
$2 + 4x = -6$
$4x = -8$
$x = -2$
The number is –2.

25. $a = 74°$. a and b are supplementary; therefore $b = 180° − a = 180° − 74° = 106°$.

26. Perimeter $= 2 \cdot \text{length} + \text{width} + \dfrac{1}{2}(\text{circumference})$

$\approx 2 \cdot (7 \text{ cm}) + 6 \text{ cm} + \dfrac{1}{2}(3.14 \cdot 6 \text{ cm})$

$= 14 \text{ cm} + 6 \text{ cm} + 9.42 \text{ cm}$

$= 29.42 \text{ cm}$

27. Area = area of rectangle + area of triangle

$= \text{length} \cdot \text{width} + \dfrac{1}{2} \cdot \text{base} \cdot \text{height}$

$= 5 \text{ in.} \cdot 4 \text{ in.} + \dfrac{1}{2} \cdot 12 \text{ in.} \cdot 5 \text{ in.}$

$= 20 \text{ in}^2 + 30 \text{ in}^2 = 50 \text{ in}^2$

28. Volume = volume of rectangular solid

$-\dfrac{1}{2} \text{ volume of cylinder}$

$= \text{length} \cdot \text{width} \cdot \text{height}$

$-\dfrac{1}{2}[\pi(\text{radius})^2 \cdot \text{height}]$

$\approx 8 \text{ in.} \cdot 4 \text{ in.} \cdot 3 \text{ in.}$

$-\dfrac{1}{2}[3.14(0.5 \text{ in.})^2 \cdot 8 \text{ in.}]$

$= 96 \text{ in}^3 - 3.14 \text{ in}^3 = 92.86 \text{ in}^3$

29. Hypotenuse $= \sqrt{(\text{leg})^2 + (\text{leg})^2}$

$\qquad\qquad\quad = \sqrt{(8 \text{ ft})^2 + (7 \text{ ft})^2}$

$\qquad\qquad\quad = \sqrt{64 \text{ ft}^2 + 49 \text{ ft}^2}$

$\qquad\qquad\quad = \sqrt{113 \text{ ft}^2} \approx 10.63 \text{ ft}$

30. Strategy   To find the perimeter of *DEF*:
- Solve a proportion to find the length of side *DF*.
- Solve a proportion to find the length of side *FE*.
- Use the formula for perimeter to find the perimeter of triangle *DEF*.

Solution

$$\frac{AB}{DE} = \frac{AC}{DF}$$

$$\frac{4 \text{ cm}}{12 \text{ cm}} = \frac{3 \text{ cm}}{DF}$$

$4 \times DF = 3 \text{ cm} \times 12$

$4 \times DF = 36 \text{ cm}$

$DF = 36 \text{ cm} \div 4 = 9 \text{ cm}$

$$\frac{AB}{DE} = \frac{BC}{FE}$$

$$\frac{4 \text{ cm}}{12 \text{ cm}} = \frac{5 \text{ cm}}{FE}$$

$4 \times FE = 5 \text{ cm} \times 12$

$4 \times FE = 60 \text{ cm}$

$FE = 60 \text{ cm} \div 4 = 15 \text{ cm}$

Perimeter $=$ side 1 $+$ side 2 $+$ side 3

$\qquad\qquad = 12 \text{ cm} + 15 \text{ cm} + 9 \text{ cm}$

$\qquad\qquad = 36 \text{ cm}$

The perimeter is 36 cm.

FINAL EXAMINATION

1. $\begin{array}{r} \overset{9\ \ \ 10}{\cancel{1}\,\overset{10}{\cancel{0}}\,\overset{8}{\cancel{0}}\,,\,\overset{0}{\cancel{9}}\,\overset{14}{\cancel{1}}\,4} \\ -\ 9\,7,6\,5\ 5 \\ \hline 3\,2\,5\,9 \end{array}$

2. $\begin{array}{r} 53 \\ 657\,\overline{)34{,}821} \\ -3285 \\ \hline 1971 \\ -1971 \\ \hline 0 \end{array}$

3. $3^2 \cdot (5-3)^2 \div 3 + 4 = 3^2 \cdot (2)^2 \div 3 + 4$

$\qquad\qquad\qquad\qquad\quad = 9 \cdot 4 \div 3 + 4$

$\qquad\qquad\qquad\qquad\quad = 36 \div 3 + 4$

$\qquad\qquad\qquad\qquad\quad = 12 + 4$

$\qquad\qquad\qquad\qquad\quad = 16$

4.

$9 = \boxed{3} \cdot 3$ (circled)

$12 = 2 \cdot 2 \cdot 3$

$16 = \boxed{2 \cdot 2 \cdot 2 \cdot 2}$

$\text{LCM} = 2 \cdot 2 \cdot 2 \cdot 2 \cdot 3 \cdot 3 = 144$

5. $\dfrac{3}{8} + \dfrac{5}{6} + \dfrac{1}{5} = \dfrac{45}{120} + \dfrac{100}{120} + \dfrac{24}{120} = \dfrac{169}{120} = 1\dfrac{49}{120}$

6. $\begin{array}{r} 7\dfrac{5}{12} = 7\dfrac{20}{48} = 6\dfrac{68}{48} \\ -\ 3\dfrac{13}{16} = 3\dfrac{39}{48} = 3\dfrac{39}{48} \\ \hline 3\dfrac{29}{48} \end{array}$

7. $3\dfrac{5}{8} \times 1\dfrac{5}{7} = \dfrac{29}{8} \times \dfrac{12}{7} = \dfrac{29 \cdot \overset{3}{\cancel{12}}}{\underset{2}{\cancel{8}} \cdot 7} = \dfrac{87}{14} = 6\dfrac{3}{14}$

8. $1\dfrac{2}{3} \div 3\dfrac{3}{4} = \dfrac{5}{3} \div \dfrac{15}{4} = \dfrac{5}{3} \times \dfrac{4}{15} = \dfrac{5 \times 4}{3 \times 15} = \dfrac{20}{45} = \dfrac{4}{9}$

9. $\left(\dfrac{2}{3}\right)^2 \left(\dfrac{3}{4}\right)^2 = \left(\dfrac{2}{3} \cdot \dfrac{2}{3}\right)\left(\dfrac{3}{4} \cdot \dfrac{3}{4}\right) = \left(\dfrac{8}{27}\right)\left(\dfrac{9}{16}\right)$

$\qquad\qquad\qquad\quad = \dfrac{72}{432} = \dfrac{1}{6}$

10. $\left(\dfrac{2}{3}\right)^2 \div \left(\dfrac{3}{4} + \dfrac{1}{3}\right) - \dfrac{1}{3} = \dfrac{4}{9} \div \left(\dfrac{9}{12} + \dfrac{4}{12}\right) - \dfrac{1}{3}$

$\qquad\qquad\qquad\qquad\qquad = \dfrac{4}{9} \div \dfrac{13}{12} - \dfrac{1}{3}$

$\qquad\qquad\qquad\qquad\qquad = \dfrac{4}{9} \times \dfrac{\overset{4}{\cancel{12}}}{13} - \dfrac{1}{3}$

$\qquad\qquad\qquad\qquad\qquad = \dfrac{16}{\underset{3}{\cancel{39}}} - \dfrac{1}{3}$

$\qquad\qquad\qquad\qquad\qquad = \dfrac{16}{39} - \dfrac{13}{39} = \dfrac{3}{39} = \dfrac{1}{13}$

11. $\begin{array}{r} \overset{2\,3\ \ 1}{4.9\,72} \\ 28.6 \\ 1.8\,8 \\ +\ 12\,8.7\,25 \\ \hline 164.177 \end{array}$

12. $\begin{array}{r} \overset{9\ \ 9\ \ 9}{\cancel{8}\,\overset{10}{\cancel{0}}\,,\,\overset{10}{\cancel{0}}\,\overset{10}{\cancel{0}}\,11} \\ 9\,0,0\,0\,1 \\ -2\,9,7\,9\,6 \\ \hline 60,2\,0\,5 \end{array}$

13. $\begin{array}{r} 2.97 \\ \times\ 0.0094 \\ \hline 1188 \\ 2673 \\ \hline 0.027918 \end{array}$

14. $\begin{array}{r} 0.687 \approx 0.69 \\ 0.062.\overline{)0.042.600} \\ -\ 372 \\ \hline 540 \\ -\ 496 \\ \hline 440 \\ -\ 434 \\ \hline 6 \end{array}$

15. $0.45 = \dfrac{45}{100} = \dfrac{9}{20}$

16. $\dfrac{323.4 \text{ mi}}{13.2 \text{ gal}} = 24.5 \text{ mi/gal}$

17. $\dfrac{12}{35} = \dfrac{n}{160}$

$12 \times 160 = n \times 35$

$1920 = n \times 35$

$1920 \div 35 = n$

$54.9 \approx n$

18. $22\dfrac{1}{2}\% = \dfrac{45}{2} \times \dfrac{1}{100} = \dfrac{45}{200} = \dfrac{9}{40}$

19. $1.35 = 1.35 \times 100\% = 135\%$

20. $\dfrac{5}{4} = \dfrac{5}{4} \times 100\% = \dfrac{500}{4}\% = 125\%$

21. Percent $\times$ base = amount

$120\% \times 30 = n$

$1.2 \times 30 = n$

$36 = n$

22. Percent $\times$ base = amount

$n \times 9 = 12$

$n = 12 \div 9 = 1.3\overline{3} = 133\dfrac{1}{3}\%$

23. Percent $\times$ base = amount

$60\% \times n = 42$

$0.60 \times n = 42$

$n = 42 \div 0.60 = 70$

24. $1\dfrac{2}{3} \text{ ft} = \dfrac{5}{3} \text{ ft} = \dfrac{5}{3} \text{ ft} \times \dfrac{12 \text{ in.}}{1 \text{ ft}} = 20 \text{ in.}$

25. $\begin{array}{r} 3 \text{ ft } 2 \text{ in.} = 2 \text{ ft } 14 \text{ in.} \\ - 1 \text{ ft } 10 \text{ in.} = \underline{1 \text{ ft } 10 \text{ in.}} \\ 1 \text{ ft } 4 \text{ in.} \end{array}$

26. $40 \text{ oz} = 40 \text{ oz} \times \dfrac{1 \text{ lb}}{16 \text{ oz}} = \dfrac{40}{16} \text{ lb} = 2.5 \text{ lb}$

27. $\begin{array}{r} 3 \text{ lb } 12 \text{ oz} \\ + \ 2 \text{ lb } 10 \text{ oz} \\ \hline 5 \text{ lb } 22 \text{ oz} = 6 \text{ lb } 6 \text{ oz} \end{array}$

28. $18 \text{ pt} = 18 \text{ pt} \times \dfrac{1 \text{ qt}}{2 \text{ pt}} \times \dfrac{1 \text{ gal}}{4 \text{ qt}} = \dfrac{18 \text{ gal}}{8} = 2.25 \text{ gal}$

29. $\begin{array}{r} 1 \text{ gal } 3 \text{ qt} \\ 3\overline{)5 \text{ gal } 1 \text{ qt}} \\ \underline{-3 \text{ gal}} \\ 2 \text{ gal} = 8 \text{ qt} \\ 9 \text{ qt} \\ \underline{-9 \text{ qt}} \\ 0 \end{array}$

30. $2.48 \text{ m} = 248 \text{ cm}$

31. $4 \text{ m } 62 \text{ cm} = 4 \text{ m} + 0.62 \text{ m} = 4.62 \text{ m}$

32. $1 \text{ kg } 614 \text{ g} = 1 \text{ kg} + 0.614 \text{ kg} = 1.614 \text{ kg}$

33. $2 \text{ L } 67 \text{ ml} = 2000 \text{ ml} + 67 \text{ ml} = 2067 \text{ ml}$

34. $55 \text{ mi} \approx 55 \text{ mi} \times \dfrac{1.61 \text{ km}}{1 \text{ mi}} \approx 88.55 \text{ km}$

35. Strategy   To find the cost:
• Find the number of watt-hours by multiplying the number of watts (2,400) by the number of hours (6).
• Convert watt-hours to kilowatt-hours.
• Multiply the product by \$.08.

Solution $2,400 \text{ W} \times 6 \text{ h} = 14,400 \text{ Wh}$
$14,400 \text{ Wh} = 14.4 \text{ kWh}$
$14.4 \times \$.08 = \$1.152$
The cost is \$1.15.

36. The number is less than 10. Move the decimal point 8 places to the right. The exponent on 10 is -8.

$0.0000000679 = 6.79 \times 10^{-8}$

37. Perimeter $= 2(\text{length}) + 2(\text{width})$

$= 2(1.2 \text{ m}) + 2(0.75 \text{ m})$

$= 2.4 \text{ m} + 1.5 \text{ m} = 3.9 \text{ m}$

38. Area $= \text{length} \times \text{width}$

$= 9 \text{ in.} \times 5 \text{ in.} = 45 \text{ in}^2$

39. Volume $= \text{length} \times \text{width} \times \text{height}$

$= 20 \text{ cm} \times 12 \text{ cm} \times 5 \text{ cm}$

$= 1200 \text{ cm}^3$

40. $-2 + 8 + (-10) = 6 + (-10) = -4$

41. $-30 - (-15) = -30 + 15 = -15$

42. $2\dfrac{1}{2} \times -\dfrac{1}{5} = \dfrac{5}{2} \times \dfrac{-1}{5} = \dfrac{-5}{10} = -\dfrac{1}{2}$

43. $-1\dfrac{3}{8} \div 5\dfrac{1}{2} = \dfrac{-11}{8} \div \dfrac{11}{2} = \dfrac{-11}{8} \times \dfrac{2}{11} = \dfrac{-1}{4} = -\dfrac{1}{4}$

44. $(-4)^2 \div (1-3)^2 - (-2) = (-4)^2 \div (-2)^2 - (-2)$

$= 16 \div 4 - (-2)$

$= 4 - (-2)$

$= 4 + 2$

$= 6$

45. $2x - 3(x-4) + 5 = 2x + (-3)[x + (-4)] + 5$

$= 2x + (-3)x + (-3)(-4) + 5$

$= 2x + (-3)x + 12 + 5$

$= -x + 12 + 5$

$= -x + 17$

46. $\dfrac{2}{3}x = -12$

$\dfrac{3}{2} \cdot \dfrac{2}{3}x = \dfrac{3}{2} \cdot (-12)$

$x = -18$

47. $3x - 5 = 10$

$3x - 5 + 5 = 10 + 5$

$3x = 15$

$\dfrac{3x}{3} = \dfrac{15}{3}$

$x = 5$

48.
$$8 - 3x = x + 4$$
$$8 - 3x - x = x - x + 4$$
$$8 - 4x = 4$$
$$8 - 8 - 4x = 4 - 8$$
$$-4x = -4$$
$$\frac{-4x}{-4} = \frac{-4}{-4}$$
$$x = 1$$

49. Strategy To find your new balance, subtract the check amounts ($321.88 and $34.23) and add the amount of the deposit ($443.56).

Solution
$$\begin{array}{r} \$872.48 \\ - \ 321.88 \\ \hline 550.60 \\ - \ \ 34.23 \\ \hline 516.37 \\ + \ 443.56 \\ \hline \$959.93 \end{array}$$
Your new balance is $959.93.

50. Strategy To find how many people will vote, solve a proportion.

Solution
$$\frac{5}{8} = \frac{n}{102,000}$$
$$5 \times 102,000 = 8 \times n$$
$$510,000 = 8 \times n$$
$$510,000 \div 8 = n$$
$$63,750 = n$$
The number of people who will vote is 63,750.

51. Strategy To find the last year's dividend, solve the basic percent equation for the base, letting *n* represent the base. The percent is 80% and the amount is $1.60.

Solution
$$\text{Percent} \times \text{base} = \text{amount}$$
$$80\% \times n = \$1.60$$
$$0.80 \times n = \$1.60$$
$$n = \$1.60 \div 0.80$$
$$n = \$2.00$$
The dividend last year was $2.00.

52. Strategy To find the mean income for the 4 months, add the incomes and divide the sum by the number of incomes (4).

Solution
$$\begin{array}{r} \$4320 \\ 3572 \\ 2864 \\ + \ \ 4420 \\ \hline \$15,176 \end{array} \qquad 4\overline{)15,176}^{\ 3794}$$
The mean income is $3794.

53. Strategy To find the simple interest due, multiply the principal ($120,000) by the interest rate by the time (in years).

Solution
$$\text{Interest} = 120,000 \times 10\% \times \frac{9}{12}$$
$$= 120,000 \times 0.10 \times \frac{9}{12}$$
$$= 9000$$
The simple interest due is $9000.

54. Strategy To calculate the probability:
• Count the number of possible outcomes.
• Count the number of favorable outcomes.
• Use the probability formula.

Solution There are 36 possible outcomes. There are 12 favorable outcomes: (1, 2), (2, 1), (1, 5), (5, 1), (2, 4), (4, 2), (3, 3), (3, 6), (6, 3), (4, 5), (5, 4), (6, 6).
$$\text{Probability} = \frac{12}{36} = \frac{1}{3}$$
The probability is $\frac{1}{3}$ that the sum of the dots on upward faces of the two dice is divisible by 3.

55. Strategy To find the percent:
• Read the graph and find the death count of China.
• Read the circle graph and find the death count of the other three countries.
• Find the sum of the death counts by adding the four death counts.
• Solve the basic percent equation for percent. The base is the sum of the four death counts and the amount is the death count of China.

Solution
$$\begin{array}{rl} \text{China:} & 1300 \text{ thousand} \\ \text{Japan:} & 1100 \text{ thousand} \\ \text{USSR:} & 13,600 \text{ thousand} \\ \text{Germany:} & + \ 3300 \text{ thousand} \\ \hline & 19,300 \text{ thousand} \end{array}$$
$$\text{Percent} \times \text{base} = \text{amount}$$
$$n \times 19,300 = 1300$$
$$n = 1300 \div 19,300$$
$$n \approx 0.067$$
China has 6.7% of the death count of the four countries.

56. Strategy To find the discount rate:
• Subtract the sale price ($226.08) from the regular price ($314) to find the amount of the discount.
• Use the basic percent equation for percent. The base is the regular price and the amount is the amount of the discount.

Solution $\begin{array}{r} \$314.00 \\ -226.08 \\ \hline \$87.92 \end{array}$

Percent × base = amount
$$n \times 314 = 87.92$$
$$n = 87.92 \div 314$$
$$n = 0.28 = 28\%$$

The discount rate for the compact disc player is 28%.

57. Strategy To find the weight of the box in pounds:
• Multiply the number of tiles in the box (144) by the weight of each tile (9 oz) to find the total weight of the box in pounds.
• Convert the weight in ounces to the weight in pounds.

Solution
$$144 \times 9 \text{ oz} = 1296 \text{ oz}$$
$$1296 \text{ oz} = 1296 \, \cancel{oz} \times \frac{1 \text{ lb}}{16 \, \cancel{oz}}$$
$$= \frac{1296}{16} \text{ lb}$$
$$= 81 \text{ lb}$$

The weight of the box is 81 lb.

58. Strategy To find the perimeter of the composite figure, add the sum of the two sides to $\frac{1}{2}$ the circumference of the circle.

Solution
$$\text{Perimeter} = 2s + \frac{1}{2}\pi(\text{diameter})$$
$$\approx 2(8 \text{ in.}) + \frac{1}{2}(3.14)(8 \text{ in.})$$
$$= 16 \text{ in.} + 12.56 \text{ in.}$$
$$= 28.56 \text{ in.}$$

The perimeter is approximately 28.56 in.

59. Strategy To find the area of the composite figure, subtract the area of the two semicircles from the area of the rectangle.

Solution
$$\text{Area} = \text{area of rectangle} - 2(\text{area of semicircle})$$
$$\text{Area} = \text{length} \times \text{width} - 2\left[\frac{1}{2}\pi(\text{radius})^2\right]$$
$$\approx 10 \text{ cm} \times 2 \text{ cm} - 2\left[\frac{1}{2}(3.14)(1 \text{ cm})^2\right]$$
$$= 20 \text{ cm}^2 - 2(1.57 \text{ cm}^2)$$
$$= 20 \text{ cm}^2 - 3.14 \text{ cm}^2$$
$$= 16.86 \text{ cm}^2$$

The area of the composite figure is approximately 16.86 cm^2.

60. The unknown number: n
$$\frac{n}{2} - 5 = 3$$
$$\frac{n}{2} - 5 + 5 = 3 + 5$$
$$\frac{n}{2} = 8$$
$$2 \cdot \frac{n}{2} = 2 \cdot 8$$
$$n = 16$$

The number is 16.